BIO 11B

Bethany Bush

REEDLEY COLLEGE

McGraw Hill create®
create.mheducation.com

over Credit: ©Darrell S. Vodopich

BN-13: 9798219036460

BN-10: 821903646M

Contents

EXERCISE 19

Human Evolution

Skull Examination

Learning Objectives

By the end of this exercise you should be able to:

1. Describe the parts of a modern human skull.
2. Distinguish between skulls of males and females.
3. Distinguish between skulls of apes and modern humans.

Please visit connect.mheducation.com to review online resources tailored to this lab.

Throughout time, no issue has interested humans more than learning about our origins. Where did we come from? What did our ancestors look like? Where did they live? Today, we are beginning to understand those questions, thanks to evidence provided by biologists and anthropologists. Nevertheless, the topic remains controversial and often elicits strong responses from people.

In this exercise, you will examine some of the information underlying recent ideas about human evolution by examining the skulls of human and humanlike organisms. Specifically, you will examine skulls of

Gorilla gorilla, the modern gorilla. Gorillas and humans share a common ancestor.

Australopithecus (fig. 19.1), a relatively small, humanlike organism that had small, pointed canine teeth and an elongated face and was adapted for bipedalism and an upright stance. The most famous australopithecine is "Lucy," discovered in 1974 in Ethiopia. *Australopithecus,* the best known of the early humanlike organisms, lived 3.5–2.5 million years ago.

Homo erectus, a human ancestor characterized by an upright posture, a sloping forehead, a large brow ridge, a thick skull, and a larger braincase (900 cm^3) than earlier humanlike organisms. *H. erectus* lived 1,800,000–27,000 years ago.

Homo sapiens, or modern humans, are characterized by a vertical forehead, small brow ridge, thin skulls, and a larger braincase (1400 cm^3) than earlier humanlike organisms. *H. sapiens* appeared 200,000 years ago.

These species represent well-documented stages in the lineage of modern humans. As you examine these skulls, think of how their traits link modern humans with other primates. Also know that anthropologists are constantly making new discoveries of early human fossils and that the specimens you will analyze represent a small subset of these species (fig. 19.2).

Dea/G Cigolini/age fotostock

Figure 19.1 *Australopithecus africanus* was the first australopithecine to be discovered. It was unearthed in southern Africa by Raymond Dart in the 1920s.

THE MODERN HUMAN SKULL

The human skull (fig. 19.3), including the lower jaw, consists of 22 bones, 8 of which are paired. All of the bones fit together at joints called **sutures,** which appear as wavy lines. Projections and raised lines are sites of muscle attachment.

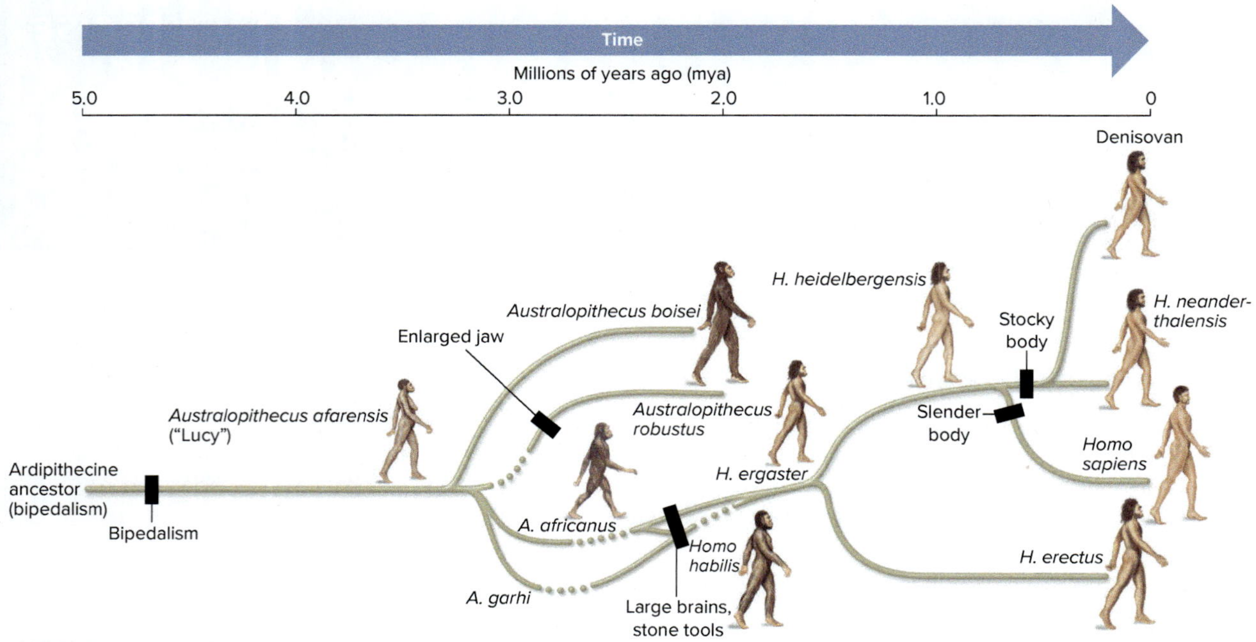

Figure 19.2 A possible scenario for human evolution. In this human family tree, several species lived contemporaneously with one another, but only one lineage gave rise to modern humans (*Homo sapiens*). Dotted lines indicate controversial time lines and branching patterns. Molecular analysis distinguishes denisovans (known from fragmented fossils discovered in Denisova Cave in Siberia) from neanderthals. Note that four species (groups) of our genus *Homo* likely occupied the world simultaneously for thousands of years.

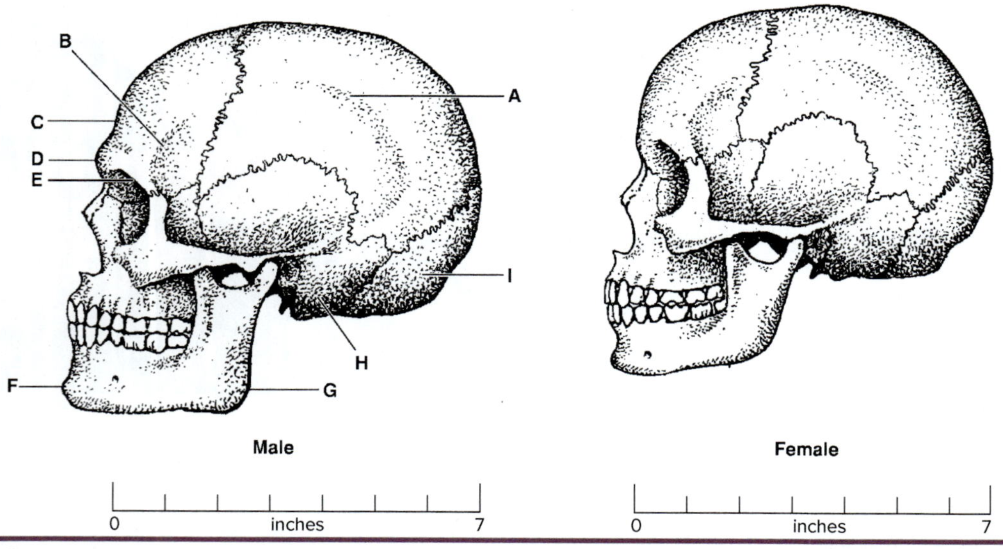

Male Skull	Feature	Female Skull
Large	**A. Braincase**	Smaller than male
Marked	**B. Muscle lines**	Slight
Retreating	**C. Forehead**	Bulging
Developed	**D. Brow ridges**	Absent
Rounded	**E. Upper rim of eye socket**	Sharp
Square	**F. Chin**	Rounded
Nearly a right angle	**G. Angle of jaw**	Angle more obtuse (over 120°)
Large	**H. Mastoid process**	Small
Present	**I. External occipital protuberance**	Absent

Figure 19.3 A comparison of male and female skulls of modern *Homo sapiens.*

Although skulls from males and females share many features, they usually can be distinguished. However, such a diagnosis—even when done by experts—is only about 90% reliable (80% if the lower jaw is missing).

Procedure 19.1 **Examine skulls of modern humans**

1. Examine the skulls of modern humans available in the lab to familiarize yourself with their structure and geometry. Be sure to examine skulls of males and females.
2. Compare these skulls with those shown in figure 19.3. The letters beside each feature in the table correspond to those shown in the diagrams.

Question 1

a. How do skulls of females differ from those of males?

b. What is the biological significance of these differences?

Procedure 19.2 **Examine the skulls of apes**

1. Use figure 19.4 and the specimens in the lab to study the following features of ape and human skulls.
2. Compare the braincases of a human skull with those of apes. A typical human brain weighs ~1400 g, compared to 400 g for chimpanzees and 550 g for gorillas.

Face

Prognathism is the extent to which the face and jaws protrude forward when viewed from the side. Their larger teeth and jaws cause apes to exhibit more prognathism than do humans.

Braincase

The **brow ridge,** the mass of bone over the eye sockets, supports the upper facial skeleton against forces produced by chewing. The brow ridge in apes is prominent. In humans, the brow ridge of modern humans is largely internalized because our frontal bone has expanded outward to a more vertical angle.

The **sagittal crest** is a thin ridge of bone atop and down the middle of the braincase. The sagittal crest is associated with having a small braincase and powerful jaws. In apes, the sagittal crest is an attachment site for the large temporalis muscle used for chewing.

The **foramen magnum** is the large opening in the base of the skull through which the spinal cord passes. The position of the foramen magnum reflects the posture of the body (and, indirectly, the pattern of movement) of **hominoids.** Humans stand erect and walk with the head directly over the vertical spinal column. Conversely, the knuckle-walking apes hold their heads forward, with the foramen magnum toward the rear. Thus, the foramen magnum is located in a more rear position in apes than in humans.

Teeth and Jaws

Adult apes and humans have the same number and types of teeth: 4 canines, 8 premolars, 12 molars, and 8 incisors. Identify these teeth on the skulls and diagrams. In apes the canine teeth are longer and more pointed than others. Notice that nonhuman primates have large lower canines (fig. 19.4). Therefore, the upper jaw must have a **diastema** (space) on each side to receive those canines when the jaw is closed. The canine teeth seldom project above the others.

In humans, the four front teeth (incisors) are smaller, more vertical, and flatter than in apes. In nonhuman primates, the canine diastema is the gap in the teeth corresponding to the canines of the opposite jaw.

Question 2

a. Between which teeth does the gap occur? Why are these gaps essential in nonhuman primates?

b. Why are they usually absent in humans?

Humans have an outward projection on the lower part of their lower jaw (i.e., chin). Apes lack this feature; instead, they have a smooth, even slant to the front part of their jaw.

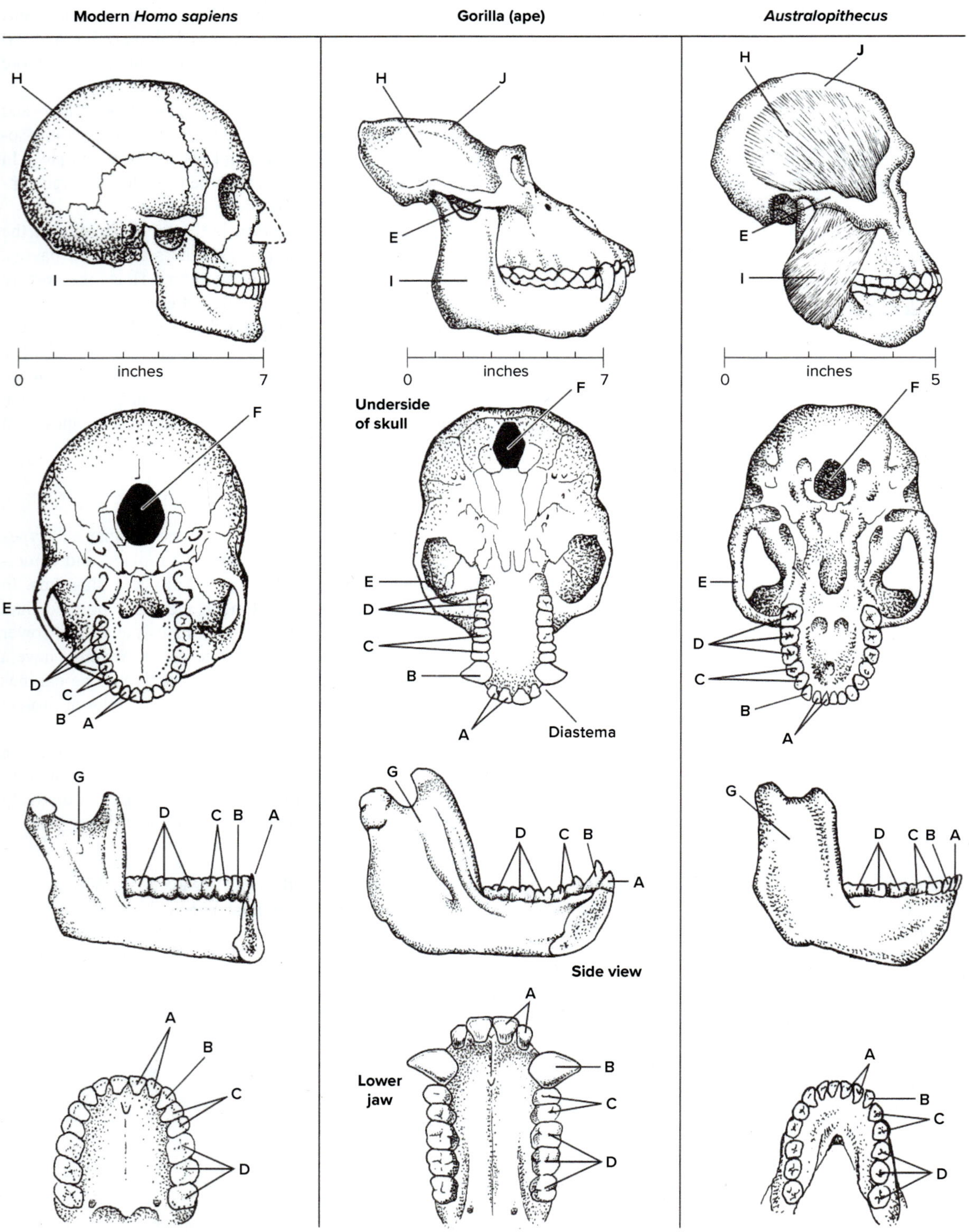

Figure 19.4 Comparison of skull features of modern *Homo sapiens,* gorilla, and *Australopithecus* skull and jaws:

A. Incisors
B. Canines
C. Premolars
D. Molars
E. Zygomatic arch
F. Foramen magnum
G. Vertical ramus for muscle attachment
H. Skull surface for muscle attachment
I. Jaw surface for muscle attachment
J. Sagittal crest for muscle attachment

In humans, teeth are arranged in a relatively continuous curve from the third molar around to the other third molar. The arrangement of teeth in apes is straighter, with a slight curve in front. This is primarily because of the larger size of the incisors and canines.

To summarize the differences between skulls of modern humans and apes, complete table 19.1.

Procedure 19.3 **Study the skulls of fossil primates**

Use the information and diagrams in figures 19.5–19.8 to learn about skulls of humans and ancient primates.

At first glance, full skeletons of humans and modern apes look quite similar. But closer examination reveals distinctive differences and adaptations, many of which are related to bipedalism, jaw shape, and brain size.

Table 19.1

Prominent Features of Skulls of Apes and Humans

Feature	Apes	Humans
Sagittal crest		
Brow ridge		
Foramen magnum		
Prognathism		
Canines		
Canine diastema		
Incisors		
Chin		
Arrangement of teeth		

Procedure 19.4 **Compare human and ape skeletal adaptations for standing**

1. Compare the skeletons illustrated in figures 19.9 and 19.10.
2. For each of the following skeletal features, record the difference between human and ape anatomy. Then speculate on the adaptive significance of those features related to posture, bipedalism, and/or food habits. Summarize your work by completing table 19.2.

Table 19.2

Comparison of Skeletal Features of Apes and Humans

Feature	Ape	Human
Tooth number and size		
Spine shape		
Braincase size		
Insertion point of head on the spinal column		
Thumb and big toe size and angle		
Arm length		
Size of heel bone (calcaneus)		
Thickness of knee joints (ends of tibia and femur)		
Length of the angled head of the femur		
Posterior exposure of sacrum		
Risk of cervical vertebrae to compression		
Size of birth canal relative to body size		

Australopithecus boisei

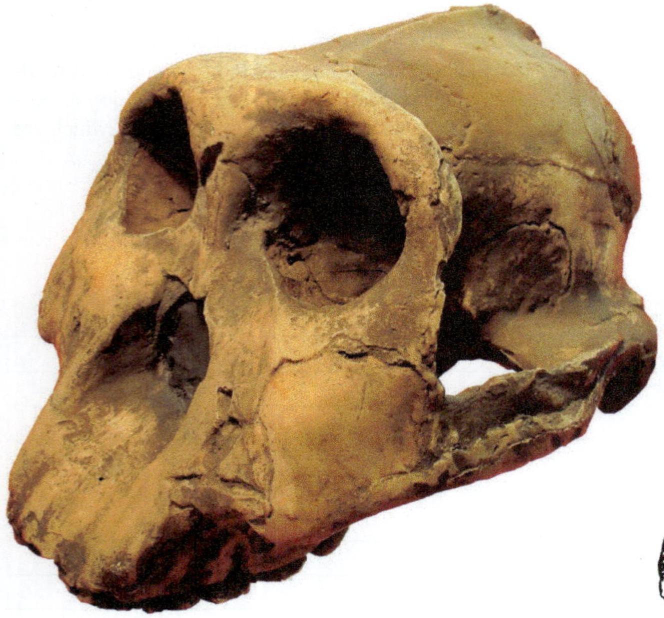

Dinoton/Shutterstock

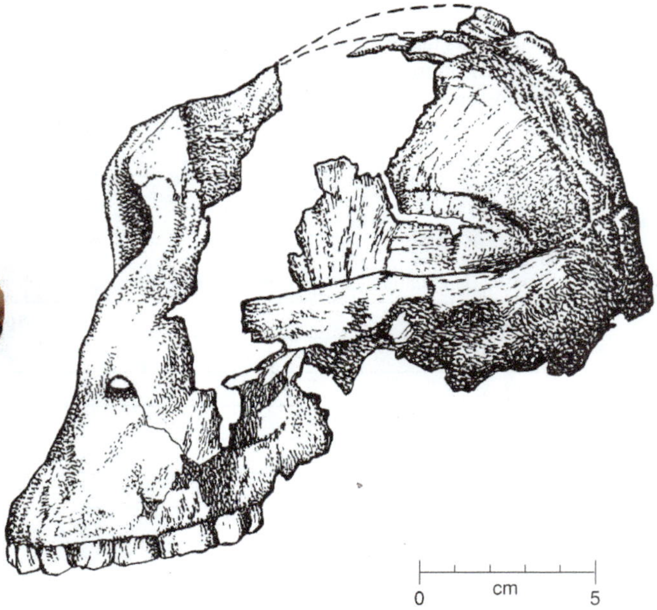

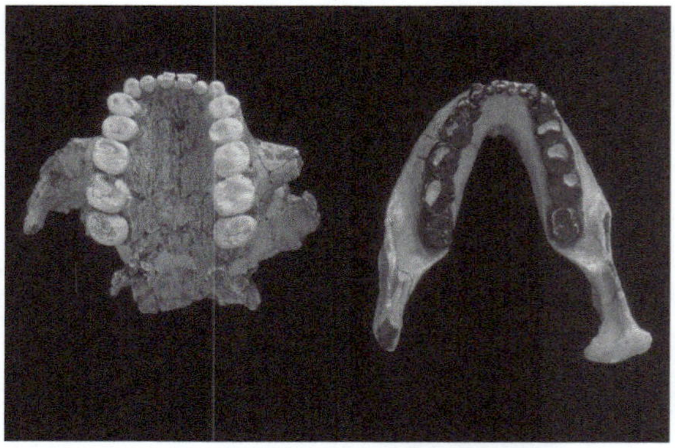

The Natural History Museum/Alamy Stock Photo

Figure 19.5 Views of the skull and teeth (upper jaw, left; lower jaw, right) of *Australopithecus boisei.* Age: 1.8 million years. This skull includes massive molar and premolar teeth (similar in size to those of gorillas) and is nearly complete except for the lower jaw. It is commonly known as "Zinj," an abbreviation of the original genus name *Zinjanthropus.* Zinj and the remains of many smaller hominins (humanlike organisms) were discovered in the Olduvai Gorge in Tanzania by Mary and Louis Leakey in 1959.

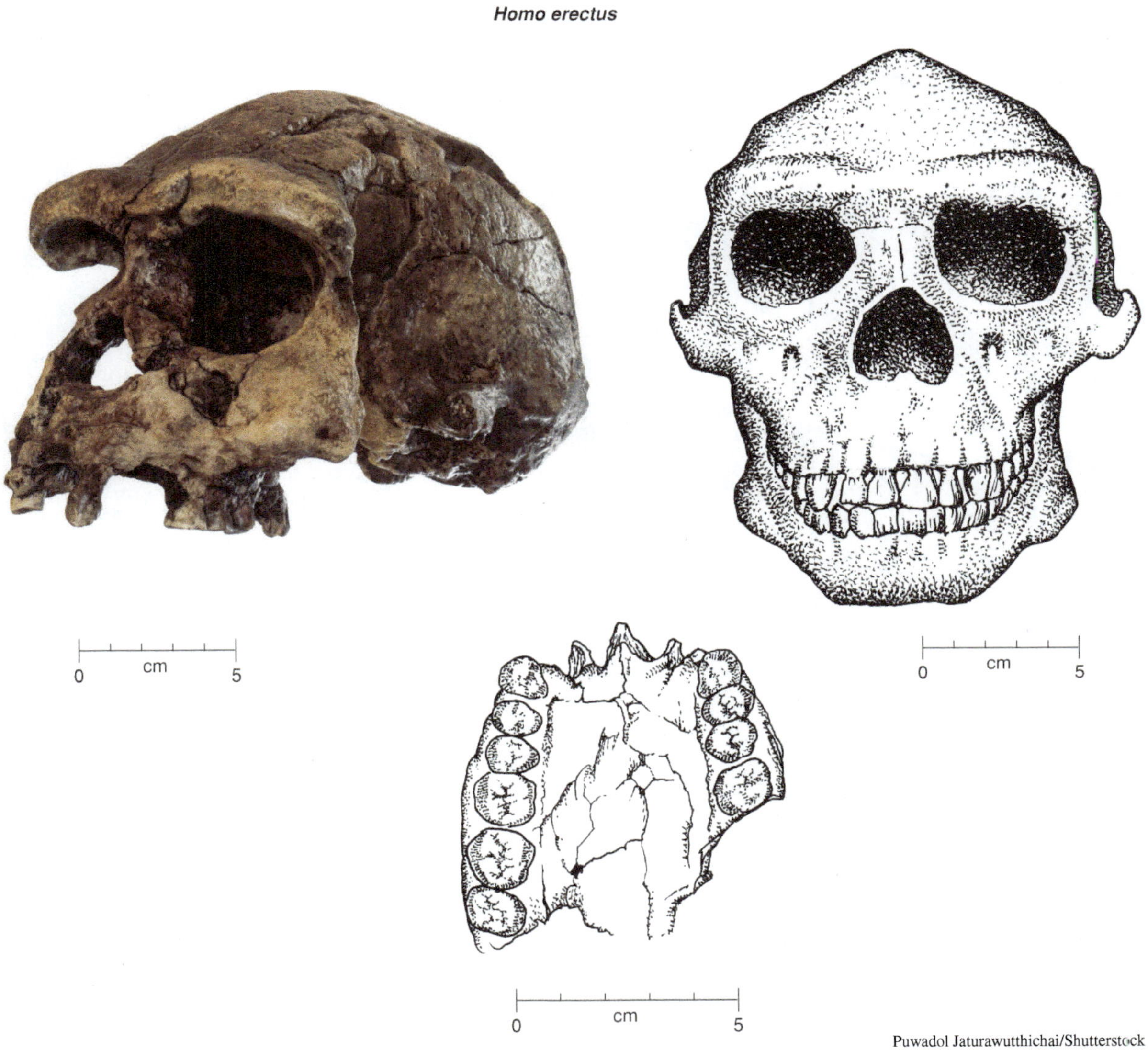

Puwadol Jaturawutthichai/Shutterstock

Figure 19.6 Views of the skull and teeth (upper jaw) of *Homo erectus*. Age: Less than 1 million years. This diagram shows a reconstruction that includes parts of skulls discovered in 1937 and 1939 in Java (Sangiran). The skull of *Homo erectus* differs from modern human skulls in that it is low vaulted and has a relatively small volume; *H. erectus* skulls have volumes of 900–1100 cm^3, whereas skulls of modern humans have volumes of approximately 1400 cm^3. Skulls of *H. erectus* also have small mastoid processes behind the ear openings, large jaws, small chins, and large molar teeth.

Homo sapiens **("Neanderthal")**

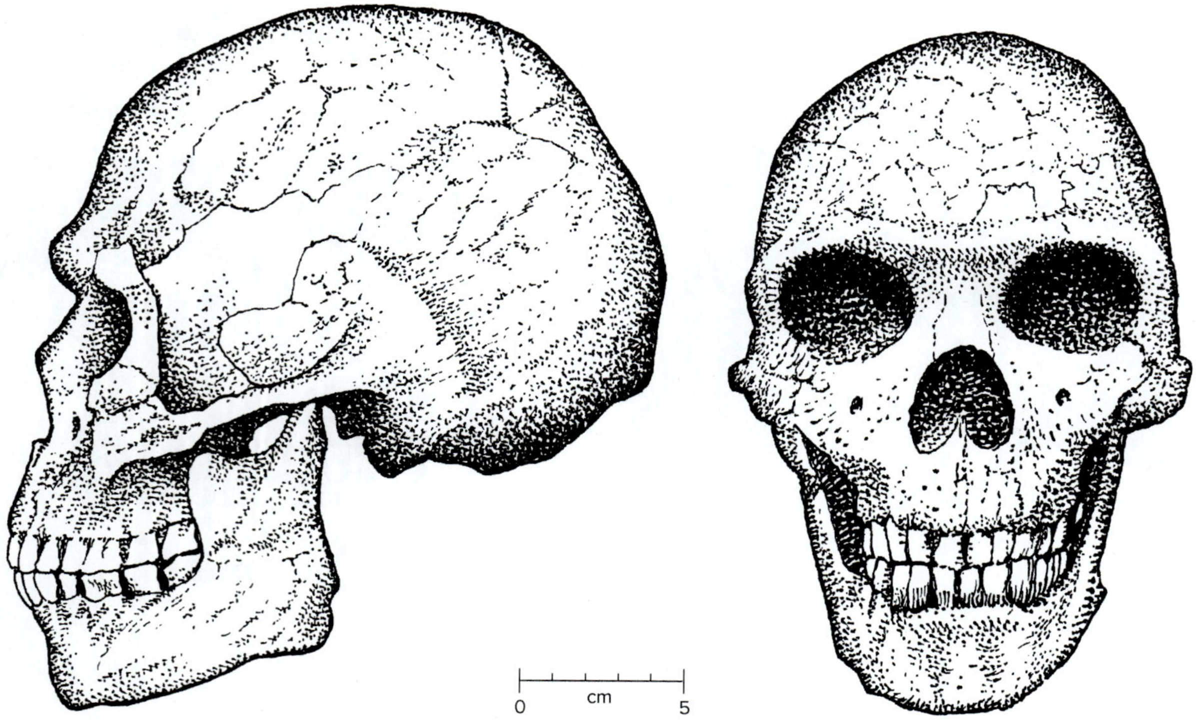

Figure 19.7 Views of the skull and teeth of *Homo sapiens.* Age: About 32,000 years. Also known as Neanderthal Man, this type was recovered in 1932 from Mugharet-es-Skhull, Wadi el-Mughara, Israel. The skull is nearly complete.

Homo sapiens **(modern)**

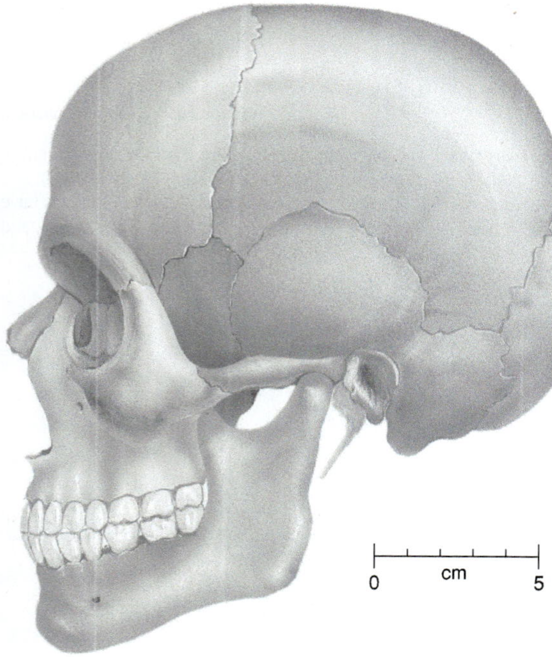

Figure 19.8 Skull of modern *Homo sapiens.* Age: About 10,000 years. This skull is from one of at least 50 skeletons recovered at Oued Agrious, Algeria, during the late 1920s. See figure 19.4 for labeled features.

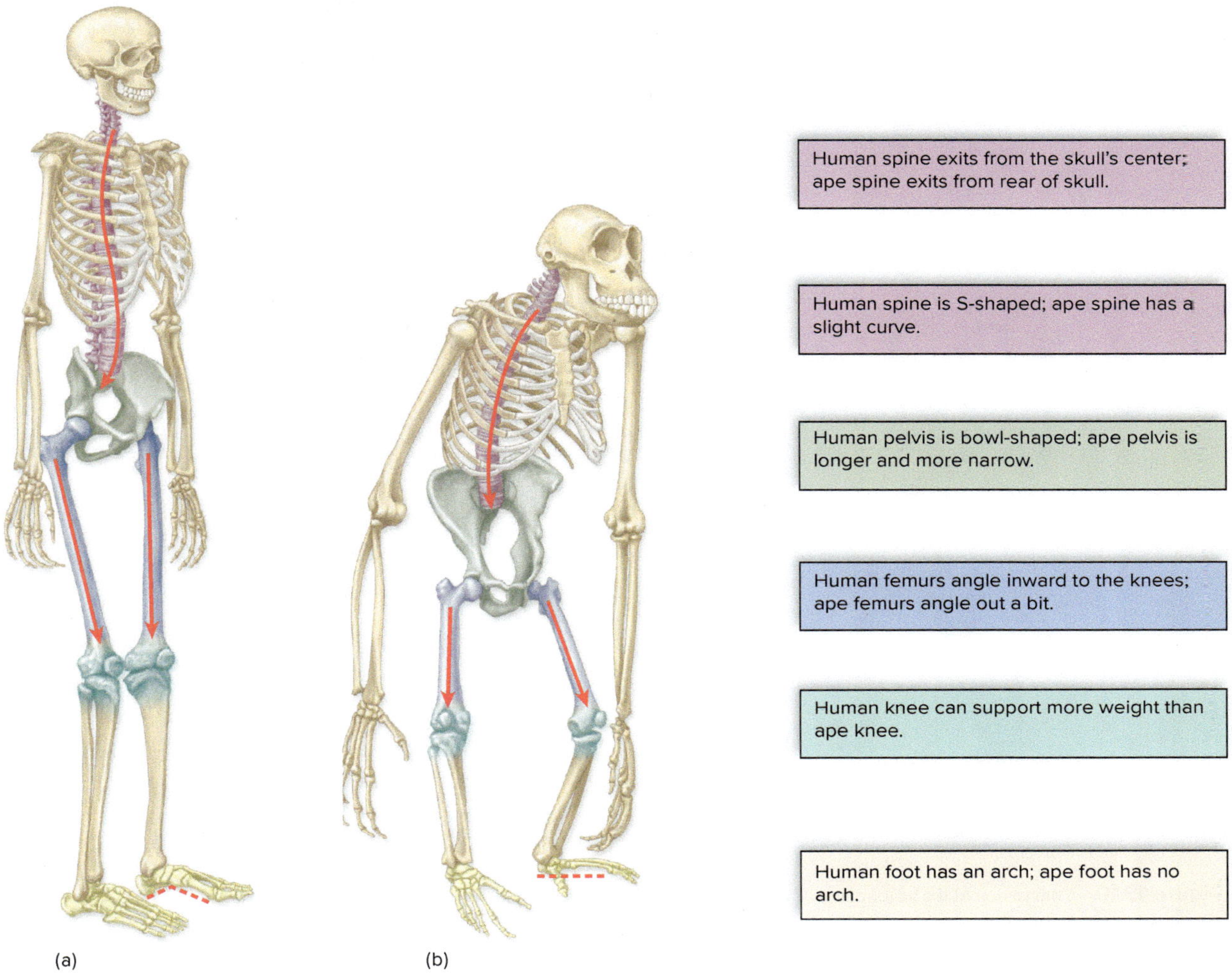

Figure 19.9 Adaptations for standing. (*a*) Human skeleton compared to (*b*) Ape (chimpanzee) skeleton.

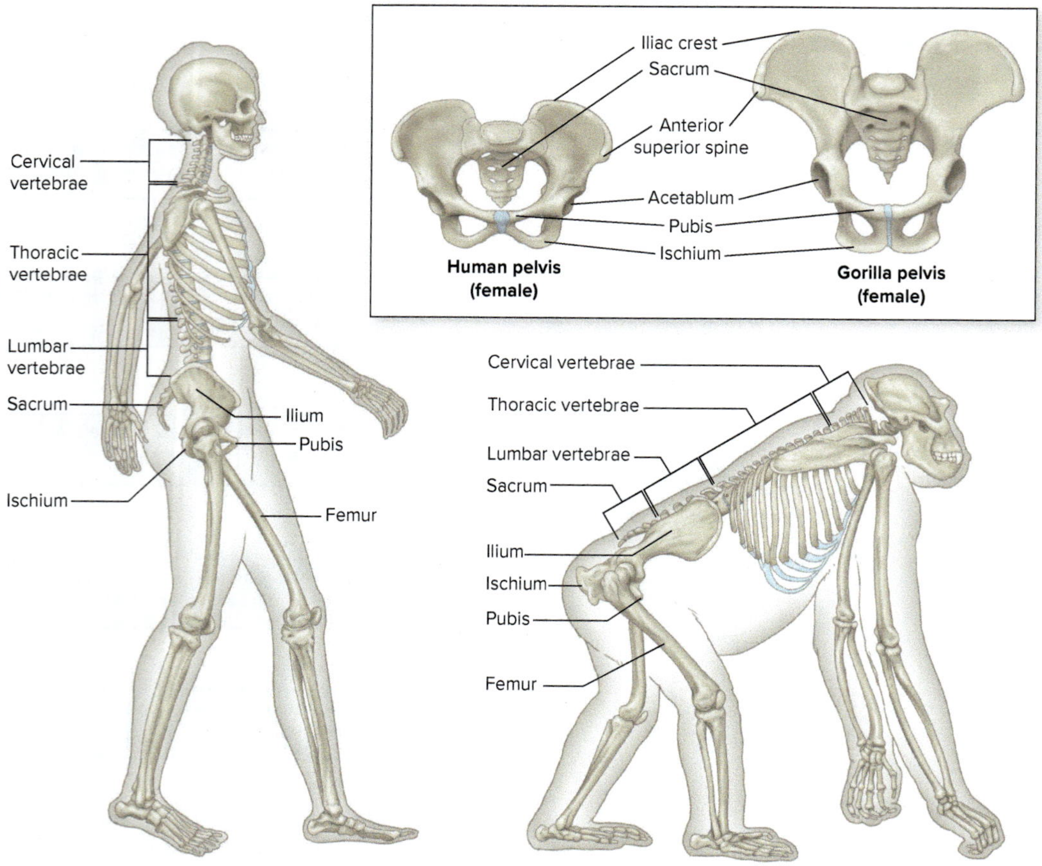

Figure 19.10 Comparison of the major bones of Human and Ape (gorilla) skeletons.

INQUIRY-BASED LEARNING

Humans don't stand still. Along what paths did our common ancestors migrate during our evolution?

Observations: Some of the oldest fossils of our human ancestors have been found in Africa. Current theories of human evolution involve the migration of ancestral primates and early hominins from Africa to Europe and back to Africa.

Question: What migration pattern of human ancestors is supported by fossil evidence?

a. Establish a working lab group and obtain Inquiry-Based Learning Worksheet 19 from your instructor.

b. Research the current literature and outline the patterns of migration by ancestral primates and early hominins away from and back to Africa.

c. Record your findings on Worksheet 19.

d. Discuss your findings with your instructor and with other students.

Questions for Further Study and Inquiry

1. What could changes in our ancestors' teeth tell us about the evolutionary history of humans?

2. Is the skull shown in figure 19.7 that of a man or a woman? Explain your answer.

3. Neanderthals were once considered to be a separate species (*Homo neanderthalensis*). However, some anthropologists consider modern humans and Neanderthal Man to be the same species, *Homo sapiens*. Do you think that they should be the same species? Why or why not? What species-related criteria are you using as a basis for your claim? What does this judgment imply about the ability of these two groups of people to interbreed when they lived together in Europe 34,000 years ago?

WRITING TO LEARN BIOLOGY

Describe what other evidence you would use to study the evolution of humans. Where would you get that evidence?

WRITING TO LEARN BIOLOGY

Read an article on a more recent fossil discovery and describe how this new early human fossil is similar to and different from the species they measured in this lab. Describe distinctive features of the skeleton and describe what it may be adapted to.

EXERCISE 20

Ecology
Diversity and Interaction in Plant Communities

Learning Objectives

By the end of this exercise you should be able to:

1. Observe the physical factors, plant dominance, and interactions among organisms in a terrestrial community and characterize the community according to these features.
2. Quantify the distribution and abundance of plants in a community.
3. Detect the experimental effects of competition and allelopathy among plants.
4. Explain four different ways to quantify the relative abundances of different plant species in a plant community, using transect data.

Please visit connect.mheducation.com to review online resources tailored to this lab.

Ecological communities are extraordinarily complex. The assemblage of plants that you observe at any time results from interactions among plants and their physical surroundings, among different species of plants, and among plants and animals. All of these interactions are driven by a flow of energy captured by green plants and passed to herbivores (plant eaters), predators, and decomposers.

It is beyond the scope of this exercise to explain all of the processes occurring in a plant community, but we can assemble some basic observations that characterize and distinguish communities.

SAFETY FIRST Before coming to lab, you were asked to read this exercise so you would know what to do and be aware of safety issues. In the space below, briefly list the safety issues associated with today's procedures. If you have questions about these issues, contact your laboratory assistant before starting work.

QUALITATIVE COMMUNITY ASSESSMENT

Procedure 20.1 **Observe and assess the ecological characteristics of a terrestrial community**

1. Locate and visit a terrestrial community designated by your instructor.

2. Characterize the community according to the criteria and questions that follow. After you've answered the questions, discuss your observations with your instructor and other groups. Be prepared to use your observations as a basis for describing your assessment of energy flow through the community, diversity of the community, and interactions among organisms.

Physical Factors

Observations

1. What levels of light intensity occur throughout the community?

2. Does the community include shade-tolerant as well as shade-intolerant plants?

3. How does light differ among different vertical levels of vegetation? Why could this be important?

4. What is the temperature 2 m above ground?

5. What is the temperature at the soil surface?

6. How does the ground slope? Why is this important?

7. How would you characterize the soil? Loam (a fertile soil of clay, sand, and organic humus)? Clay (a sticky mix of ultrafine mineral particles)? Sand (a granular mix of hard mineral grains)?

8. What is the nature of the ground cover? Grasses? Bare soil?

9. Is there a layer of leaf litter on the ground?

10. Is the community generally a moist, moderate, or dry environment?

Interpretations

1. How might shade affect the temperature of the community?

2. How might different amounts of light at different vertical levels within the community be important?

3. What parts of the community might be cooler than others? Why is this important?

4. Why would ground slope be important?

5. Based on your observations of slope and soil type, would you expect the soil to retain moisture?

6. How long has this community been left to develop naturally? That is, what is the age of the community?

Plant Dominance

Observations

1. Which plant species are most abundant in numbers?

2. Which plant species are most abundant in biomass?

3. What general categories of plant types (shrubs, trees, etc.) are apparent?

4. What is the vertical distribution of vegetation?

Interpretations

1. Would you describe this community as diverse? Why or why not?

2. What comparison community in your local area would you consider to be more diverse? Less diverse?

3. What observations led you to your conclusion for the previous question?

4. What specific factors make your comparison community more or less diverse? Human impact? Stressful environmental factors? Geology?

Interactions among Organisms

Observations

1. What evidence do you see of resident vertebrates?

2. What evidence do you see of resident invertebrates?

3. What evidence do you see of plant–animal interactions?

4. What evidence do you see of plant–plant interactions?

5. What adaptations do the plants have to discourage herbivores?

6. Do you see any obvious or subtle evidence of competition by plants for available resources?

Interpretations

1. If you don't see any vertebrates, does that mean they are not around? Explain your answer.

2. Reexamine the observations that you just listed. How would each observation affect the type and growth of plants in the community that you studied?

3. What types of competitive interactions are apparent in the community?

4. What types of mutually beneficial interactions are apparent in the community?

QUANTITATIVE COMMUNITY ASSESSMENT

Many techniques have been developed to measure the numbers, densities, and distributions of organisms in terrestrial plant communities. One widely used technique is to count organisms within randomly distributed **quadrats** (sometimes called **plots**) of uniform size. Other techniques involve measuring distances between plants or the distance from randomly chosen points to nearest plants. In the **line-intercept method,** a **transect,** or line, is established and laid out within the community. Organisms that touch this line are counted and measured. Calculations based on these measurements of quadrats and transects reveal the relative abundances, frequencies, and distributions of the plant species that compose the community.

Procedure 20.2 **Assessing a community with the line-intercept method**

1. With the help of your instructor, locate a suitable field site with a plant community to be examined.

2. Obtain a measuring tape 15–20 m long, a meterstick, and a notepad. If a measuring tape is unavailable, use a measured piece of string or rope.

3. Assess the general layout of the community to be sampled. With the aid of your instructor, decide on a reasonable set of criteria to govern the placement of a transect for each group of students.

Table 20.1

Summary of Raw Data for Species Occurring along a Transect

Total transect length = ________ **Total number of intervals** ________

Species *i*	**f_i = Number of Intervals in Which Species *i* Occurs**	**n_i = Number of Individuals Encountered**	**c_i = Length of Transect Intercepted**
	F = Total of all frequencies = ________	N = Total of all individuals = ________	C = Total length of transect intercepted = ________

Question 1

What concepts or ideas should govern the placement of your transect to obtain a representative sample of the community? Are there any "wrong" places to put a transect? Why or why not?

4. You and your lab partners will work on a single transect. Stretch the measuring tape on the ground to establish a transect.
5. Divide the transect into 1-, 2-, or 3-m intervals. Each interval represents a separate unit of the transect.
6. At one end of the transect begin counting plants that touch, overlie, or underlie the transect line. For each plant encountered record the type of plant (species) and the length of the line that the plant intercepts. For plants that overhang the line, record the length of the imaginary vertical plane of the line that the plant would intercept. Record these raw data for each interval in your field notepad. Also record any uncovered (bare) lengths within the transects.
7. When all plants from all intervals have been recorded, summarize your data in table 20.1.
8. Sum the values in each of the three data columns of table 20.1 to calculate F, N, and C. Record the calculations at the bottom of each column.
9. Use the data in table 20.1 to calculate the following four parameters for each species within the community. Record your results in table 20.2.

Relative frequency = f_i/F

Relative density = n_i/N

Relative coverage = c_i/C

Importance value of species *i* = Relative frequency + Relative density + Relative coverage

Question 2

What is the meaning of an importance value? Why would we calculate this in addition to density, coverage, and frequency?

Table 20.2

Relative Values of Each Species in a Selected Community Using Parameters of the Line-Intercept Method

Species	Relative Frequency	Relative Density	Relative Coverage	Importance Value

PLANT INTERACTIONS

Competition

Competition is the interaction among individuals seeking a scarce common resource (fig. 20.1). The interaction is usually negative (disadvantageous) for both competitors. The intensity of competition for a resource such as light, food, water, space, and nutrients depends on the amount of resource, the number of individuals competing, and the needs of each individual for that resource. Competition among members of the same species is **intraspecific competition,** whereas competition among members of different species is **interspecific competition.**

Procedure 20.3 **Examine competition by sunflower seedlings**

1. Obtain five pots containing enough potting soil to plant seeds. Measure the area at the surface of the potting soil.
2. Follow your lab instructor's directions to plant 2, 4, 8, 16, and 32 sunflower seeds in each of the five pots, respectively. In each of the pots plant at least 20% more seeds so you can later reduce the number of seedlings to the numbers previously listed. Label each pot with the number of seeds, the date, and your name.
3. Water each pot gently with a consistent amount of water.
4. Place the pots in the greenhouse or growth area so that each pot has the same environmental conditions of light, temperature, and so on. After the seedlings are established, remove excess seedlings so the treatments will have the correct number of seedlings listed in step 2. Use that number of plants in later calculations even if plants die before the experiment ends. For calculations of mean weight and mean height, use the treatment's original number of plants.
5. Examine the pots at regular intervals as directed by your instructor. Record measurements of the parameters called for in table 20.3. Your instructor may ask you to plot your results.

Dom Hammond//PunchStock/Design Pics, Inc.

Figure 20.1 Competition in a forest. Each plant in this forest competes with all the individuals around it for light, soil nutrients, and moisture.

Question 3

a. In which pot was competition greatest?

b. Which parameters best showed the effects of competition?

Table 20.3

Effects of Competition on Sunflower Seedlings

	Plants per Pot				
	2	4	8	16	32
General observations:					
Date ________					
Date ________					
Date ________					
Date ________					
Mean height of individuals:					
Date ________					
Date ________					
Date ________					
Date ________					
Range of height of individuals:					
Date ________					
Date ________					
Date ________					
Date ________					
Mean width of 10 widest leaves:					
Date ________					
Date ________					
Date ________					
Date ________					
Mean fresh weight of aboveground biomass (measured at the end of the experiment):					
Date ________					

c. What other characteristics of the competitors might you have measured?

d. Did competition more noticeably affect the number of individuals or the biomass of each individual?

e. In what environments would you expect that competition among sunflowers would be most intense?

f. Would you expect different results if a different type of potting soil was used? Why?

g. Would plants and animals compete for the same resources? What might be some differences?

Allelopathy

Allelopathy is a form of competition involving the inhibition of a plant's germination or growth by exposure to compounds produced by another plant. These compounds may be airborne or leach from various plant parts. Rainfall, runoff, and diffusion typically distribute inhibitory compounds in the immediate area surrounding the producing plant (fig. 20.2).

Question 4

a. What are the benefits of producing allelopathic compounds?

b. What are the possible disadvantages of producing allelopathic compounds?

Procedure 20.4 Demonstrate allelopathy

1. Determine from your instructor the overall experimental design for the class—how many plants your group will test and how many replicates you will set up for each plant.

2. Obtain tissue (stems and leaves) from the variety of plants provided by your instructor. Some of these plants are suspected to produce allelopathic compounds.

3. For each plant: Mix 10 g of tissue with 100 mL of water in a blender and homogenize. Let the slurry soak for 5–10 min to leach chemicals from the disrupted tissue.

4. Filter or strain the slurry to remove large particulates. Drain the filtrate into a beaker.

5. Obtain a petri dish (7.6 cm diameter) and line its bottom with a circular piece of filter paper.

6. Label the petri dish appropriately for the plant extract and replicate being tested in that dish.

7. Saturate the filter paper with a measured amount (5–8 mL) of the extract.

8. Repeat steps 3–7 for each replicate and each plant being tested.

9. Obtain seeds of radish, lettuce, or oat. Distribute 50 seeds uniformly on the filter paper in each dish.

10. Your instructor may enhance the experimental design by asking you to set up replicate dishes for each extract and to test the effects on different kinds of seeds. Follow the directions given by your instructor.

11. You may also be asked to design and implement an experimental control of a water-only treatment that is blended and then filtered.

12. Incubate the covered dishes at room temperature in the laboratory or in the greenhouse. After 24 and 48 h, count the number of seeds that have germinated and calculate the percent germinated.

13. After 72 h (or the length of time specified by your instructor) measure the length of the radicle and make relevant observations about the apparent rate of growth of the emerged embryos. Extend your observations for as many days as specified by your instructor.

14. Record your results in table 20.4.

Arto Hakola/Shutterstock

Figure 20.2 Allelopathic inhibition of nearby plant competitors for nutrients and water often results in barren areas surrounding the established plant producing the inhibitors.

Question 5

a. Did your observations reveal any differences in allelopathy among tested plants?

b. Which plant species demonstrated the most intense allelopathy?

Table 20.4

Effects of Allelopathy on Germination and Growth Rate of Seeds

Plant Extract ________ Seed Species ________ Total No. of Seeds ________

	Time	% Germinated	Length of Radicle	Observations
Rep 1				
Rep 2				
Rep 3				

Plant Extract ________ Seed Species ________ Total No. of Seeds ________

	Time	% Germinated	Length of Radicle	Observations
Rep 1				
Rep 2				
Rep 3				

Plant Extract ________ Seed Species ________ Total No. of Seeds ________

	Time	% Germinated	Length of Radicle	Observations
Rep 1				
Rep 2				
Rep 3				

Plant Extract ________ Seed Species ________ Total No. of Seeds ________

	Time	% Germinated	Length of Radicle	Observations
Rep 1				
Rep 2				
Rep 3				

INQUIRY-BASED LEARNING

Do leaves, roots, and stems all produce allelopathic compounds?

Observation: Many plants produce chemicals that inhibit the growth of other plants and/or the germination of other seeds. These allelopathic chemicals are adaptive because they reduce competition.

Question: Do all organs (e.g., roots, stems, leaves) of allelopathic plants produce the same amounts of allelopathic chemicals?

a. Establish a working lab group and obtain Inquiry-Based Learning Worksheet 20 from your instructor.

b. Discuss with your group well-defined questions relevant to the preceding observation and question. Choose and record your group's best question for investigation.

c. Translate your question into a testable hypothesis and record it.

d. Outline on Worksheet 20 your experimental design and supplies needed to test your hypothesis. Ask your instructor to review your proposed investigation.

e. Conduct your procedures, record your data, answer your question, and make relevant comments.

f. Discuss with your instructor any revisions to your questions, hypothesis, or procedures. Repeat your work as needed.

Questions for Further Study and Inquiry

1. Diverse plant communities have many species representing a variety of plant types such as grasses, shrubs, succulents, hardwood trees, softwood trees, vines, and ferns. What factors increase a community's diversity? Age of the community? Energy input? Moisture? Nutrients? Disturbance? Human activity? How do they do so?

2. What characteristics of a community make it more resilient than other communities to a particular disturbance? Describe both the type of disturbance and the type of community.

3. What characteristics would indicate that a community has not been disturbed for a few years?

4. How does competition influence natural selection? Is the presence of competitors a selective force?

Pollution

The Effects of Chemical, Thermal, and Acidic Pollution

EXERCISE 23

Learning Objectives

By the end of this exercise you should be able to:

1. Describe how chemical pollution, thermal pollution, and acid rain affect the growth and reproduction of selected organisms.
2. Determine what factors can lead to excessive growth of algae.

Please visit connect.mheducation.com to review online resources tailored to this lab.

Seldom a day passes in which we don't hear a pollution-related story in the news. Oil spills, leaks of toxic chemicals, and noise pollution are a few of the pollution-related problems that affect our lives. The effects of pollution can be disastrous, as exemplified by oil spills along the coasts of Alaska and Spain; the leakage of 40 tons of toxic gas in 1984 in India (which killed thousands and injured hundreds of thousands of people); and the dumping of toxic wastes into our rivers, lakes, and oceans.

A **pollutant** is any physical or chemical agent that decreases the aesthetic value, economic productivity, or health of the biosphere. There are many types of pollutants, including noise, chemicals, radiation, and heat. Whether or not it makes headlines, pollution of any type affects fundamental aspects of life, namely our health and how organisms grow and behave. Because populations of organisms interact among themselves and with their environment, pollution always affects more than one organism. For example, consider the following food chain:

algae → zooplankton → small fish → large fish → humans

A pollutant that reduces a population at any step of this food chain will affect all other levels of the food chain because all steps of the chain are linked. Thus, water polluted with chemicals, such as herbicides that kill algae, will decrease populations of zooplankton that eat the algae; this ultimately affects populations of fish and other organisms that are part of the food chain.

In this exercise, you will study several types of pollution. Specifically, you will

- Simulate the effects of **acid precipitation** by examining seed germination and survival of organisms at acidic pH.
- Study chemical pollution by examining the effects of differing concentrations of nutrients and **pesticides** on growth of algae and shrimp.
- Study **thermal pollution** by examining how organisms survive at unnaturally high temperatures.
- Examine water polluted with an overabundance of algae.
- Use the Allium Test to assay the effects of a variety of pollutants on plant growth.

SIMULATING THE EFFECTS OF ACID RAIN

Acid rain is a worldwide problem caused by atmospheric pollution (fig. 23.1*a*). Compounds such as nitrates and sulfates released into the atmosphere by automobiles and industries combine with water to form nitric and sulfuric acids that are deposited across the countryside in precipitation having a pH as low as 4.2. There, the acid decreases the pH of the soil, affects the availability of nutrients and metals, and usually diminishes plant growth. The effects of acid rain are often subtle but significant; for example, the cumulative pollution of decades of acid rain can destroy landmarks and reduce forests and crop yields (fig. 23.1*b*).

In this exercise, you will study the influence of acid rain on (1) seed germination and (2) growth and reproduction of brine shrimp. You will simulate acid rain with dilute solutions of sulfuric acid.

⚠ Be careful while you handle the cultures and do not get the sulfuric acid solutions on yourself. If you do, wash yourself immediately and thoroughly.

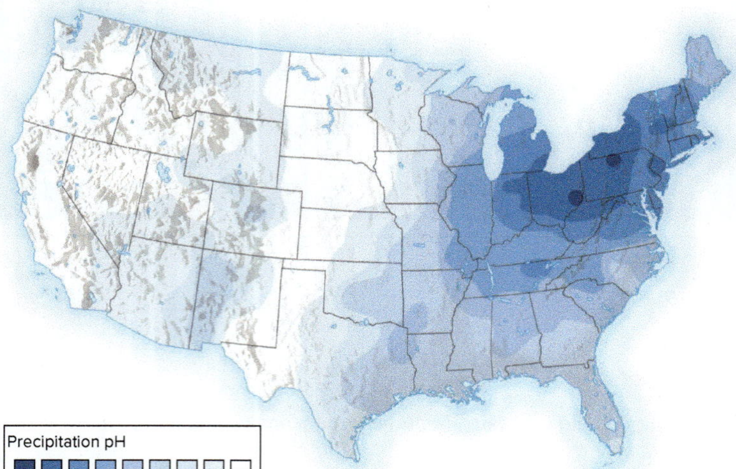

(a)

Marek Mnich/E+/Getty Images
(b)

Figure 23.1 Acid rain and its effects. (*a*) The pH of snow and rain throughout the United States. (*b*) Acid rain often kills trees and damages landmarks such as sculptures and buildings.

SAFETY FIRST Before coming to lab, you were asked to read this exercise so you would know what to do and be aware of safety issues. In the space below, briefly list the safety issues associated with today's procedures. If you have questions about these issues, contact your laboratory assistant before starting work.

Procedure 23.1 Observe seed germination

1. Examine the set of petri dishes labeled "Acid Rain and Seed Germination." Each of the dishes was inoculated several days ago with 50 seeds of corn (*Zea mays*). Dish 1 contains seeds soaked in a solution having a neutral pH (i.e., pH 7, which is near the pH at which corn seeds germinate in their typical environments), dish 2 contains seeds soaked in a solution having a pH of 4, and dish 3 contains seeds soaked in a solution having a pH of 2.

Question 1

a. Examine dish 1 (pH 7). What percentage of the seeds germinated? A germinated seed is one in which the root or shoot has punctured the seed coat. Use the following formula to calculate the percentage germination:

$$\text{\% Germination} = \frac{\text{number of germinated seeds}}{\text{total number of seeds}} \times 100$$

b. How do you think that lowering the pH will affect seed germination? Formulate a hypothesis for how lowering the pH will affect seed germination.

2. Compare the seeds in dish 1 with those in dishes 2 (pH 4) and 3 (pH 2). Calculate the percentage germination for each treatment. Record your results in table 23.1.

3. Do your data in table 23.1 support your hypothesis? Explain.

Question 2

a. How did acidity affect seed germination? How accurate were your predictions in Question 1?

b. What does this tell you about the effect of acid rain on seed germination in natural environments?

c. How would the decreased growth of plants in response to acid rain affect animals in the same environment?

Table 23.1

Germination of Seeds in Environments of Different pH

Treatment	Total Number of Seeds	Number Germinated	% Germination
pH = 7	______	______	______
pH = 4	______	______	______
pH = 2	______	______	______

Table 23.2

Growth of Brine Shrimp in Environments of Different pH

	Counts of Number of Living Shrimp						
Treatment	1	2	3	4	5	Total	Mean
pH = 7	______	______	______	______	______	______	______
pH = 4	______	______	______	______	______	______	______
pH = 2	______	______	______	______	______	______	______

Procedure 23.2 Observe growth of brine shrimp

Brine shrimp (*Artemia salina*) is a small crustacean that lives in salty aquatic environments such as the Great Salt Lake. Brine shrimp are often used as live food in aquariums.

1. Examine the three cultures labeled "Acid Rain and Brine Shrimp." Several days ago all of the cultures were inoculated with the same number of brine shrimp eggs. Culture 1 contains shrimp growing in a solution having a pH of 4, and culture 2 contains shrimp growing in a solution having a pH of 2. The pH of typical environments for brine shrimp ranges from 7 to 8.5.

2. Compare these cultures with that in dish 3 at neutral pH (pH = 7). Use your dissecting microscope to count the number of living shrimp in five randomly selected fields of view of each culture.

3. Record your results in table 23.2.

Question 3

a. How does acidity affect hatching and growth of brine shrimp in your experiment? What does this tell you about the effect of acid rain on survival of brine shrimp?

b. Consider the food chain described in the introduction of this exercise. How would acid rain affect fish and humans in this food chain?

c. Comment on the validity of extrapolating the results of your experiments with seed germination and brine shrimp survival to the natural world.

NUTRIENT ENRICHMENT: EUTROPHICATION

As naturally occurring nutrients are washed from the soil by rain, bodies of water such as lakes and ponds undergo a nutrient-enrichment process called **eutrophication** (fig. 23.2). Some instances of eutrophication occur naturally, whereas others are produced by human activities such as pollution. Most nutrients can be absorbed by an ecosystem without damaging the ecosystem's integrity. However, if the critical load of nutrients is exceeded, eutrophication occurs.

Nagel Photography/Shutterstock

Figure 23.2 A eutrophic pond. The surface bloom of green algae indicates the abundance of nutrients in the water.

Table 23.3

Effects of Nutrient Enrichment on Algal Cultures

Culture	Observations
1 Eutrophication	
2 Eutrophication, anaerobic	
3 Control	

Nitrogen and phosphorous are important nutrients that can limit plant growth. These nutrients are also major sources of anthropogenic eutrophication. These nutrients can enter aquatic systems from untreated sewage, agricultural runoff, excess fertilizer from lawns and gardens, and some detergents. In extreme situations, this excessive input of nutrients can cause rapid growth of algae and aquatic plants, thereby leading to the accelerated filling of lakes and ponds. In this exercise, you will study eutrophication by observing the growth rates of several organisms in both nutrient-enriched and unenriched media.

Several days ago algal cultures were started with similar amounts of algae. Culture 1 contains an excess of nitrogen, a condition similar to that of lakes enriched with raw sewage. Culture 2 contains large amounts of nitrogen and minimal oxygen to simulate anaerobic conditions that occur naturally when bacterial populations in water increase rapidly in response to added sewage. As the bacteria grow and reproduce, they rapidly deplete the oxygen supply in the water, thus producing anaerobic conditions. Culture 3 is an aerobically grown control containing normal amounts of nitrogen. This experiment relates the well-publicized and frequent occurrences of fish-kills in overly eutrophic lakes and ponds to nutrient stimulation of rapid bacterial growth that depletes the dissolved oxygen.

Procedure 23.3 **Examine the effects of simulated eutrophication**

1. Examine the two algal cultures labeled "Eutrophication" and the culture labeled "Control."
2. Make at least three observations of each culture. Record your observations in table 23.3.

Question 4

a. Which culture has the most algal growth?

b. Which has the least?

Fred Ward Productions

Figure 23.3 Huge amounts of pesticides and herbicides, such as those being applied by this crop duster, are used in agriculture each year. One of the most important tasks facing agriculturists and other biologists is the development of integrated pest management and systems involving genetically engineered organisms that will avoid the need for such chemical excesses.

c. What does this tell you about the influence of excessive amounts of nutrients on algal growth?

d. How could you reverse the effects of eutrophication?

e. Can eutrophication ever be beneficial? If so, when?

PESTICIDE POLLUTION

Pesticides are common pollutants, and the increasing demands for food by our expanding population have caused farmers to increase their reliance on pesticides to increase crop yields (fig. 23.3). These pesticides are often effective, but residues washed from the soil by rain contaminate lakes, ponds, and the underground water supply. These pollutants can harm organisms that drink and live in these waters.

Table 23.4

Effects of Pesticide on Survival of Brine Shrimp

Culture	Observations			
	Replicate 1	Replicate 2	Replicate 3	Mean
Pesticide culture 1	_______	_______	_______	_______
Pesticide culture 2	_______	_______	_______	_______
Pesticide culture 3	_______	_______	_______	_______
Control culture 4	_______	_______	_______	_______

Table 23.5

Effects of Temperature on Survival of Brine Shrimp

Culture	Observations			
	Replicate 1	Replicate 2	Replicate 3	Mean
Culture 1 (room temperature)	_______	_______	_______	_______
Culture 2 (35°C)	_______	_______	_______	_______

Procedure 23.4 **Examine the effect of a pesticide on the survival of brine shrimp**

Do not open the pesticide cultures or spill the liquids on yourself; the pesticides are toxic.

1. Examine the four cultures labeled "Pesticides." Several days ago these cultures were started with the same number of brine shrimp and kept at the same temperature.

2. Culture 1 contains the recommended dosage of the pesticide, culture 2 contains 1/10 the recommended dosage, and culture 3 contains 10-times the recommended dosage. Culture 4 is a control that lacks pesticide.

3. Examine three samples from each culture with your dissecting microscope. Record your observations of each culture in table 23.4.

Question 5

a. Which culture contains the most living shrimp?

b. Which culture contains the fewest living shrimp?

c. What does this tell you about how pesticides affect the growth of aquatic organisms?

d. How might this pollution affect terrestrial organisms that depend on the water for drinking water?

THERMAL POLLUTION

Excessive heat is a common pollutant. Many factories use water from lakes, reservoirs, or rivers to cool heat-generating equipment and release the hot water to reservoirs or ponds, where it raises the water temperature. Thermal pollution:

- Speeds biochemical reactions, thereby altering the growth of organisms and composition of communities.
- Speeds the evaporation of water, thereby concentrating other pollutants in the water.
- Decreases the oxygen supply in the water because warm water holds less oxygen than does cool water.

Procedure 23.5 **Examine the effect of temperature on the survival of brine shrimp**

1. Use your dissecting microscope to examine the two cultures labeled "Thermal Pollution." Both cultures were started several days ago with the same number of brine shrimp.

2. Culture 1 was grown at room temperature (20–25°C), and culture 2 was grown at 35°C in oxygen-depleted water. Count the number of living brine shrimp in three fields of view. Record your observations of each culture in table 23.5.

Question 6

a. Which culture contains more living shrimp?

b. Which culture contains fewer living shrimp?

c. What does this tell you about the influence of thermal pollution on growth of aquatic organisms?

d. How might thermal pollution of a lake or pond affect nearby terrestrial organisms?

THE ALLIUM TEST

Many pollutants directly affect the growth and development of organisms. Thus, the most informative assay for these pollutants involves measuring growth of living organisms exposed to controlled treatments of suspected pollutants. The **Allium Test** provides this kind of assay. In the Allium Test, bulbs of the common onion, *Allium,* are subjected to solutions of water being investigated for pollutant toxicity. Roots develop quickly, and their number and length after five days estimate the effect of the potential pollutant on a common aspect of plant growth.

Do not open the pesticide cultures or spill the liquids on yourself. Pesticides are toxic.

Table 23.6

Growth of *Allium* Roots Exposed to Solutions Containing Different Pollutants

	Mean Number of Roots		Mean Length of Roots	
Solutions Assayed	**Treatment**	**Control**	**Treatment**	**Control**

Table 23.7

Root Growth of Replicate *Allium* Bulbs Exposed to a Specific Treatment Solution

Treatment: ______		Control: ______	
Number of Roots	**Length of Roots**	**Number of Roots**	**Length of Roots**
Mean =	Mean =	Mean =	Mean =

Procedure 23.6 **Use the Allium Test to assay a variety of pollutants**

1. Examine the solutions provided by your instructor for your class to assay with the Allium Test. Record the solutions in table 23.6.
2. You may need to work in small groups to assay all of the solutions. Determine which solutions or treatments are your group's responsibility. Some of the treatments may involve different temperatures and some may involve solutions with unknown contents. Record in table 23.7 the treatment and control solutions tested by your group.
3. Obtain onion bulbs. You will need 5 to 10 bulbs for each treatment and 5 to 10 bulbs for controls for each treatment.
4. Trim the outer, loose layers of each bulb. Use a razor blade to trim exposed tissue from the root crown. Your instructor will demonstrate how to do this.
5. Obtain a beaker (or test tube) wide enough to support the onion but that will allow the root crown to protrude into the solution.
6. Fill 5 to 10 replicate beakers with the solution being tested.
7. Fill 5 to 10 beakers with the appropriate control solutions.

Question 7

What is the appropriate control solution for your treatment? Explain your answer.

8. Put a trimmed onion bulb on your beaker so the root crown is completely submerged (fig. 23.4).
9. Incubate the treatments and controls in the dark at room temperature for three to five days.

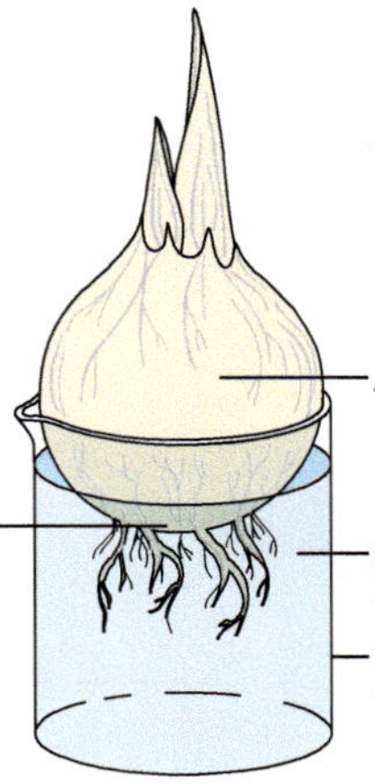

Figure 23.4 Allium Test setup. The number and length of the growing roots estimate the toxicity of the solution being assayed.

10. Check and supplement the solution levels in the beakers periodically if needed.
11. After four to seven days, determine the mean root length and number of roots for each bulb. Record the values in table 23.7. These are the results for the treatment solution that your group tested.
12. Record your mean values from table 23.7 in table 23.6.
13. In table 23.6 record the results for the solutions tested by the other groups.
14. Compare the number and length of roots on treated bulbs with those of controls.

Question 8

a. Why did each group need to run controls, rather than one set of controls for everyone?

INQUIRY-BASED LEARNING

Using the Allium Test to Detect Variation in Water Quality

Observations: The presence and concentration of biologically active compounds can be measured directly or by their indirect effect on a quantifiable biological process, such as a bioassay. Root production by *Allium* is sensitive to a variety of compounds. The number of biologically active compounds that humans introduce each year into our environment is immense.

Question: How does root growth by *Allium* respond to environmental contaminants?

a. Establish a working lab group and obtain Inquiry-Based Learning Worksheet 23 from your instructor.

b. Discuss with your group and instructor a specific contaminant and question relevant to the preceding observation and question. Record it on Worksheet 23.

c. Translate your question into a testable hypothesis and record it.

d. Review the Allium Test procedure 23.6. Then outline on Worksheet 23 your experimental design and supplies needed to test your hypothesis. You may choose to use a bioassay other than the Allium Test. Ask your instructor to review your proposed investigation.

e. Conduct your procedures, record your data, answer your question, and make relevant comments.

f. Discuss with your instructor any revisions to your questions, hypotheses, or procedures. Repeat your work as needed.

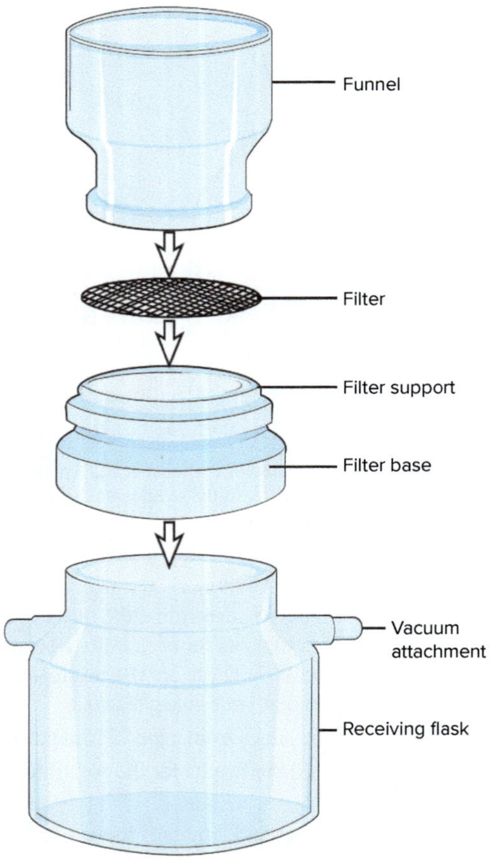

Figure 23.5 Diagram of filtration apparatus used to collect organisms from a water sample.

b. The Allium Test is a common test in environmental science. What are some disadvantages to this test?

c. The Allium Test for thermal pollution may show that higher temperatures accelerate growth. Is this bad? Why or why not?

ALGAL POLLUTION OF WATER SUPPLIES

Growing human populations have created great demands on groundwater supplies. To circumvent these demands, many areas have started using surface waters such as ponds and lakes to meet their water needs. Unlike groundwater, which is usually free of algae and cyanobacteria, surface water usually contains large amounts of algae and cyanobacteria that produce tastes and odors (*Anabaena, Nitella, Pandorina, Hydrodictyon*), clog pipes (*Synedra, Anacystis*), produce toxins (*Anabaena*), and form mats covering a lake or reservoir (*Oscillatoria, Spirogyra*).

Work in small groups to assess algal pollution in the water samples available in the lab.

Procedure 23.7 Determine algal pollution of water samples

1. Dip a filter base and funnel (fig. 23.5) into 70% ethanol for 1 min. Shake off the excess alcohol and let the apparatus air dry.

2. Use sterilized forceps to place a filter having pores 0.45 μm in diameter on the filter base. Screw the funnel onto the support (fig. 23.5) so that the filter is held securely in place.

3. Add 200 mL of the water sample to the funnel. If the water sample is heavily polluted, decrease the sample to 100, 50, or 25 mL.

4. Use a hand vacuum pump or water aspirator (fig. 23.6) to evacuate the receiving flask. Algae will be trapped on the filter.

5. Remove and dry the filter in an oven at 45°C for 8 min (or overnight at room temperature).

6. Float the dried filter on approximately 5 mL of immersion oil in a small petri dish. The dry filter will become transparent in the oil. If the filter remains opaque after a few minutes, place the petri dish on a warm surface until the filter clears.

7. Remove the filter from the petri dish by dragging it across the side of the dish.

8. Cut the filter into fourths and place all of these pieces of filter on one clean microscope slide.

9. Use a dissecting microscope to examine the filter with reflected light at low total magnification (10–14×). You will see much algae and cyanobacteria on the filter.

10. Count the algae in each of 10 randomly selected fields of view and record the number in the following table:

No. of Algae in Each Field of View

Field	1	2	3	4	5	6	7	8	9	10	Total
Number of Algae	___	___	___	___	___	___	___	___	___	___	_____

11. Use the following formula to calculate the total number of algae on your filter:

$$\text{Total no. of algae on filter} = \frac{1380\ \text{mm}^2}{\text{area of one field (mm}^2\text{)}} \times \frac{\text{total no. of algae counted}}{\text{no. of fields counted}}$$

Total no. of algae on filter = ______________

Calculate the area of the field of view with a stage micrometer or ruler (see Exercise 3). The value 1380 mm^2 is the area of the 47-mm filter that you used to collect the algae.

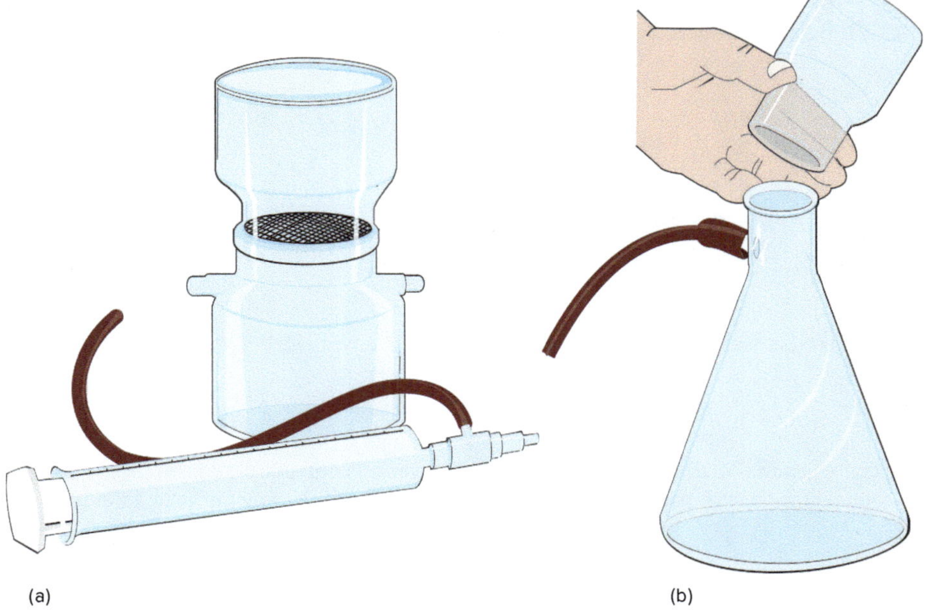

Figure 23.6 Vacuum filtration using (*a*) a hand vacuum pump or (*b*) a water aspirator. Applying a vacuum speeds the filtration of samples.

Area of field of view = ______________

12. Use the following formula to calculate the number of algae per milliliter of sample:

$$\text{No. of algae mL}^{-1} = \frac{\text{no. of algae on filter}}{\text{volume of sample (mL)}}$$

No. of algae mL^{-1} = ______________

Algal populations exceeding 1000 organisms mL^{-1} indicate that the water is over-enriched by sewage or other nutrients.

Question 9

a. How many algae per milliliter are in your water sample?

b. Is your sample polluted?

c. In what ways might these algae benefit other organisms in the water?

d. In what ways might they harm other organisms?

e. Suppose you are responsible for "cleaning up" a lake polluted with algae. What would you do, and why?

BIOREMEDIATION

Bioremediation is the use of living organisms (usually microbes) to remove pollutants from an environment. For example, bacteria that metabolize hydrocarbons are often used to minimize the impact of oil spills. Similarly, several bacteria have been genetically engineered (Exercise 16) to metabolize ionic mercury and other dangerous chemicals from nuclear waste.

Question 10

Bacteria that remove dangerous chemicals have an obvious value for remedying pollution. Are there any drawbacks or limitations of bioremediation? If so, what are they?

Questions for Further Study and Inquiry

1. What causes pollution?

2. Is all pollution bad? Why or why not?

3. What are some ways that you could "purify" water polluted with a pesticide?

4. What are the consequences of using water from ponds and rivers to cool industrial processes? What are the consequences of stopping the thermal pollution (forcing an industry to stop releasing the heated water)?

5. Polluting lakes with laundry water and sewage often produces an algal "bloom." Why don't populations of predators (i.e., zooplankton and fish) increase to offset this bloom and keep the algal population in check?

6. Diseases such as typhoid fever (caused by *Salmonella typhosa*) and dysentery are often associated with lakes polluted by sewage. Suppose that you live near a large lake in which sewage is dumped and that as a local health official you are in charge of reducing the incidence of typhoid in your area. How would you do this, assuming that you had the complete cooperation of city and other local officials? How would you do this if you couldn't prevent sewage from being dumped in the lake? What would be the consequences of your actions?

7. Define pollution. Can something be a pollutant in some situations but not in others? Explain your answer.

8. Does pollution always result from the demands of expanding populations? Why or why not?

DOING BIOLOGY YOURSELF

Design your own assay similar in concept to the Allium Test. Use a living plant and test potential pollutants of your choice.

WRITING TO LEARN BIOLOGY

Rachel Carson devoted much of her life to studying how organisms are affected by pollution. Her masterpiece, *Silent Spring*, raised public awareness of pollution and made the word *ecology* a household word. Go to the library and read an article about this remarkable woman. What was her message, and why is it still important?

EXERCISE 24

Survey of Prokaryotes

Domains Archaea and Bacteria

Learning Objectives

By the end of this exercise you should be able to:

1. Describe distinguishing features of Domains Archaea and Bacteria.
2. Describe the major differences between bacteria and cyanobacteria.
3. Identify representative examples of archaea, bacteria, and cyanobacteria.
4. Perform a Gram stain.

Please visit connect.mheducation.com to review online resources tailored to this lab.

Cellular organisms have evolved along two lines. Species whose cells lack membrane-bound organelles are **prokaryotes** (table. 24.1); bacteria are examples of prokaryotes. Species whose cells have membrane-bound organelles are **eukaryotes** and include plants, animals, fungi, and protists. About 5000 species of prokaryotes have been described, and many more await identification and description.

Prokaryotes were long thought to be a unified group commonly called bacteria. However, analyses as recently as 1996 of prokaryotic cell structure and genetics revealed two groups with surprisingly different DNA sequences. The sequences within both groups were strikingly different from each other and from those of eukaryotes. This new information has revealed three distinctive domains of organisms (fig. 24.1).

Species of **Domain Archaea** are prokaryotic and often inhabit, but are not restricted to, extreme and stressful environments. Species of **domain Bacteria** are also prokaryotic and the most abundant organisms on Earth (fig. 24.2). Species of **domain Eukarya** are all eukaryotic and include **kingdoms Fungi, Plantae, Animalia,** and the polyphyletic (multiple origins) **protists** (see Exercise 25 for the current status of protistan classification). Organisms in domain Eukarya are described in Exercises 25–31 and 36–40. This classification of living organisms into three domains is now widely accepted, but much phylogenetic information remains to be revealed. Discovering the relationships among life forms is an exciting and ongoing process.

DOMAIN ARCHAEA

Domain Archaea likely includes the oldest forms of life on earth, and domains Bacteria and Eukarya probably diverged from Archaea independently. In domain Archaea, a single cell is called an **archaeon**. Archaeons are diverse prokaryotes that share ribosomal RNA sequences as well as several important biochemical characteristics that distinguish them from all other types of organisms. Archaeons have distinctive membranes, unusual cell walls, and unique metabolic cofactors.

Today's archaea are probably survivors of ancient lines that persist in habitats similar to those present when archaeons first evolved. These habitats are often extremely acidic, hot, or salty. This is why many archaea are called **extremophiles**. Many archaea can live in an anaerobic atmosphere rich in carbon dioxide and hydrogen as well as the more benign environments typical of bacteria and eukaryotes. Exercise 16 includes a procedure involving *Halobacterium salinarum,* a common archaea that inhabits salty environments.

DOMAIN BACTERIA

A single cell of organisms in domain Bacteria is called a **bacterium** (pl. bacteria). Bacteria are distributed more widely than any other group of organisms, and individual cells are microscopic (1 µm or less in diameter, figs. 24.2 and 24.3). A single gram of soil may contain over a billion bacteria. Bacteria have cell walls, which give them three characteristic shapes (fig. 24.4).

- Bacillus (rod-shaped)
- Coccus (spherical)
- Spirillum (spiral)

Most bacteria are **heterotrophic,** meaning that they derive their energy from organic molecules made by other organisms.

Table 24.1

Prokaryotes Compared to Eukaryotes

Feature	Example
Unicellularity. All prokaryotes are basically single-celled. Even though some bacteria may adhere together or form filaments, their cytoplasm is not directly interconnected, and their activities are not as integrated and coordinated as cells of multicellular eukaryotes.	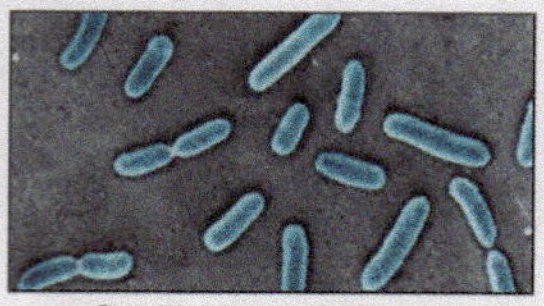 Cavallini James/BSIP SA/Alamy Stock Photo Prokaryotic cell
Cell Size. Most prokaryotic cells are 1–10 µm in diameter, while most eukaryotic cells are over 10 times that size.	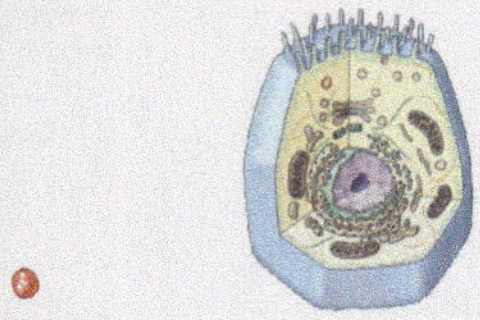 Bacterial cell — Eukaryotic cell
Chromosomes. Prokaryotic DNA exists as a single circle in the cytoplasm, while in eukaryotes, proteins are complexed with the DNA into multiple chromosomes.	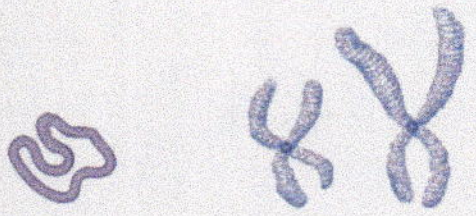 Bacterial genome — Eukaryotic chromosomes
Cell Division. Prokaryotic cells divide by binary fission. The cells pinch in two. In eukaryotes, microtubules pull chromosomes to opposite poles during the cell division process, called mitosis.	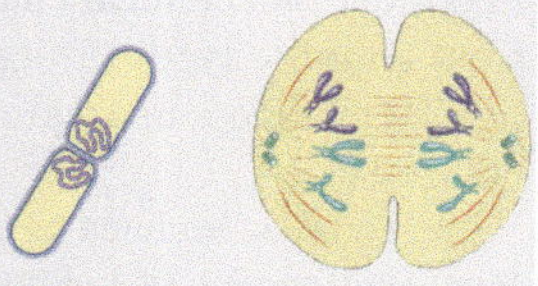 Binary fission in bacteria — Mitosis in eukaryotes
Internal Compartmentalization. Unlike eukaryotic cells, bacterial cells contain little internal compartmentalization, no internal membrane systems such as Golgi or ER, and no cell nucleus.	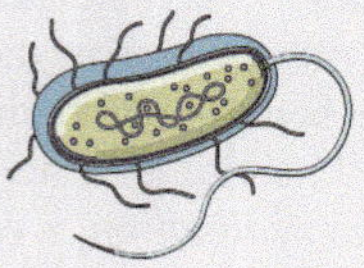 Bacterial cell
Flagella. Prokaryotic flagella are simple, composed of a single fiber of protein that spins like a propeller. Flagella in eukaryotes are complex structures that whip back and forth, rather than rotating.	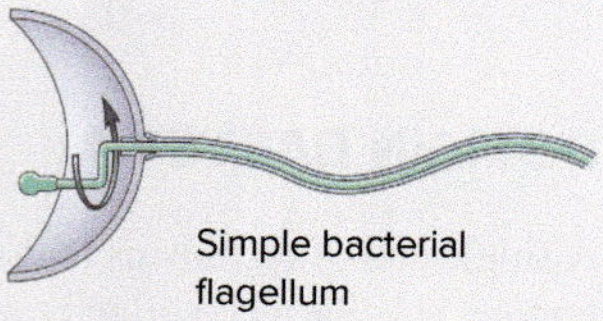 Simple bacterial flagellum
Metabolic Diversity. Prokaryotes possess many metabolic abilities not found among eukaryotes; some prokaryotes can perform several different kinds of anaerobic and aerobic photosynthesis, obtain their energy from oxidizing inorganic compounds, or fix atmospheric nitrogen.	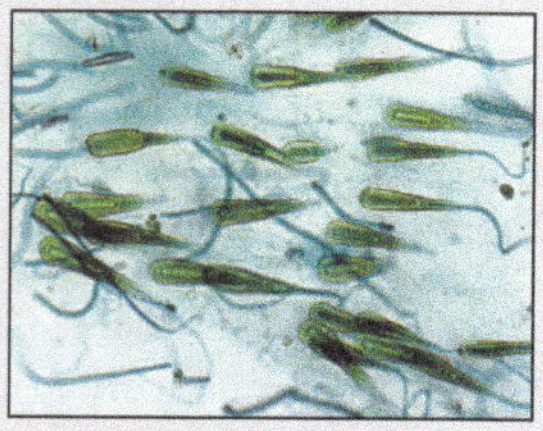 ©BiologyImaging.com Some prokaryotes, such as *Rivularia*, can fix nitrogen from the atmosphere.

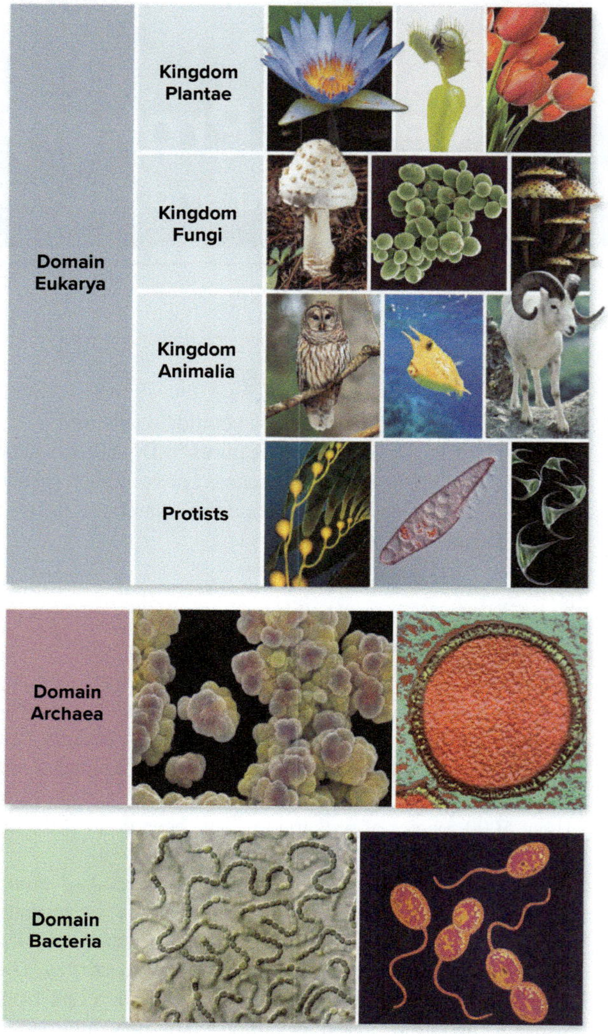

(a) Emma Grigor/EyeEm/Getty Images; (b) Don Farrall/Photodisc/Getty Images; (c) Alina Pavlova/Taiga/123RF; (d)Ro-ma Stock Photography/Corbis/Getty Images; (e) Steve Gschmeissner/Science Photo Library/Alamy Stock Photo; (f) Russell Illig/Getty Images; (g) Gary W. Carter/Corbis Super/Alamy Stock Photo; (h) Khoroshunova Olga/Shutterstock; (i) Alan and Sandy Carey/Getty Images; (j) Joebelanger/123RF; (k) Science History Images/Alamy Stock Photo; (l) MichaelTaylor/Shutterstock; (m) Power and Syred/Science Source; (n) Wolfgang Baumeister/Science Source; (o) Don Rubbelke/McGraw Hill; (p) Alfred Pasieka/Science Source

Figure 24.1 The diversity of life. Biologists categorize all living things into three overarching groups called domains: Bacteria, Archaea, and Eukarya. Domain Eukarya is composed of protists and three kingdoms: Plantae, Fungi, and Animalia.

Heterotrophic bacteria are **decomposers** because they feed on dead organic matter and release nutrients locked in dead tissue.

Question 1

a. Why is it important that decomposers such as bacteria release nutrients?

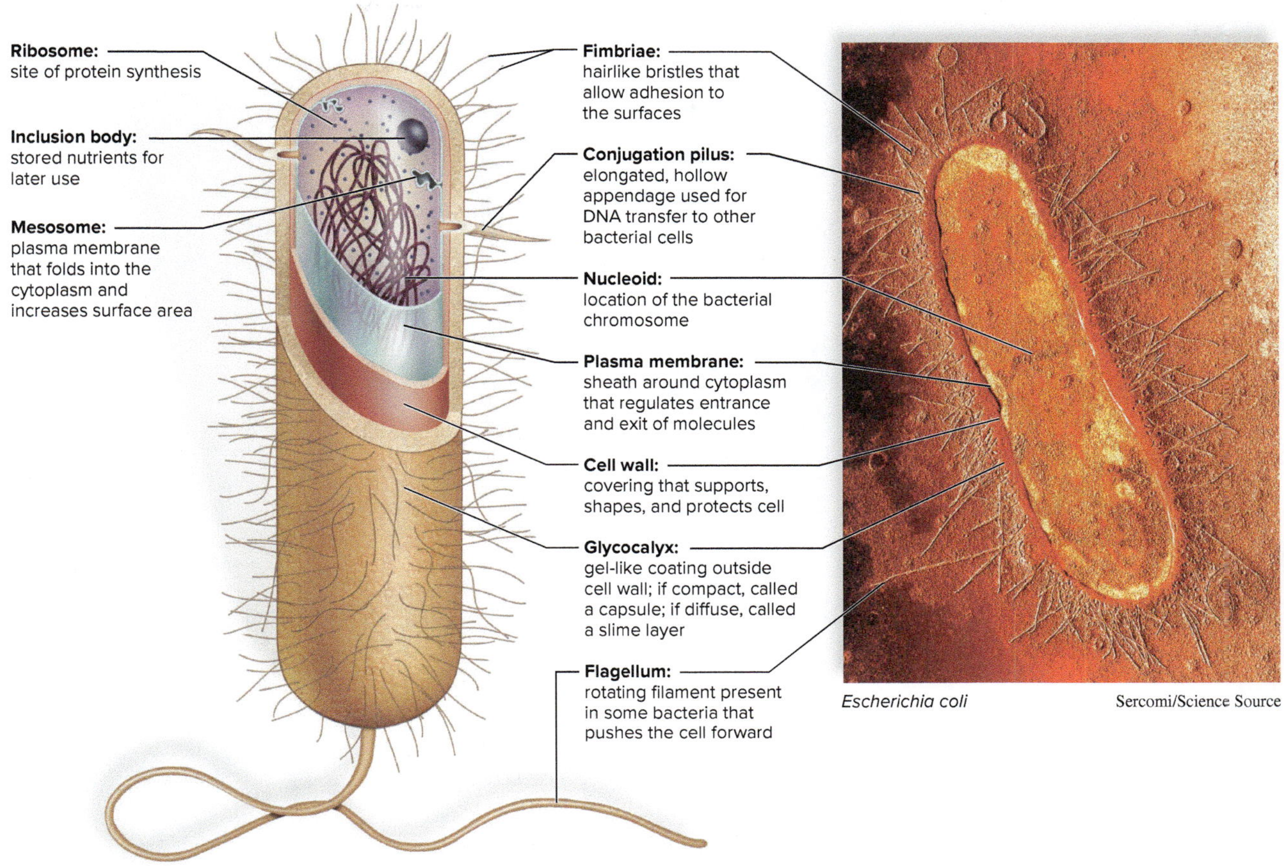

Escherichia coli Sercomi/Science Source

Figure 24.2 The structure of a bacterial cell.

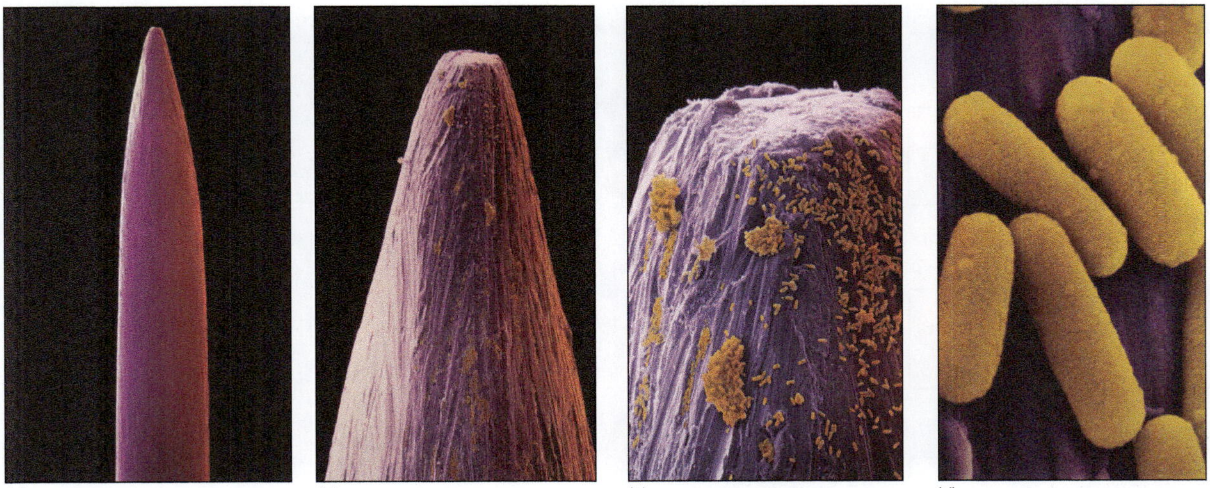

(a) Dr. Tony Brain/David Parker/Science Photo Library/Science Source (b) Dr. Tony Brain/David Parker/Science Photo Library/Science Source (c) Dr. Tony Brain/David Parker/Science Photo Library/Science Source (d) Dr. Tony Brain/David Parker/Science Photo Library/Science Source

Figure 24.3 Four views of a contaminated pin, which would seem an unlikely site for bacteria to grow. (*a*) The tip of the pin, magnified 7×. When scanning electron micrographs are shown at increasing magnifications—(*b*) 35×, (*c*) 178×, and (*d*) 4375×—you see rod-shaped bacteria growing there.

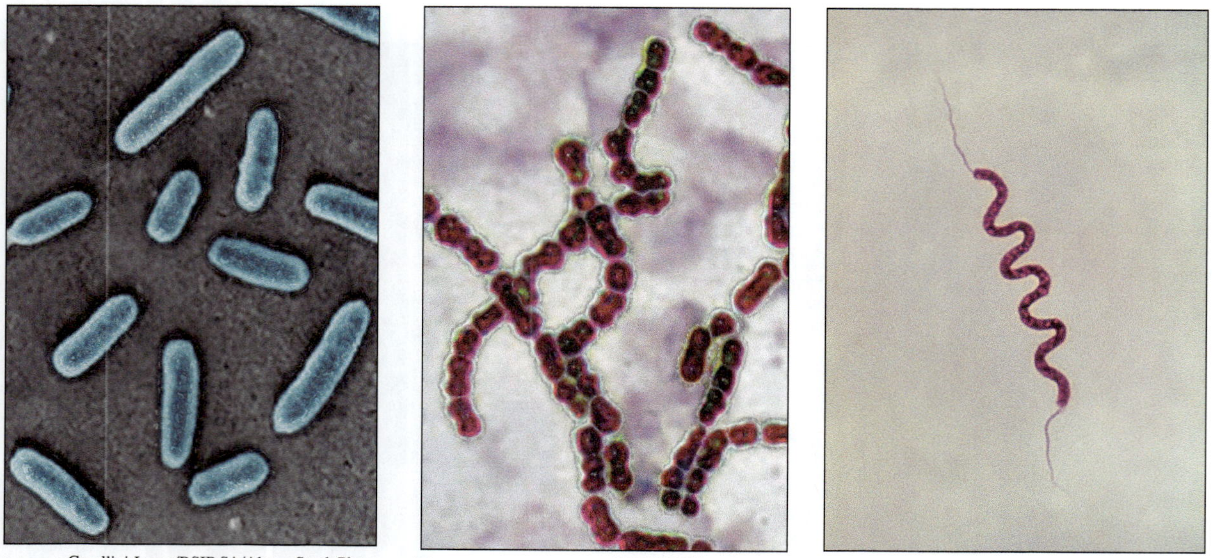

(a) Cavallini James/BSIP SA/Alamy Stock Photo (b) BSIP/UIG/Getty Images (c) Ed Reschke/Stone/Getty Images

Figure 24.4 The three basic shapes of bacteria: (*a*) bacillus (*Pseudomonas*); (*b*) coccus (*Streptococcus*); and (*c*) spirillum (*Spirilla*), 400×.

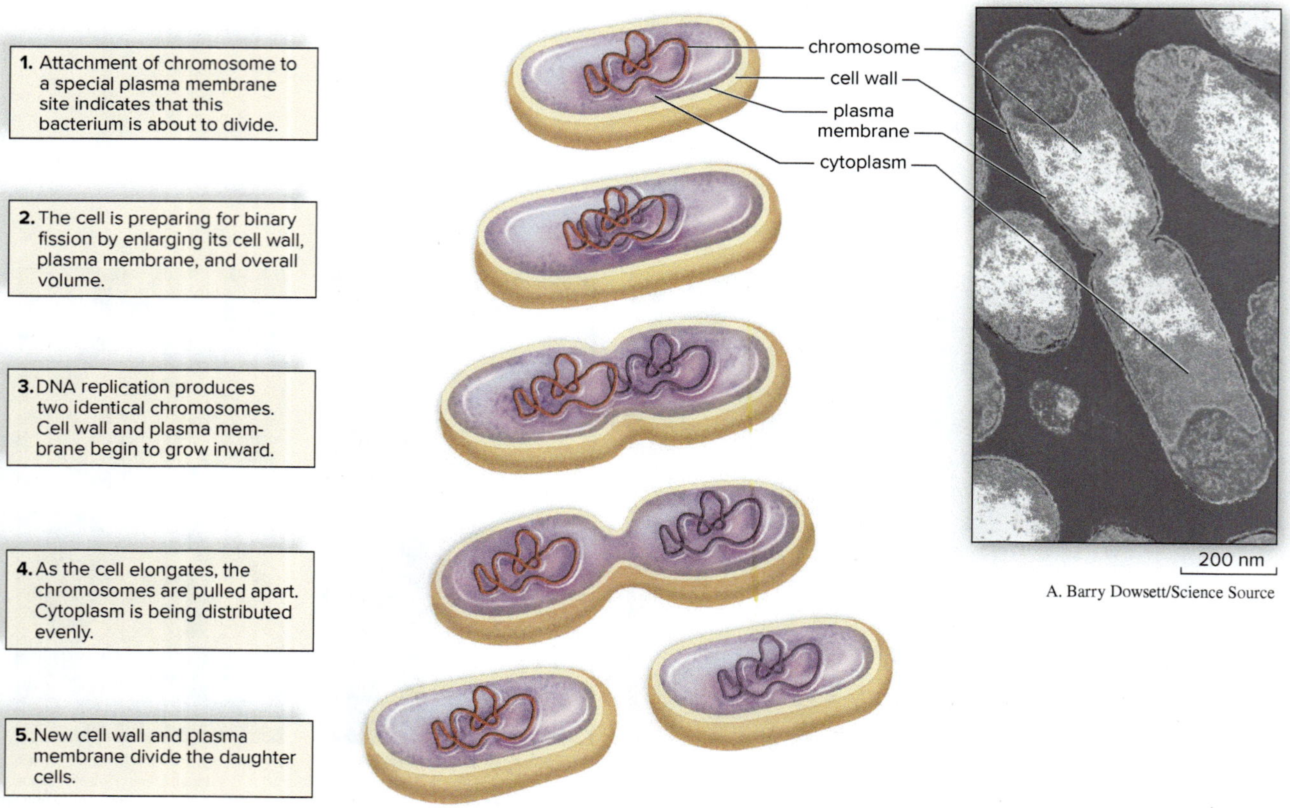

A. Barry Dowsett/Science Source

Figure 24.5 Binary Fission. First, DNA replicates, and as the cell lengthens, the two chromosomes separate, and the cells divide. The two resulting bacteria are identical. In step 3 a protein called FtsZ forms a ring at the site of cell separation. This protein is similar to that forming the mitotic spindle of eukaryotes.

Table 24.2

Important Bacterial Diseases That Affect Humans

Disease	Pathogen	Vector/Reservoir	Epidemiology
Anthrax	*Bacillus anthracis*	Animals, including processed skins	Bacterial infection that can be transmitted through contact or ingested. Rare except in sporadic outbreaks. May be fatal.
Botulism	*Clostridium botulinum*	Improperly prepared food	Contracted through ingestion or contact with wound. Produces acute toxic poison; can be fatal.
Chlamydia	*Chlamydia trachomatis*	Humans, Sexually Transmitted Infection	Urogenital infections with possible spread to eyes and respiratory tract. Occurs worldwide; increasingly common over past 30 years.
Cholera	*Vibrio cholerae*	Human feces, plankton	Causes severe diarrhea that can lead to death by dehydration; 50% peak mortality if the disease goes untreated. A major killer in times of crowding and poor sanitation; over 100,000 died in Rwanda in 1994 during a cholera outbreak.
Dental caries (tooth decay)	*Streptococcus*	Humans	A dense collection of this bacteria on the surface of teeth leads to secretion of acids that destroy minerals in tooth enamel—sugar alone will not cause cavities.
Gonorrhea	*Neisseria gonorrhoeae*	Humans only	Sexually Transmitted Infection on the increase worldwide. Usually not fatal.
Hansen's disease (leprosy)	*Mycobacterium leprae*	Humans, feral armadillos	Chronic infection of the skin; worldwide incidence about 10–12 million, especially in Southeast Asia. Spread through contact with infected individuals.
Lyme disease	*Borrelia bergdorferi*	Ticks, deer, small rodents	Spread through bite of infected tick. Lesion followed by malaise, fever, fatigue, pain, stiff neck, and headache.
Peptic ulcers	*Helicobacter pylori*	Humans	Infects the stomach, where it causes ulcers. About 40% of the world's population harbors *H. pylori.* Barry Marshall, who isolated the bacterium in 1982, drank a culture of *H. pylori*; he got an ulcer.
Plague	*Yersinia pestis*	Fleas of wild rodents: rats and squirrels	Killed ¼ of the population of Europe in the 14th century; endemic in wild rodent populations of the western United States today.
Pneumonia	*Streptococcus, Mycoplasma, Chlamydia*	Humans	Acute infection of the lungs, often fatal without treatment
Tuberculosis	*Mycobacterium tuberculosis*	Humans	An acute bacterial infection of the lungs, lymph, and meninges. Its incidence is on the rise, complicated by the development of new strains of the bacteria that are resistant to antibiotics.
Typhus	*Rickettsia typhi*	Lice, rat fleas, humans	Historically a major killer in times of crowding and poor sanitation; transmitted from human to human through the bite of infected lice and fleas. Typhus has a peak untreated mortality rate of 70%.

b. What term best describes heterotrophic bacteria that feed on living tissue?

Bacteria that derive their energy from photosynthesis or the oxidation of inorganic molecules are **autotrophic.** However, photosynthesis in bacteria is often different from that in eukaryotes because molecular sulfur rather than oxygen is sometimes produced as a by-product.

A laboratory culture of bacteria usually consists of a tube of liquid nutrients (broth) containing growing bacteria or a tube or plate of solidified agar with bacteria growing on the surface.[1] The jellylike agar is melted, mixed with nutrients, and poured into tubes or plates to solidify. Many species of bacteria can be cultured in nutrient broth or on a layer of nutrient-rich agar. It may surprise you to know that most species of bacteria are *not* culturable in vitro. We just don't know enough about the nutrient and environmental requirements of these bacteria to grow them in the lab.

Bacteria reproduce asexually via **binary fission,** in which a cell's DNA replicates and the cell pinches in half without the nuclear and chromosomal events associated with mitosis (see Exercise 14) (fig. 24.5). Some bacteria have genetic recombination via **conjugation,** in which all or part of the genetic material of one bacterium is transferred to another bacterium and a new set of genes is assembled.

*1 **Agar is a gelatinous polysaccharide used in culture media for microbiology labs. You'll learn more about agar and the red algae it comes from in Exercise 25.***

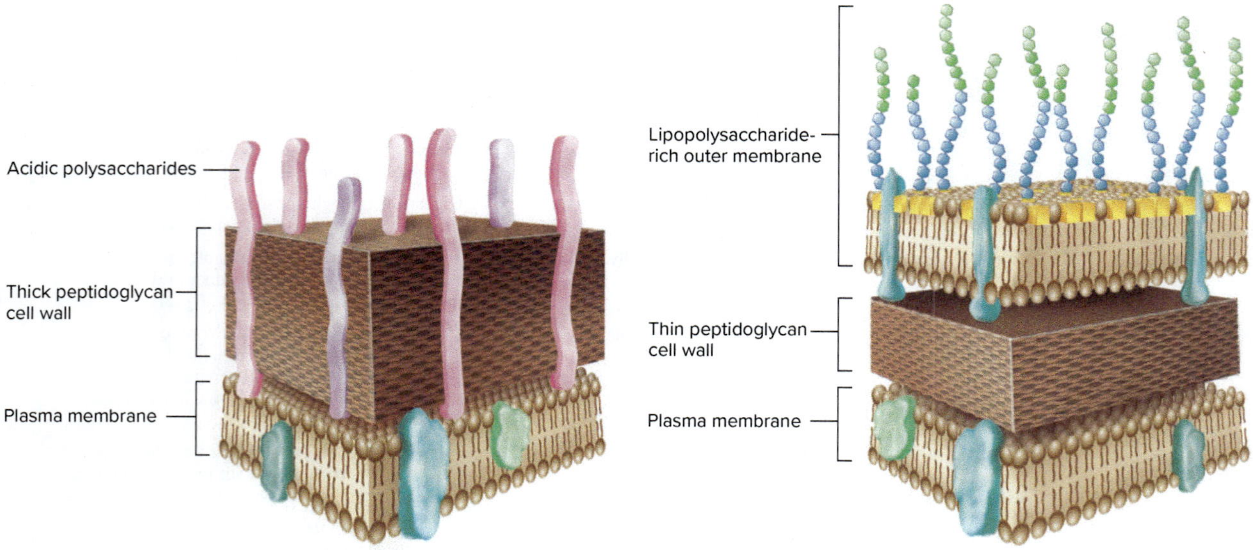

(a) Gram-positive: thick peptidoglycan cell wall layer, no outer membrane

(b) Gram-negative: thinner peptidoglycan cell wall layer, with outer membrane

Figure 24.6 Cell walls of Gram-positive and Gram-negative bacteria. (*a*) Cell walls of Gram-positive bacteria have a thick peptidoglycan layer and no outer membrane. Gram stains stick readily to this peptidoglycan layer. (*b*) Cell walls of Gram-negative bacteria have an outer membrane of lipopolysaccharides. Gram stains don't stick to this outer membrane.

Some bacteria are pathogenic (table 24.2); that is, they cause diseases such as pneumonia and tuberculosis. However, most bacteria are harmless to humans. Indeed, many beneficial bacteria live in and on your body. Nevertheless, you should handle all bacterial cultures with care. The preparation of wet mounts of bacterial cultures requires proper use of a transfer loop and sterilizing flame. Your instructor will demonstrate this aseptic technique.

SAFETY FIRST Before coming to lab, you were asked to read this exercise so you would know what to do and be aware of safety issues. In the space below, briefly list the safety issues associated with today's procedures. If you have questions about these issues, contact your laboratory assistant before starting work.

Procedure 24.1 **Culture common bacteria**

1. Obtain a sterile cotton swab and a closed petri dish containing sterile nutrient agar.
2. Open the packaged swab and drag the tip over a surface such as your teeth, face, or tabletop.
3. Open the petri dish and drag the exposed swab over the surface of the agar in the manner demonstrated by your instructor.
4. Close the lid and tape it shut. Label the dish with a wax pencil.
5. Turn the dish upside down and place it in the incubator or in a warm area.
6. After 24–48 h examine the agar for bacterial growth. Record your observations.

Question 2
What are the shape and size of each bacterial colony?

Gram Stain

One of the most important techniques to classify bacteria is the **Gram stain,** based on the different structural and chemical compositions of bacterial cell walls (fig. 24.6). Gram staining is important because it often correlates with the sensitivity of a bacterium to antibiotics. **Gram-positive** bacteria (e.g., *Streptococcus, Micrococcus*) have a thick cell wall that retains a purple dye, whereas **Gram-negative** (e.g., *Escherichia coli, Serratia*) bacteria have a much thinner cell wall that does not retain the dye.

During the Gram stain technique, crystal violet and iodine are applied to stain all of the bacteria purple. Then alcohol is used to remove the stain from the surface of the Gram-negative cell walls that do not bind the stain. Finally, safranin is used to counterstain the Gram-negative cells with a red color contrasting with purple Gram-positive cells (fig. 24.7). In the following procedure, you will Gram-stain some of your bacteria to see the difference between Gram-negative and Gram-positive organisms.

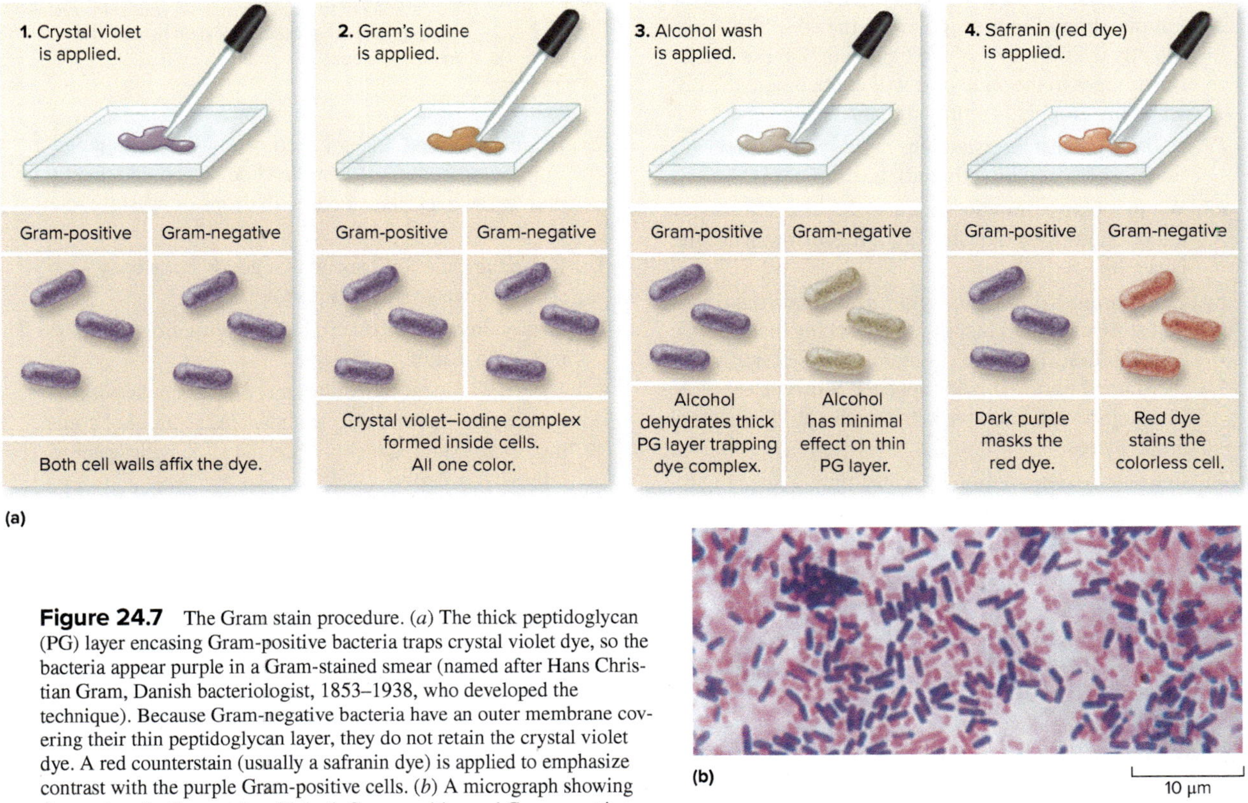

De Agostini Picture Library/Getty Images

Figure 24.7 The Gram stain procedure. (*a*) The thick peptidoglycan (PG) layer encasing Gram-positive bacteria traps crystal violet dye, so the bacteria appear purple in a Gram-stained smear (named after Hans Christian Gram, Danish bacteriologist, 1853–1938, who developed the technique). Because Gram-negative bacteria have an outer membrane covering their thin peptidoglycan layer, they do not retain the crystal violet dye. A red counterstain (usually a safranin dye) is applied to emphasize contrast with the purple Gram-positive cells. (*b*) A micrograph showing the results of a Gram stain with both Gram-positive and Gram-negative cells.

Procedure 24.2 Observe stained bacteria with oil-immersion magnification

1. Obtain a microscope and a small bottle of immersion oil. Recall from Exercise 3 that the resolving power of a lens depends, among other things, on the amount of light that it gathers. More light improves resolution, and light is scattered when it passes through air. If a drop of immersion oil, a fluid with the same refractive index (ability to bend light) as glass, is placed between the objective lens and the specimen, then the lens can gather more light.
2. Examine the microscope, and verify with your instructor that the microscope is equipped with an oil-immersion objective. This objective can resolve micrometer-sized particles such as bacteria.
3. Rotate the low-power objective into observation position.
4. Obtain some prepared slides of stained bacteria from your instructor. These may be commercially prepared slides or slides with bacteria that your instructor has stained to help you practice with your microscope.
5. Place a slide on the stage with the specimen centered over the light path through the hole in the stage.
6. While watching from the side, slowly rotate the low-power objective as close as possible to the slide without the objective touching the slide. Adjust the diaphragm for medium light-intensity.
7. Look through the oculars and slowly adjust the coarse adjustment to increase the working distance. Stop when you see the color of the stained bacteria and are roughly focused on the smear of bacteria.
8. Improve the illumination and sharpen the image as much as possible with the fine-adjustment knob. At this low magnification you will only see small dots, at best.
9. Rotate a higher-power objective into position and refocus.
10. Rotate the nosepiece so that the alignment is halfway between the oil-immersion objective and the next lowest-power objective. There should not be an objective in correct position for observation. This position will allow you to place a drop of oil on the slide.

11. Put one drop of immersion oil on the coverslip directly over the spot of the light path. Do not touch the dropper to the slide or it will contaminate the oil when the dropper is returned to the bottle.

12. Rotate the oil-immersion lens directly into observation position and directly into the drop of oil.

13. While looking from the side, use the fine-adjustment knob to lower the objective until it *gently* touches the coverslip.

14. Look through the oculars and *slowly* rotate the fine-adjustment knob to increase the working distance. This rotation should be counterclockwise. Stop when the stained bacterial color appears. Slowly rotate the fine-adjustment knob back and forth until the bacteria are in focus.

15. Improve your resolution by adjusting the diaphragm.

16. Examine the sizes, shapes, and stains of the bacteria on the slide.

17. Repeat this entire procedure for each of the slides offered by your instructor.

18. When you finish your work, clean the oil from the slides and objectives with the lens paper provided.

Procedure 24.3 **Use known bacterial cultures to prepare and observe a Gram stain**

1. Obtain a slide, coverslip, transfer loop, alcohol burner, and a culture of living bacteria.

2. Available cultures should include the following bacteria, among others:

Bacillus megaterium—a large bacterium resistant to radiation, desiccation, and heat.

Rhodospirillum rubrum—a photosynthetic purple bacterium.

Escherichia coli—found in the human intestine; the most intensively studied of all bacteria.

Staphylococcus epidermidis—a coccus found among the normal flora of skin.

3. Apply a loop of bacteria to a drop of water on a slide. If your cultures are in liquid broth, then add one drop of culture medium to your slide. Your instructor will demonstrate how to use sterile technique to open, sample, and close the culture of bacteria. Do not add a coverslip to the slide.

4. Heat the slide gently by holding it with a clothespin and passing it over the top of a flame three to four times. Drying time is critical to success. Check with your instructor to avoid heating the slide too little or too much. Be careful not to break the slide. If hot plates are available, hold the slide to the hot surface for 10 sec. This heat will adhere the bacteria to the slide.

Be careful not to inhale or spill on your skin crystal violet or any other biological stain used in the following steps.

5. Examine figure 24.7 and become familiar with the staining procedure. When the slide has cooled, gently drench the bacterial smear with drops of crystal violet for 20 sec.

6. Rinse the slide for 2 sec with a gentle but steady stream of water from a squirt bottle.

7. Gently drench the bacterial smear with drops of iodine for 1 min.

8. Drop 95% alcohol (decolorizer) on the smear with an eyedropper until no purple shows in the alcohol coming off the slide. Quickly rinse the slide with water to remove the alcohol.

9. Gently drench the bacterial smear with drops of safranin for 20 sec.

10. Gently rinse the slide with water. Air dry the slide, or blot gently if necessary. Add a coverslip.

11. Observe the smear with your microscope using low power and then high power and/or the oil-immersion objective. You will not need a coverslip for heat-fixed slides with oil immersion.

12. Determine if the bacteria are Gram positive or Gram negative.

13. Repeat the Gram stain procedure using other known bacteria that you obtain from culture tubes.

14. Record your observations in table 24.3.

Procedure 24.4 **Use a Gram stain to observe living bacteria from your teeth**

1. Obtain a slide, coverslip, transfer loop, and alcohol burner.

2. Use the wide end of a toothpick to scrape your teeth near the gum line.

3. Thoroughly mix what's on the tip of the toothpick in a small drop of water on a microscope slide.

4. Allow this bacterial smear to dry.

5. Repeat steps 4–13 of procedure 24.3.

Question 3

a. Which type of bacteria is most prevalent in the sample from your teeth?

Deadly Food! Beware!

The bacterium *Clostridium botulinum* can grow in food products and produces a toxin called botulinum, the most toxic substance known. Microbiologists estimate that 1 gram of this toxin can kill 14 million adults! The good news is that *C. botulinum* requires anaerobic conditions for growth, which limits its prevalence. The bad news is that *C. botulinum* is extremely tolerant to stress; it can withstand boiling water (100°C) for short periods, but is killed at 120°C in 5 min. This tolerance makes *C. botulinum* a serious concern when people can vegetables. If home canning is not done properly, this bacterium will grow in the anaerobic conditions of the sealed container and be extremely poisonous. Several adults and infants die every year from botulism in the United States.

Tolerance to stress is enhanced in *C. botulinum* and many other bacteria by the formation of thick-walled **endospores** that surround their chromosome and a small portion of the surrounding cytoplasm. These highly resistant endospores (fig. 24.8) may later germinate and grow after decades or even centuries of inactivity. The endospores of *Clostridium botulinum* can germinate in poorly prepared canned goods, so never eat food from a swollen (gas-filled) can of food; you risk contracting botulism leading to nerve paralysis, severe vomiting, and death.

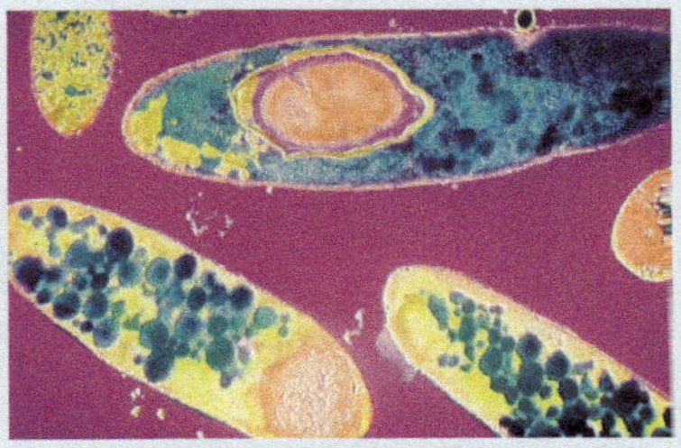

PTP/Phototake/Medical Images

Figure 24.8 Endospores. The round orange circle in the upper cell is an endospore forming within a cell of *Clostridium botulinum*, the bacterium that causes the disease botulism. These resistant endospores enable the bacterium to survive in improperly sterilized canned and bottled foods.

Table 24.3

The Relative Size and Shape of Some Common Bacteria

Bacterial Species	Gram Stain (+/−)	Relative Size	Shape

b. Is *Bacillus megaterium* Gram positive or Gram negative? How do you know?

Bacterial Colony Morphology

A bacterial **colony** is a visible speck or patch of millions of bacterial cells that are typically the progeny of a single cell that reproduced on the agar's surface. A bacterial colony growing on the surface of nutrient agar often has distinctive characteristics depending on the species. Careful observation of the shape, color, size, texture, and margins of a colony is important for any bacterial identification. Figure 24.9 illustrates the most common features of bacterial colony morphology.

Procedure 24.5 **Evaluate the colony morphology of bacterial species**

1. Obtain 48-h growth plates of 4–6 bacterial species provided by your instructor for your evaluation.

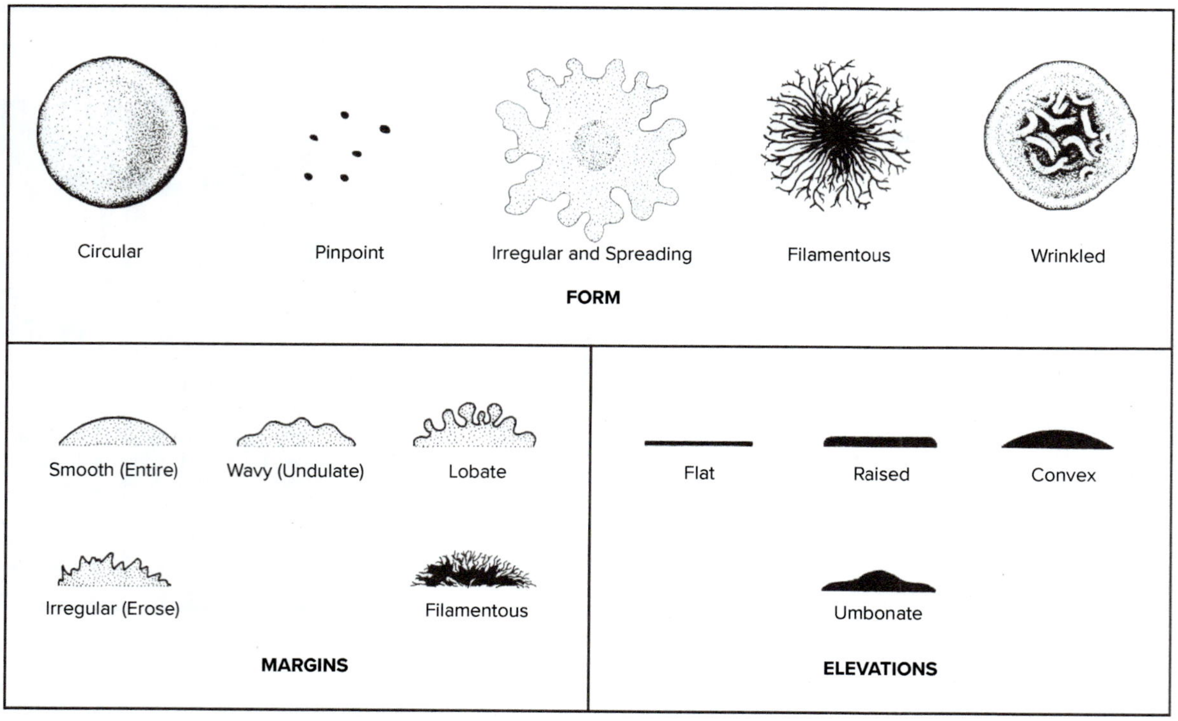

Figure 24.9 Colony characteristics. Colonies are classified by form, margin, and elevation as well as color and diameter.

2. Familiarize yourself with the colony morphologies shown in figure 24.9.
3. Use a stereomicroscope or hand lens to evaluate a representative colony on each of the plates. Record your observations in table 24.4.
4. When you have completed your analysis, your instructor will provide the species names associated with each of the plates and the accepted colony descriptions for that species. Use this information to evaluate the accuracy of your observations.

Table 24.4

An Evaluation of Bacterial Colony Morphology

Plate ID	Species	Colony Diameter (mm)	Color	Form	Elevation	Margin
A						
B						
C						
D						
E						
F						

Nitrogen Fixation by Bacteria

Certain bacteria and cyanobacteria transform atmospheric nitrogen (N_2) into other nitrogenous compounds that can be used as nutrients by plants. This process is called **nitrogen fixation.** All organisms need nitrogen as a component of their nucleic acids, proteins, and amino acids. However, chemical reactions capable of breaking the strong triple bond between atoms of atmospheric nitrogen are limited to certain bacteria and cyanobacteria. This process uses an enzyme called nitrogenase along with ATP, energized electrons, and water to convert N_2 to ammonia (NH_3). Ammonia can be absorbed by plants and used to make proteins and other macromolecules.

Rhizobium is a bacterium that can fix nitrogen and can grow intimately with roots of some plants called legumes (e.g., clover, alfalfa, and soybeans). Such associations between *Rhizobium* and host roots form **nodules** on the roots (fig. 24.10). These resident nitrogen fixers provide ammonia to the plant while the plant provides sugars and other nutrients to the bacteria.

Procedure 24.6 **Observe root nodules**

1. Observe the root systems on display and note the nodules.
2. Examine a prepared slide of a cross section of a nodule.

Nigel Cattlin/Alamy Stock Photo

Figure 24.10 Root nodules containing nitrogen-fixing bacteria.

Question 4

a. Where are the bacteria? Are they between cells or inside cells?

b. Why is this relationship between a plant and bacterium called a mutualism?

c. How does *Rhizobium* benefit from this association?

d. How does the host plant benefit from the association?

Bacterial Sensitivity to Inhibitors

Growth of some bacterial species is more sensitive to inhibitors such as antibiotics than is growth of other species. For example, an antibiotic may be more effective against *Staphylococcus* than against *Streptococcus.* This is important information to a physician who must select one of many available antibiotics to treat a bacterial infection.

To determine the effectiveness of an antibiotic, a medical laboratory may set up a **sensitivity plate.** A sensitivity plate is a petri dish of solid medium that has been uniformly inoculated on its entire surface with a known bacterium or an unknown sample from an infected patient. After inoculation, four to eight small paper disks—each soaked in a different antibiotic—are placed equidistant from each other on the culture surface. After 24 h, an effective antibiotic will produce a visible halo of clear surface around the disks where it inhibited growth of the bacteria—the larger the halo, the more effective the antibiotic (fig. 24.11). If the antibiotic was totally ineffective, the bacteria will grow to the edge of the paper disk.

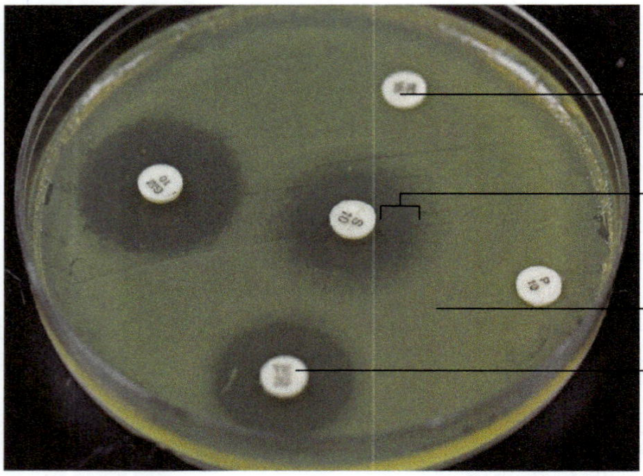

Lisa Burgess/McGraw Hill

Figure 24.11 A sensitivity plate is used to determine the effectiveness of one or more antibiotics, each of which is on the surface of a paper disk. Any disk containing an effective antibacterial agent inhibits nearby bacterial growth on the agar. If the agent is ineffective, then bacteria will grow up to the disk.

You've probably seen television commercials for products such as mouthwash or disinfectant that "kills germs on contact." Many of these products are developed and tested using sensitivity plates. The mouthwash is effective if no bacteria grow around a paper disk soaked in the mouthwash.

Procedure 24.7 Examine sensitivity plates

1. Obtain from your instructor one of each type of prepared sensitivity plate.
2. Examine each plate, note the bacteria used to inoculate the plate, and note the types of disks distributed on the plate.
3. Determine which disks inhibited bacterial growth strongly, weakly, or not at all.
4. Record the bacterial species and your observations in table 24.5.

Table 24.5

Inhibition of Four Bacterial Species by Various Growth Inhibitors

Antibiotic/Antiseptic	Plate 1 ________	Plate 2 ________	Plate 3 ________	Plate 4 ________

Question 5
Based on their appearance, which drugs or chemicals inhibit the growth of bacteria?

Cyanobacteria (Blue-Green Algae)

Cyanobacteria are a major group of photosynthetic bacteria that grow in many environments. Most cyanobacteria are free-living, whereas others live symbiotically with plants and other organisms. Cyanobacteria are photosynthetic, and their pigments include **chlorophyll *a*** and the accessory pigments **phycocyanin** (blue) and **phycoerythrin** (red). Because of various proportions of these pigments, only about half of the cyanobacteria are blue-green in color; other cyanobacteria range in color from brown to olive green.

Cyanobacteria reproduce by fission and are often surrounded by a jellylike **sheath.** Because cyanobacteria are prokaryotes, they are not related to other algae, all eukaryotic. Cyanobacteria such as *Oscillatoria* often cause many of the disagreeable tastes, colors, and odors in water.

Procedure 24.8 Examine cyanobacteria

1. With your microscope, examine living material and prepared slides of *Oscillatoria* (fig. 24.12*a*). *Oscillatoria* grows as long chains of cells called **trichomes.**

Question 6
Do all cells of a trichome of *Oscillatoria* appear similar?

2. Examine living material and a prepared slide of *Anabaena* or *Nostoc* (commonly called witch's butter or starjelly) (fig. 24.12*b*). Filaments of these cyanobacteria consist of small grapelike colonies. Trichomes of *Nostoc* consist of small vegetative cells and larger, thick-walled **heterocysts,** in which nitrogen fixation occurs.
3. Examine a wet mount of living *Gloeocapsa,* characterized by a thick, gelatinous sheath (fig. 24.12*c*).
4. Add a drop of dilute India ink to the slide of *Gloeocapsa* so that the sheath will stand out against the dark background. Often *Gloeocapsa* forms clusters of cells and therefore has a colonial body form. Locate one of these colonies.
5. Add a drop of methylene blue to a fresh slide of *Gloeocapsa* and determine if it enhances your observation more than does India ink.

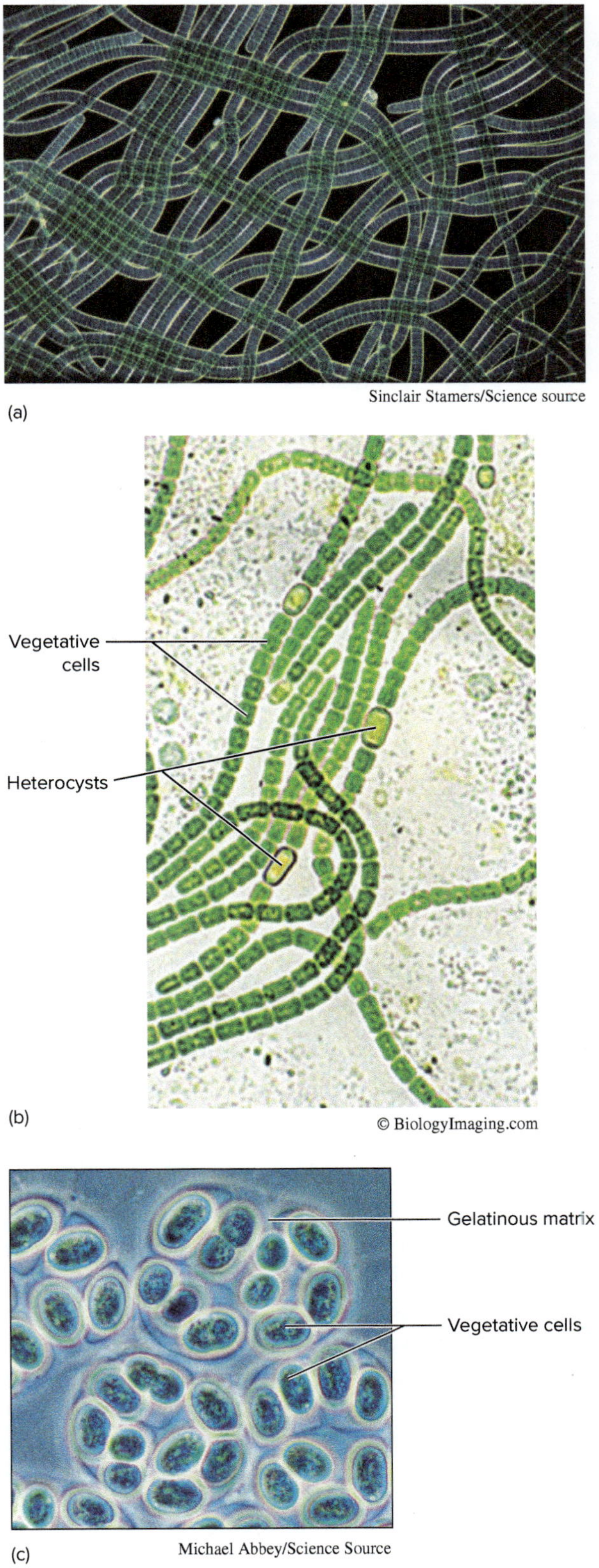

Figure 24.12 Cyanobacteria. (*a*) *Oscillatoria* (200×); (*b*) *Nostoc* (400×); and (*c*) *Gloeocapsa* (400×).

Question 7

a. Do adjacent cells of *Gleocapsa* share a common sheath?

b. What do you suppose is the function of the sheath?

c. Do clusters of *Gloeocapsa* represent multicellular organisms? Why or why not?

d. What is the best stain for *Gloeocapsa,* India ink or methylene blue?

6. Examine some living *Merismopedia,* which also form colonies. Sketch these cyanobacteria in the following space.

Question 8

a. How is the shape of *Merismopedia* different from other cyanobacteria you studied in this exercise?

b. How would a colony attain this shape?

INQUIRY-BASED LEARNING

Do we reach every microenvironment with our household disinfectants?

Observations: Bacteria grow virtually everywhere. Some common household products effectively inhibit bacterial growth on floors and tabletops but may leave behind resistant species. Some microenvironments, however, are not exposed to the selection pressure of disinfectants.

Question: Do some places in our surroundings harbor bacteria that resist household disinfectants especially well?

a. Establish a working lab group and obtain Inquiry-Based Learning Worksheet 24 from your instructor.
b. Discuss with your group a well-defined question relevant to the preceding observation and question. Record your question on Worksheet 24.
c. Translate your question into a testable hypothesis and record it.
d. Review the discussion of sensitivity plates and procedure 24.7. Outline on Worksheet 24 your experimental design and supplies needed to test your hypothesis. Ask your instructor to review your proposed investigation.
e. Conduct your procedures, record your data, answer your question, and make relevant comments.
f. Discuss with your instructor any revisions to your questions, hypotheses, or procedures. Repeat your work as needed.

Questions for Further Study and Inquiry

1. Why do Gram-positive and Gram-negative bacteria stain differently?

2. What happens when milk is pasteurized?

3. What causes milk to sour?

4. What ecological roles are performed by cyanobacteria?

5. How do antibiotics kill bacteria? Why do they not affect viruses?

6. How could bacteria become resistant to an antibiotic?

7. Antibiotic resistance is promoted by overprescription of antibiotics. How do we stop this trend? What problems are involved? Is antibiotic resistance inevitable?

DOING BIOLOGY YOURSELF

Obtain two nutrient agar plates. Touch and drag the tip of your finger on the agar surface of one plate. Wash your hand and repeat the procedure on the other plate. Incubate the plates for 24–48 h. Then compare colony appearances and number. What can you conclude about the presence and diversity of bacteria on your plates? Is the appearance of a colony a good way to distinguish species of bacteria? Why or why not?

DOING BIOLOGY YOURSELF

Doorknobs, sinks, tables, and so on are often laden with bacteria. Devise a protocol to swab areas of your workplace to determine which microenvironments harbor the most bacteria.

WRITING TO LEARN BIOLOGY

There is a great diversity of roles in a typical ecosystem, some shared by a variety of organisms. What ecological roles are performed by cyanobacteria?

EXERCISE 25

Survey of Protists

Algal Autotrophs

Learning Objectives

By the end of this exercise you should be able to:

1. Discuss the distinguishing features of different groups of algae.
2. Appreciate the economic importance of algae.
3. Outline the events of "alternation of generations" in green algae.

Please visit connect.mheducation.com to review online resources tailored to this lab.

This exercise begins your study of **domain Eukarya. Eukaryotes** are organisms composed of cells having membrane-bound nuclei. This domain is commonly divided into three well-defined kingdoms—Fungi, Animalia, Plantae—along with a group of organisms with diverse origins called protists. In one sense, **protists** include all eukaryotes that lack the distinguishing characteristics of fungi, animals, or plants. Protists live in moist habitats and include simple eukaryotes such as amoebas, as well as multicellular organisms such as the brown alga, kelp.

Fungi have cell walls and are heterotrophic. **Heterotrophic** organisms feed on organic matter produced by other organisms because they cannot make their required organic compounds from inorganic substances; that is, heterotrophs require organic nutrients produced by other organisms.

Animals are ingestive-feeding heterotrophs that lack cell walls and can respond rapidly to external stimuli. Animals are multicellular.

Plants are multicellular, **autotrophic** organisms, meaning that they can synthesize all required organic compounds from inorganic substances using external energy, usually sunlight.

Protists share two common characteristics: Most protists are microscopic, and most are successful in moist habitats. Their high surface-to-volume ratio causes them to dry out quickly. Despite shared features, modern phylogenetic analyses of DNA sequences and cellular features reveal that protists are not a well-defined, monophyletic (single origin) group. Protists probably share common ancestry with fungi, plants, and animals. The diversity of protists clearly makes understanding their phylogeny and taxonomy a challenge.

Currently, we classify protists into several eukaryotic **supergroups** that each displays distinctive features that represent a separate evolutionary lineage. Notice that one supergroup, Archaeplastida, includes a lineage giving rise to land plants (table 25.1). Review information in your textbook to thoroughly examine the phylogenetic relationships among protists.

To recognize two major ecological roles of protists, this lab exercise examines autotrophic species commonly referred to as **algae**, Latin for "seaweeds." Exercise 26 examines heterotrophic species referred to as **protozoa** (fig. 25.1). Notice that some supergroups of protistans include both autotrophic species and heterotrophic species. Despite having photosynthesis in common, algae do not form a monophyletic group descended from a single common ancestor. They are polyphyletic (i.e., they have more than one common ancestor), and the details of their phylogenies are still being discovered. Diversity of protists is immense (fig. 25.2).

INTRODUCTION TO ALGAE

One way to determine the importance of something is to remove it and see what happens. If we did that with algae, most people would be shocked by the result. Global oxygen production would immediately decline. A major food source, perhaps *the* major food source for the world's ecosystems,

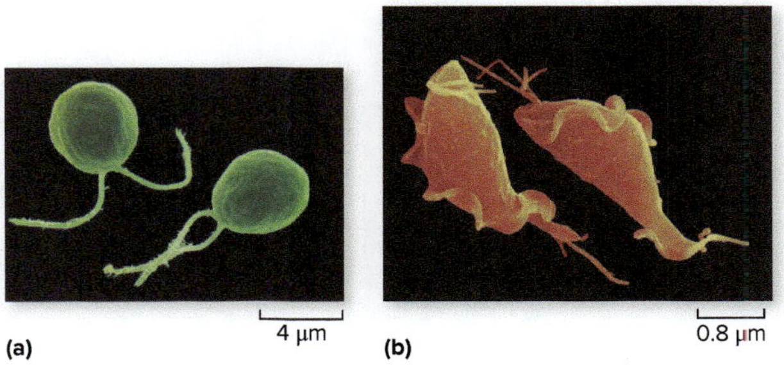

(a) (b)

(a) Aaron J. Bell/Science Source; (b) David M. Phillips/Science Source

Figure 25.1 Representative algal and protozoan protists. (*a*) *Chlamydomonas* sp., a common unicellular alga. (*b*) *Trichomonas vaginalis,* a protozoan causing trichomoniasis, a common sexually transmitted disease in humans.

DOMAIN: Eukarya
Protists

CHARACTERISTICS
Usually a complex single cell; photosynthesize, ingest, or absorb food; haploid life cycle

Supergroup	Members	Distinguishing Features
Archaeplastida	Green algae, red algae, charophytes	Plastids; unicellular, colonial, and multicellular
Chromalveolata	Stramenopiles: brown algae, diatoms, golden brown algae, water molds Alveolates: ciliates, apicomplexans, dinoflagellates	Most with plastids; unicellular and multicellular Alveoli support plasma membrane; unicellular
Excavata	Euglenoids, kinetoplastids, parabasalids, diplomonads	Feeding groove; unique flagella; unicellular
Amoebozoa	Amoeboids, plasmodial and cellular slime molds	Pseudopods; unicellular
Rhizaria	Foraminiferans, radiolarians	Thin pseudopods; some with tests; unicellular
Opisthokonta	Choanoflagellates, animals, nucleariids, fungi	Some with flagella; unicellular and colonial

Assorted fossilized diatoms

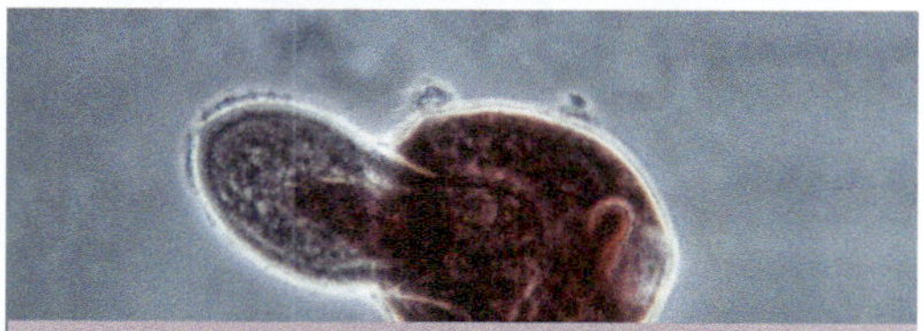
Onychodromus, a giant ciliate ingesting one of its own kind

Plasmodium, a slime mold

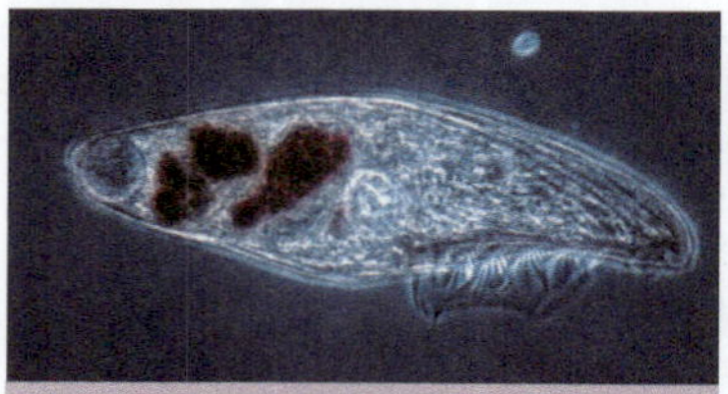
Blepharisma, a ciliate with visible vacuoles

Licmorpha, a stalked diatom

Acetabularia, a single-celled green alga (chlorophyte)

Nonionina, a foraminiferan

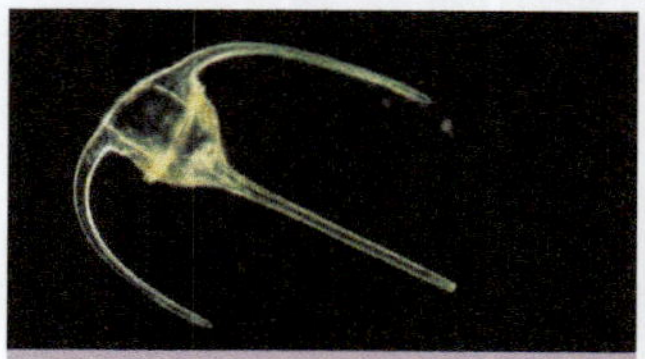
Ceratium, an armored dinoflagellate

Bossiella, a coralline red alga

Synura, a colony-forming golden brown alga

Amoeba proteus, a protozoan

(a) Science History Images/Alamy Stock Photo; (b) Eric Grave/Science History Images/Alamy Stock Photo; (c) MichaelTaylor/Shutterstock; (d) Franz Neidl/Alamy Stock Photo; (e) Jeff Rotman/Alamy Stock Photo; (f) Borut Furlan/Getty Images; (g) Stephen Durr; (h) Roland Birke/Getty Images; (i) Scenics&Science/Alamy Stock Photo; (j) Eric Grave/Science History Images/Alamy Stock Photo; (k) Patrick W. Grace/Science Source

Figure 25.2 Diversity of algal protists.

Table 25.1

Eukaryotic Supergroups and Examples of Included Protists

Supergroup	Groups	Unifying Characteristics
EXCAVATA	euglenoids flagellates	Unicellular flagellates, often with feeding groove; mitochondria highly modified in specialized parasites; secondary plastids (when present) derived from endosymbiotic green algae
ARCHAEPLASTIDA	red algae green algae	Primary plastids having only two envelope membranes
CHROMALVEOLATA	dinoflagellates ciliates apicomplexans diatoms brown algae	Peripheral membrane sacs (alveoli); some dinoflagellates have secondary plastids derived from red algae, some have secondary plastids derived from green apicomplexan algae, and some have tertiary plastids derived from diatoms; some apicomplexans have secondary plastids derived from red or green algae. Straw-like flagellar hairs; fucoxanthin accessory pigment common in autotrophic forms
RHIZARIA	radiolarians foraminiferans	Thin, cytoplasmic projections; secondary plastids (when present) derived from endosymbiotic green algae
AMOEBOZOA	slime molds amoebas	Amoeboid movement by pseudopodia
OPISTHOKONTA	choanoflagellates	Swimming cells possess a single posterior flagellum.

Table 25.2

The Common Pigments in Addition to Chlorophyll *a*, Storage Products, and Cell Wall Components Characteristic of Common Algae

Groups	Predominant Organization	Pigment	Storage Product	Cell Wall
Green algae	Unicellular, filamentous, colonial	Chlorophyll *a,b*	Starch	Mainly cellulose
Brown algae	Filamentous, multicellular	Chlorophyll *a,c,* fucoxanthin	Laminarin, mannitol, lipids	Cellulose, alginates
Red algae	Multicellular	Chlorophyll *a,d,* phycobilins	Modified starch	Cellulose, agar, carrageenan
Diatoms	Unicellular	Chlorophyll *a,c,* fucoxanthin	Chrysolaminarin, lipids	Silica
Dinoflagellates	Unicellular	Chlorophyll *a,c*	Starch, lipids	Pectin, cellulose plates
Euglenoids	Unicellular	Chlorophyll *a,b*	Paramylon, lipids	Proteinaceous pellicle

would be gone. Tens of thousands of irreplaceable algal species would be lost, along with their unique diversity of potentially useful chemicals, many with pharmaceutical value. The absence of algae would lead to rapid extinction of many hundreds of thousands of invertebrate animal species. Ecosystems would collapse. Fortunately we won't lose algae anytime soon; they have thrived for 1.5 billion years.

Algae are photosynthetic, eukaryotic organisms typically lacking multicellular sex organs. The major groups of algae are distinguished in part by their energy storage products, cell walls, and color, resulting from the type and abundance of colored **pigments** (substances that absorb light) in their plastids (table 25.2). Biologists often group algae by the colors of these pigments—for example, green algae, brown algae, and red algae.

Algae are also distinguished by their cellular organization (fig. 25.3). **Unicellular** algal species occur as single, unattached cells that may or may not be motile. **Filamentous** algal species occur as chains of cells attached end to end. These filaments may be few to many cells long and may be unbranched or branched in various patterns. **Colonial** algae occur as groups of cells attached to each other in a nonfilamentous manner. For example, a colony may include several to many cells adhering to each other as a sphere, flat sheet, or other three-dimensional shape. Colonies are clonal—that is, their cells are identical. Multicellular organization is not typical of protists but describes algae of more complex design than simple colonies. Multicellular species have cells of different types and functions and show significant interdependence.

GREEN ALGAE (SUPERGROUP ARCHAEPLASTIDA)

Green algae, collectively referred to as chlorophytes, are the most diverse and familiar algae in fresh water (fig. 25.4). However, a few genera live in salt water. Although

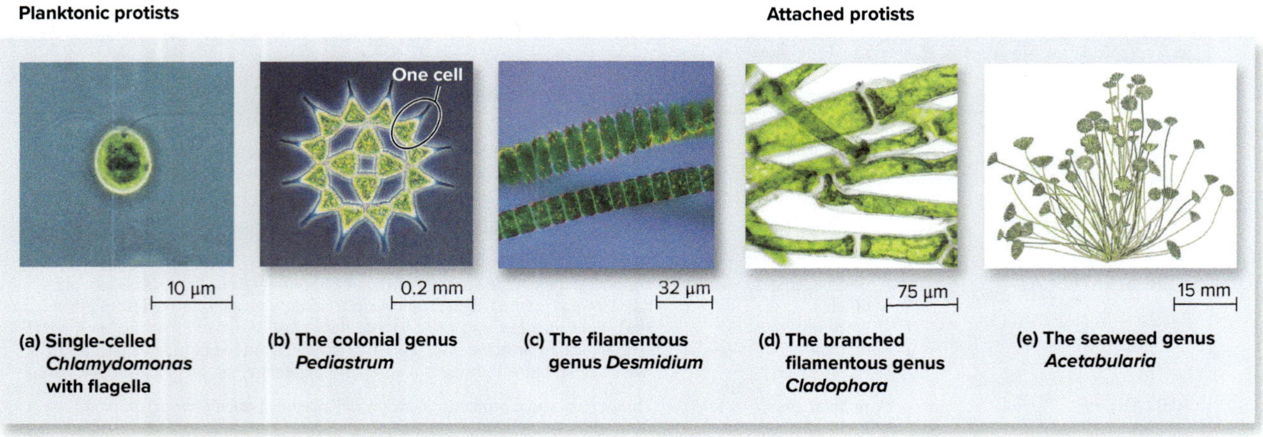

(a) DE AGOSTINI PICTURE LIBRARY; (b) Roland Birke/Phototake/Medical Images; (c) Marek Mis/Science Source; (d) Buccaneer/123RF; (e) Free_styler/Shutterstock

Figure 25.3 The diversity of algal body-types reflects their habitats. (*a*) The single-celled flagellate genus *Chlamydomonas* occurs in the phytoplankton of lakes. (*b*) The colonial genus *Pediastrum* consists of several cells arranged in a lacy star shape that helps to keep this alga afloat in water. (*c*) The filamentous genus *Desmidium* occurs as a twisted row of cells. (*d*) The branched filamentous genus *Cladophora* that grows attached to nearshore surfaces is large enough to see with the unaided eye. (*e*) The relatively large seaweed genus *Acetabularia* lives on rocks and coral rubble in shallow tropical oceans.

the common name chlorophyte means "green chlorophyll plants" (*chloro* + *phyta*), algae are not classified as plants. Green algae may be ancestral to land plants and share many characteristics with land plants, such as

- Chlorophyll *a,* which occurs in algae and green plants
- Chlorophyll *b,* which occurs in land plants and in the green and euglenoid algae
- Starch as the carbohydrate storage material
- Cell walls made of cellulose

Let's examine a few representatives of green algae.

Unicellular Green Alga: *Chlamydomonas*

Chlamydomonas is a motile, unicellular alga found in soil, lakes, and ditches (fig. 25.4). It probably has the simplest structure and type of reproduction among green algae. The egg-shaped cells of *Chlamydomonas* contain a large chloroplast and a pyrenoid involved in the production and storage of starch.

M. I. Walker/Science source

Figure 25.4 *Chlamydomonas* is a common alga that is rich in chlorophylls *a* and *b*. It is a single-celled green alga less than 100 µm long.

Procedure 25.1 Observe *Chlamydomonas*

1. Observe a drop of water containing living *Chlamydomonas* and note the movement of the cells. If necessary, review in Exercise 3 the proper use of a compound microscope and the associated videos at the lab manual's website.
2. If the movement is too fast, make a new preparation by placing one or two drops of methylcellulose on a slide and adding a drop of water containing *Chlamydomonas.*
3. Mix gently and add a coverslip.
4. Note the **stigma,** which appears as a reddish, light-absorbing spot at the anterior end of the cell.
5. Examine prepared slides of *Chlamydomonas.*

Question 1

a. Is the movement of *Chlamydomonas* smooth or does it appear jerky?

b. Can you see both flagella? (You may need to reduce the light intensity to see flagella.)

c. How does methylcellulose affect the movement of *Chlamydomonas*?

d. How does the stigma help *Chlamydomonas* survive?

Asexual and Sexual Reproduction in *Chlamydomonas*

Chlamydomonas usually reproduces asexually via mitosis. Each mitosis produces two haploid daughter cells. As many as 16 daughter cells can form inside the cell wall of the original parent cell before being released.

Sexual reproduction in *Chlamydomonas* is a response to unfavorable environmental conditions. For sexual reproduction, vegetative cells of *Chlamydomonas* undergo mitosis to produce gametes. The gametes fuse to form a diploid **zygote,** the resting stage of the life cycle. In most species of *Chlamydomonas,* the gametes of two strains are **isogamous,** meaning they have an identical shape and appearance. For convenience, isogametes of *Chlamydomonas* are referred to as + or −. Gametes unite to form a diploid zygote. **Syngamy** is the pairing and fusion of morphologically similar haploid gametes to form a diploid zygote. Syngamy is similar to fertilization, but fertilization typically refers to fusion of unequal gametes such as egg and sperm cells. The zygote surrounds itself with a resistant surface and is called a **zygospore.** Under favorable conditions the zygote undergoes meiosis to produce haploid individuals called **spores.** Spores are reproductive cells capable of developing into an adult without fusing with another cell. Study the life cycle of *Chlamydomonas* shown in figure 25.5.

Procedure 25.2 Observe syngamy

1. Place drops of + and − gametes of *Chlamydomonas* provided by your instructor next to each other on a microscope slide, being careful not to mix the two drops. Do not add a coverslip.
2. While you observe the drops through low power of the microscope, mix the two drops of isogametes.
3. Switch to high magnification and note the clumping gametes. Try to locate cells that have paired.

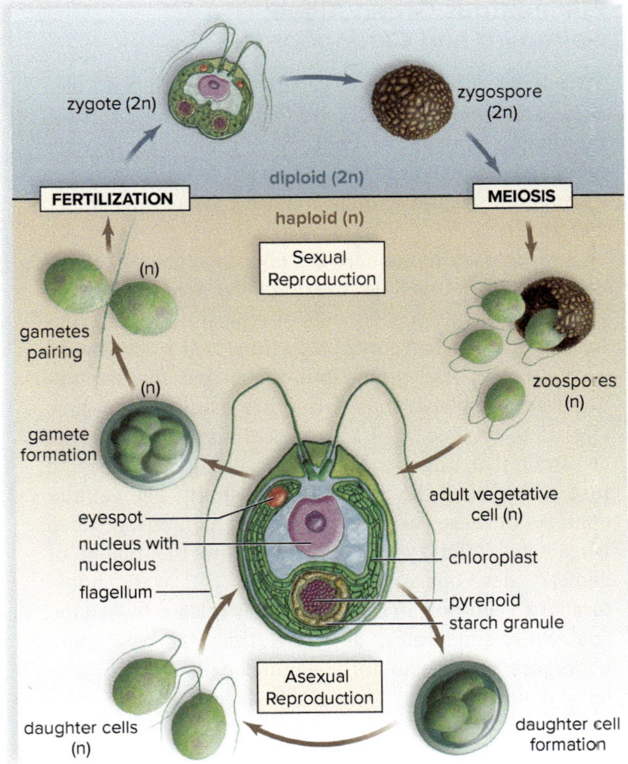

Figure 25.5 *Chlamydomonas* life cycle. Individual cells of *Chlamydomonas* (a microscopic, biflagellated alga) are haploid and divide asexually, producing copies of themselves. At times, such haploid cells act as gametes, fusing to produce a zygote. The zygote develops a thick, resistant wall, becoming a zygospore (as shown in the lower right-hand side of the diagram). Meiosis of the zygospore then produces four haploid individuals. Only + and − strains can mate with one another, although both may also divide asexually and reproduce themselves.

Question 2

a. Under what environmental conditions would a zygote not undergo meiosis immediately?

b. Are spores of *Chlamydomonas* haploid or diploid?

c. Which portions of the life cycle of *Chlamydomonas* are haploid?

d. Which are diploid?

Filamentous Green Algae: *Spirogyra* and *Cladophora*

Two of the most common genera of filamentous green algae are *Spirogyra* and *Cladophora. Spirogyra* grows in running streams of cool fresh water and secretes mucilage that makes it feel slippery. *Cladophora* is also common in streams and has a much coarser appearance and texture.

Spirogyra reproduces sexually by a process called **conjugation** as well as asexually by **fragmentation.** During conjugation, filaments of opposite mating types lie side by side and form projections that grow toward each other. These projections touch and the separating wall dissolves, thus forming a **conjugation** tube (fig. 25.6, also see fig. 4.15). The cellular contents of the (−) strain then migrate through the conjugation tube and fuse with that of the nonmotile (−) strain. The cellular contents of these two strains function as nonflagellated isogametes. The zygote resulting from the fusion of gametes develops a thick, resistant cell wall and is termed a zygospore. The zygospore is released when the filament disintegrates, at which time the zygospore undergoes meiosis to form haploid cells that become new filaments.

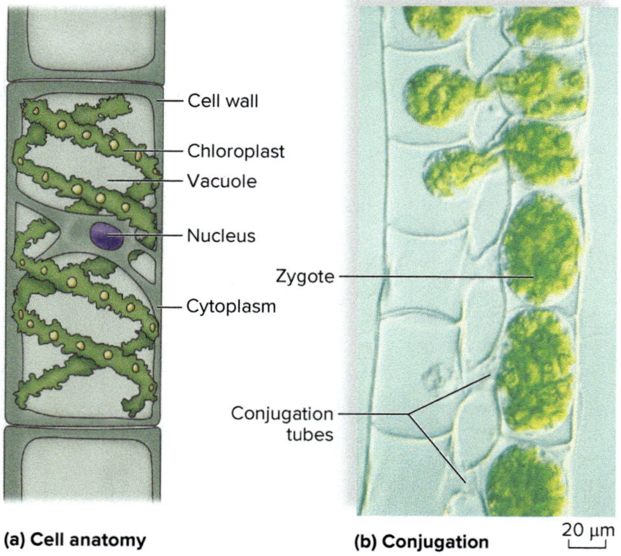

M. I. Walker/Science Source

Figure 25.6 *Spirogyra* (watersilk). (*a*) Spiral, ribbonlike chloroplasts occur in each cell. (*b*) During conjugation, the cell contents of one filament (+strain) enter the cells of another filament (−strain) through a conjugation tube (200×).

Procedure 25.3 **Examine *Spirogyra* and *Cladophora***

1. Obtain and examine a living culture of *Spirogyra.*
2. Prepare a wet mount with a small amount of living *Spirogyra.* Examine it with low and then high magnification. Notice the arrangement of the chloroplasts and whether filaments are branched.
3. Examine a prepared slide of *Spirogyra.*
4. In the following space, sketch a filament of *Spirogyra* and note the arrangement of its chloroplasts.
5. Examine a prepared slide of filaments of *Spirogyra* undergoing conjugation. Locate the conjugation tubes, gametes, and zygotes.
6. Prepare and examine a wet mount of living *Cladophora* (fig. 25.7). Then examine a prepared slide of *Cladophora*.
7. In the following space, draw a few cells of *Cladophora* showing their general shape and the filament's branching pattern.

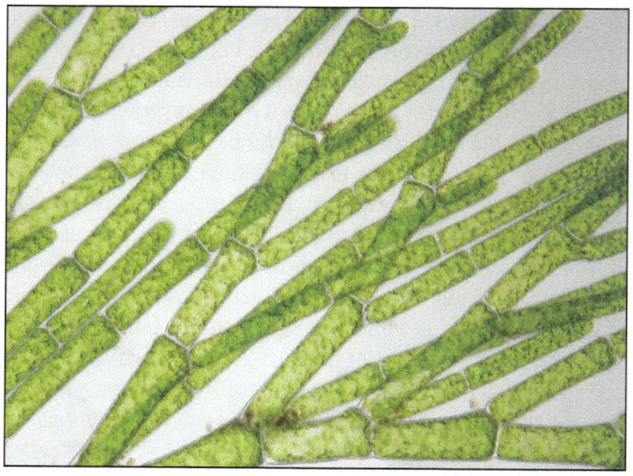

Blickwinkel/Alamy Stock Photo

Figure 25.7 The green alga *Cladophora* forms branched filaments consisting of multinucleate cells (100×).

Question 3

a. Are filaments of *Spirogyra* branched?

b. What is the arrangement of chloroplasts of *Spirogyra?*

c. Can you see any conjugation tubes? If you can't, examine the prepared slides that show these structures.

d. How do you think *Spirogyra* reproduces asexually?

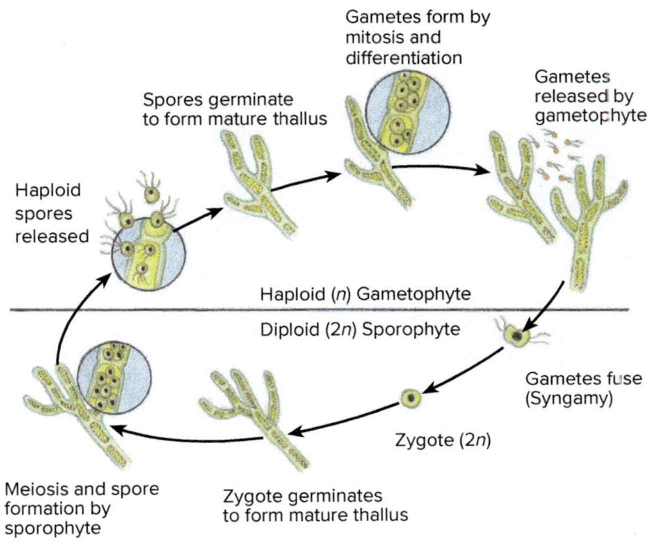

Figure 25.8 Alternation of generations in *Cladophora.* During alternation of generations, the haploid (*n*) gametophytes alternate with diploid (2*n*) sporophytes. Gametophytes produce haploid gametes that fuse to form a diploid zygote, the first cell of the sporophyte generation. The zygote germinates and undergoes meiosis to form haploid spores, the first cells of the gametophyte generation. The gametophyte then uses mitosis to produce gametes, thereby completing the life cycle.

Question 4

a. How is *Cladophora* morphologically similar to *Spirogyra?* How is it different?

b. What is the shape of its chloroplasts?

Mature *Cladophora* exists in diploid and haploid forms. The diploid stage of the life cycle produces spores and is called the **sporophyte.** The haploid stage of the life cycle produces gametes and is called the **gametophyte.** This phenomenon of alternating haploid and diploid stages of a life cycle is called **alternation of generations.** Alternation of generations is a reproductive cycle in which the haploid gametophyte produces gametes that fuse to form a zygote that germinates to produce a diploid sporophyte. Within the sporophyte, meiosis produces spores that germinate into gametophytes, thus completing the cycle (fig. 25.8). Alternation of generations occurs in many green algae, including *Cladophora,* and in all land plants.

You should become familiar with the concept of alternation of generations because it occurs frequently in the plant kingdom and we will refer to it repeatedly in this and future exercises. Refer to your textbook or instructor for more information.

Colonial Green Alga: *Volvox*

Volvox consists of many *Chlamydomonas*-like cells bound in a common spherical matrix (fig. 25.9). Each cell in the sphere has two flagella extending outward from the surface of the colony. Synchronized beating of the flagella spins the colony through the water like a globe on its axis. *Volvox* is one of the most structurally advanced colonial forms of algae, so much so that some biologists consider *Volvox* to be multicellular. Some of the cells of a *Volvox* colony are functionally differentiated; a few specialized cells can produce new colonies, and eggs and sperm are formed by different cells in the colony.

Volvox reproduces by **oogamy.** Motile sperm swim to and fuse with the large nonmotile eggs to form a diploid zygote. The zygote enlarges and develops into a thick-walled zygospore released when the parent colony disintegrates. The zygospore then undergoes meiosis to produce haploid cells that subsequently undergo mitosis and become a new colony. During asexual reproduction, some cells of *Volvox* divide, bulge inward, and produce new colonies called **daughter colonies** that initially are held within the parent colony.

Procedure 25.4 Observe *Volvox*

1. Examine a prepared slide of *Volvox.*

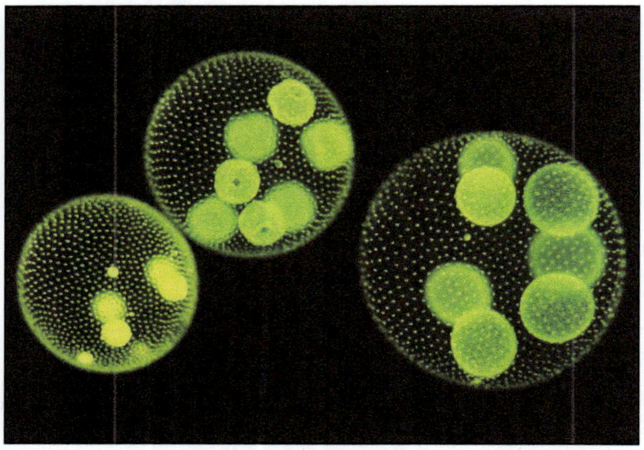

Lebendkulturen.de/Shutterstock

Figure 25.9 Colonies of *Volvox*. Many parent colonies contain asexually produced daughter colonies (400×).

2. Obtain a living culture of *Volvox*. Place a drop of living *Volvox* on a depression slide.
3. Under low magnification, observe the large, hollow, spherical colonies for a few minutes to appreciate their elegance and beauty.
4. Search for flagella on the surface.
5. Complement your observations of this alga by re-examining prepared slides of *Volvox*.

Question 5

a. What is oogamy?

b. What are the tiny spheres inside the larger sphere of *Volvox*?

c. How do you suppose they get out?

d. How do you think the number of cells in a young *Volvox* colony compares to the number in a mature colony?

RED ALGAE (SUPERGROUP ARCHAEPLASTIDA)

Red algae obtain their color from the presence of red **phycobilins** in their plastids. Red algae typically live in warm marine waters. The thallus of a red alga can be attached or free-floating, filamentous, or fleshy (figs. 25.10, 25.11).

Procedure 25.5 Examine *Polysiphonia*, *Porphyra*, and commercial products of red algae

1. Obtain prepared slides of *Polysiphonia* and *Porphyra* and any living cultures that are available in lab.
2. Examine a prepared slide of *Polysiphonia*. Notice the thickness of the filaments compared with that of filamentous green algae.
3. Examine living *Polysiphonia*. This genus is highly branched and filamentous. As with other red algae, their life cycles can be quite complex. Gametophytes of these organisms are dioecious (i.e., they are either male or female).
4. Examine some prepared slides and living *Porphyra* if available. Compare the structure of *Polysiphonia* with that of *Porphyra*. "Blades" of *Porphyra* consist of two layers of cells separated by colloidal material.
5. Study the display of carrageenan, agar, and other products derived from red algae.
6. Pick up and feel a piece of agar, noting its texture. Agar is a gelatinous polysaccharide from red algae used as a solidifying agent in culture media for microbiology labs (fig. 25.11*b*). A 1% suspension of hot agar remains liquid until it cools to about body temperature.

A. Martin UW Photography/Getty Images

Premaphotos/Alamy Stock Photo

Figure 25.10 Red algae come in many forms and sizes.

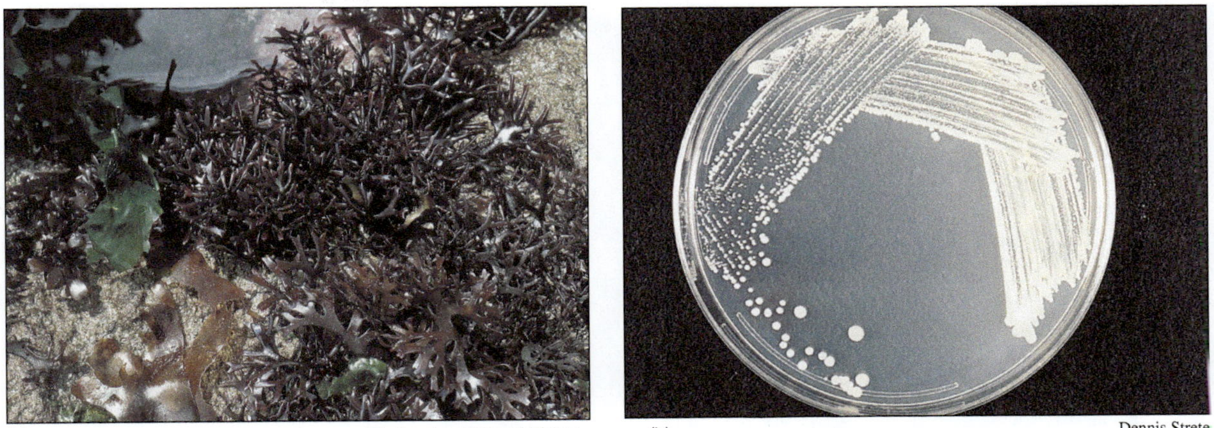

(a) Heather Angel/Natural Visions (b) Dennis Strete

Figure 25.11 A red alga and a common extract, agar. (*a*) Irish moss (*Chondrus crispus*) is a red alga that is commercially important as a source of carrageenan. Carrageenan is used as a stabilizer in paints and cosmetics, as well as in foods such as salad dressings and dairy products. (*b*) Microbiologists grow a variety of organisms on media solidified with agar (shown here) extracted from seaweeds such as *Gracilaria*. The bacteria shown here and the fungus shown in figure 27.10 are growing on agar. Agar is also used to make drug capsules, cosmetics, and gelatin desserts.

BROWN ALGAE (SUPERGROUP CHROMALVEOLATA)

Brown algae are primarily marine and structurally complex; there are no unicellular or colonial brown algae. Brown algae usually grow in cool water and obtain their name from the presence of a brown pigment called **fucoxanthin.** Brown algae range in size from microscopic forms to kelps over 50 m long. Thalli of some brown algae are similar to those of land plants. Review table 25.2 for the characteristics of brown algae.

Among the larger brown algae is *Macrocystis,* a kelp reaching 100 m in length (fig. 25.12). The flat blades of this kelp float on the surface of the water, while the base is anchored far below the surface. Another ecologically important brown alga is sargasso weed (*Sargassum;* fig. 25.13), which forms huge floating masses that dominate the vast Sargasso Sea in the Atlantic Ocean northeast of the Caribbean. These mats are microhabitats for a variety of highly adapted and cryptically colored animals.

Fucus

Fucus is another common genus of brown algae (fig. 25.14). *Fucus* (rockweed) typically attaches to rocks in the intertidal zone via a specialized structure called a holdfast. The outer surface of *Fucus* is covered by a gelatinous sheath. Tips of *Fucus* branches, called **conceptacles,** may be swollen and contain reproductive structures called **oogonia** (female) and **antheridia** (male), as shown in figure 25.14. Oogonia are multicellular sex organs that produce eggs. Antheridia are multicellular sex organs that produce sperm. Most protists do not have multicellular reproductive organs.

The life cycle of *Fucus* is similar to the common life cycle of animals. The mature thallus is diploid, and cells within reproductive structures undergo meiosis to produce gametes, thereby skipping the multicellular haploid stage common to many protists, plants, and fungi.

Karen Gowlett-Holmes/Getty Images

Figure 25.12 Brown alga. The giant kelp, *Macrocystis pyrifera,* grows in relatively shallow water along coasts throughout the world and provides food and shelter for many different kinds of organisms.

Kingsley Stern

Figure 25.13 *Sargassum,* a floating brown alga from which the Sargasso Sea got its name. *Sargassum* also lives in other oceans.

Procedure 25.6 Examine *Fucus*

1. Refer to figure 25.14 as you examine *Fucus* in the lab. Use your dissecting microscope to examine a cross section of the flattened, dichotomously branched thallus of *Fucus.*
2. Notice the presence of swollen areas on the thallus of *Fucus.*
3. Work in a group to dissect one of these structures.
4. Examine prepared slides of antheridia and oogonia of *Fucus.* Sketch what you see.

Question 6

a. How does the structure of *Fucus* differ from that of the green algae that you examined earlier in this exercise?

b. Are all portions of the thallus photosynthetic? How can you tell?

c. Considering where *Fucus* lives, what do you think is the function of its gelatinous sheath?

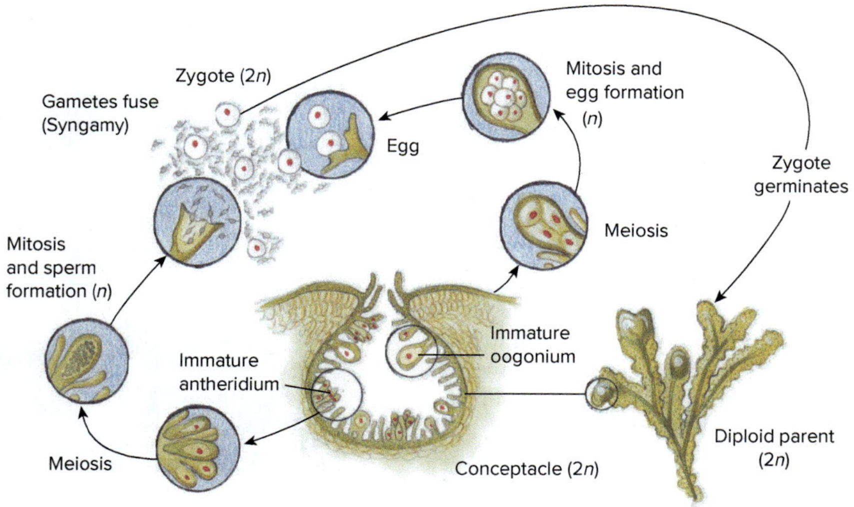

Figure 25.14 Sexual reproduction in a monoecious species of *Fucus.* Some species of *Fucus* have separate male thalli and female thalli containing conceptacles with only antheridia, and only oogonia, respectively.

d. Are the swollen structures solid masses or are they hollow?

Question 7

a. Are the gametes of *Fucus* isogamous or oogamous?

b. How does the structure of tissue surrounding the reproductive structures com.pare with that of green algae?

Economic Importance of Brown Algae

In the Orient, some brown algae are used as food. One of these algae is *Laminaria,* a kelp marketed as "kombu." Brown algae are also important sources of **alginic acid,** a hydrophilic substance (i.e., it absorbs large quantities of water). Alginic acid is used as an emulsifier (an additive used to stabilize processed food and other products) in dripless paint, ice cream, pudding mixes, and cosmetics.

Procedure 25.7 **Examine some commercial products of brown algae**

1. Taste a small piece of kelp packaged as a food product. How would you describe its taste and texture?

2. Observe the products on display and, in the case of foods, read their contents labels.

DIATOMS (SUPERGROUP CHROMALVEOLATA)

Diatoms are unicellular algae containing chlorophylls *a* and *c* and xanthophyll pigments that give them their golden-brown color. Although diatoms are tiny, their great numbers, rapid rates of reproduction, and photosynthetic capacity make them vitally important as a primary link in the food chain of the oceans.

Diatoms have a hard cell wall made of silicon dioxide (glass) (fig. 25.15). These walls are arranged in overlapping halves, much like the halves of a petri dish. The glass walls of diatoms persist long after the remainder of the cell disintegrates (fig. 25.16) and may accumulate in layers of **diatomaceous earth** several hundred meters deep. This depth indicates how many diatoms have existed through the ages.

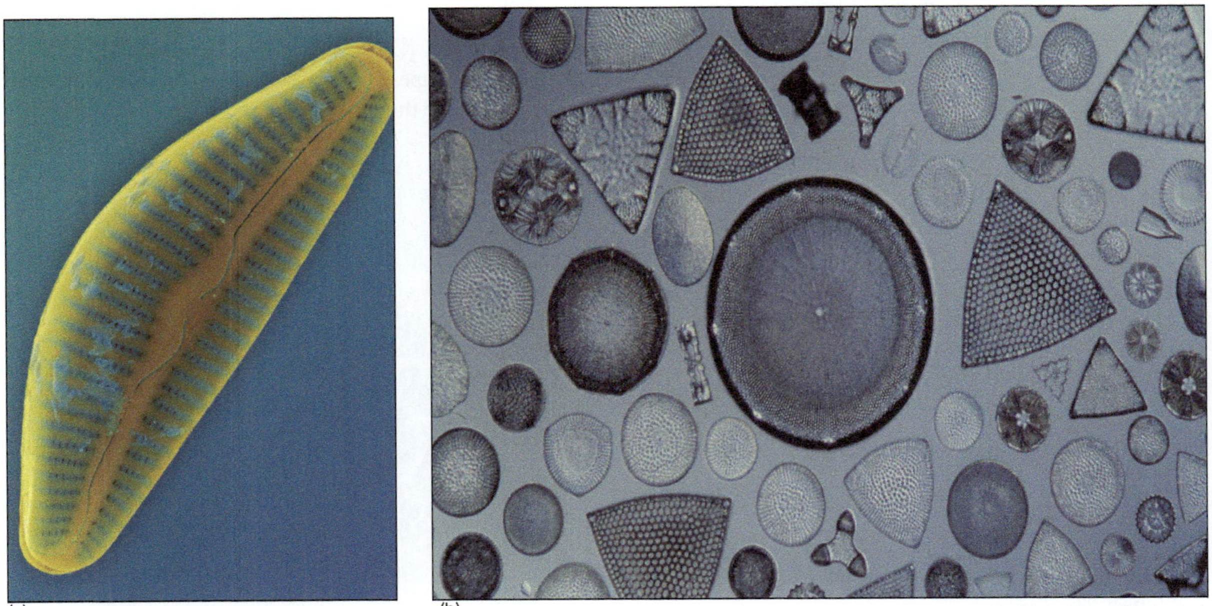

(a) DENNIS KUNKEL MICROSCOPY/Science Photo Library/Alamy Stock Photo (b) Dr. Norbert Lange/Shutterstock

Figure 25.15 Diatoms. (*a*) A pennate (bilaterally symmetrical) diatom. (*b*) Several different kinds of diatoms, including some centric (round) species (600×).

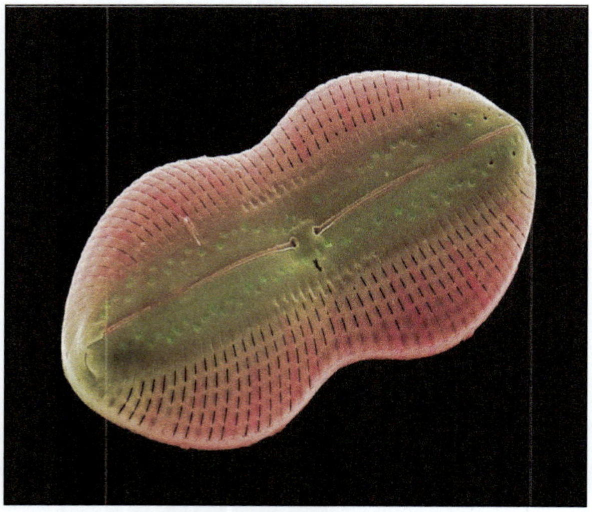

STEVE GSCHMEISSNER/Science Photo Library/Alamy Stock Photo

Figure 25.16 A pennate diatom showing its ornate, silicon cell wall (1000×).

Procedure 25.8 **Examine prepared slides of diatoms, living diatoms, and diatomaceous earth**

1. Examine a prepared slide of diatoms. Sketch some of the cells here. Some cells are long and thin, whereas others are disklike.
2. Prepare a wet-mount slide from a culture of living diatoms. Compare the shapes of the cells with those on the prepared slide.
3. Mount a small amount of diatomaceous earth in water on a microscope slide. Examine the diatomaceous earth with your microscope. Note the variety of shells, some broken and others intact. A mass of these shells is clean, insoluble, and porous.

Question 8

a. Can you see any pores in the walls of diatoms?

b. Are any of the diatoms moving?

c. If diatoms lack flagella, how do you explain their motility?

d. How would diatomaceous earth compare to sand as a material for swimming pool filters? Which would be better and why?

DINOFLAGELLATES (SUPERGROUP CHROMALVEOLATA)

Dinoflagellates are all unicellular and characterized by the bizarre appearance of their cellulose plates and by the presence of two flagella located in perpendicular grooves (fig. 25.17). Blooms of a red-pigmented dinoflagellate called *Ptychodiscus brevis* produce a "red tide." Toxin production and oxygen depletion by these blooms of algae can kill massive numbers of fish. Dinoflagellates are important primary producers in oceans (second only to diatoms) and include many autotrophic and heterotrophic forms. Some dinoflagellates are bioluminescent, others live symbiotically with corals, and some are heterotrophic.

Procedure 25.9 **Examine dinoflagellates**

1. Examine a prepared slide of *Peridinium* or *Ceratium.* Look for longitudinal and transverse flagella and flagellar grooves.
2. Prepare a wet-mount slide from a living culture of dinoflagellates. Dinoflagellates are quite small, so be patient while searching for organisms.

Question 9

How do the shapes of dinoflagellates compare with other unicellular algae that you have observed in this exercise?

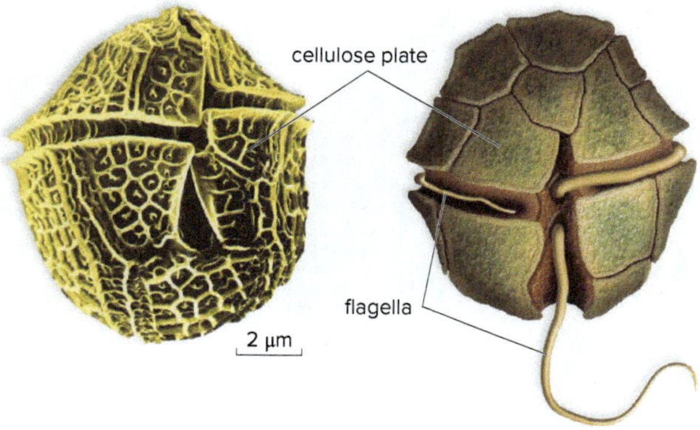

Figure 25.17 Dinoflagellate *Gonyaulax*. Most dinoflagellates have rigid cellulose plates and pair of flagella in perpendicular grooves.

EUGLENOIDS (SUPERGROUP EXCAVATA)

Euglenoids include mostly freshwater unicellular forms. Although plastids of euglenoids contain chlorophylls *a* and *b* (like the green algae), euglenoids are distinctive because they lack a cell wall, and instead have a pellicle made largely of protein. The protein makes the cell more flexible. Euglenoids are motile and have two flagella (fig. 25.18).

Procedure 25.10 Observe *Euglena*

1. Observe living and prepared slides of *Euglena* available in the lab while referring to figure 25.18.
2. You may want to add a drop of methylcellulose in your preparation to slow the *Euglena.*
3. Note the colored **eyespot** near the base of the flagella.
4. Observe the movement and changing shapes of the organisms.

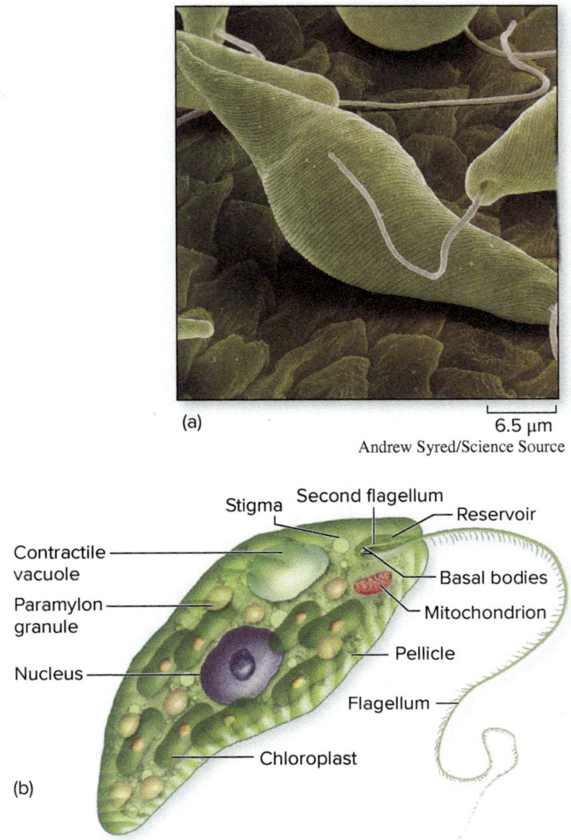

Figure 25.18 Euglenoids. (*a*) Micrograph of individuals of the genus *Euglena* (400×). (*b*) Diagram of *Euglena.* Paramylon granules store food reserves.

Chloroplasts of *Euglena* may contain a single pyrenoid, which appears as a clear, circular area within the plastid. *Euglena* is best known for its ability to be autotrophic, heterotrophic, and saprophytic. Its specific mode of nutrition is determined by current environmental conditions. This phenomenon illustrates why it is often impossible to distinguish plant from animal at the cellular level and why classification of protists seems so unwieldy. Our classification schemes for these and other organisms will improve as we learn more about them.

Question 10
What is the function of the eyespot of *Euglena?*

INQUIRY-BASED LEARNING

How do algae respond to changing environmental stimuli?

Observation: You learned in Exercise 23 how algae are affected by environmental changes involving nutrients, pollutants, and temperature changes. As a result, algal growth can be used to monitor environmental changes (e.g., eutrophication).

Question: How can algae be used to monitor environmental conditions?

a. Establish a working lab group and obtain Inquiry-Based Learning Worksheet 25 from your instructor.
b. Discuss with your group well-defined questions relevant to the preceding observation and question. Choose and record your group's best question for investigation.
c. Translate your question into a testable hypothesis and record it.
d. Outline on Worksheet 25 your experimental design and supplies needed to test your hypothesis. Ask your instructor to review your proposed investigation.
e. Conduct your procedures, record your data, answer your question, and make relevant comments.
f. Discuss with your instructor any revisions to your questions, hypotheses, or procedures. Repeat your work as needed.

Questions for Further Study and Inquiry

1. What are examples of unicellular, filamentous, and colonial green algae?

2. How are green algae different from cyanobacteria?

3. What is meant by "alternation of generations"?

4. What is meant by the term "kelp"?

5. Are the stem, holdfast, and blade of brown algae the same as stems, roots, and leaves of land plants? Why or why not?

6. Brown algae contain chlorophyll. Why, then, do they appear brown and not green?

7. What are the main differences and similarities among the major groups of algae? How are these groups related to each other evolutionarily?

8. Describe three ways that algae affect your life.

9. Why are green algae considered to be ancestral to plants?

DOING BIOLOGY YOURSELF

Would you expect environmental conditions to influence syngamy? Design two experiments to investigate the effects of two environmental conditions on the frequency of syngamy. How would you measure syngamy? Its frequency?

WRITING TO LEARN BIOLOGY

Describe the plant-like and animal-like characteristics of *Euglena.* Which characteristics conclusively define a plant? Which ones define an animal?

EXERCISE 26

Survey of Protists

Protozoan Heterotrophs

Learning Objectives

By the end of this exercise you should be able to:

1. Describe the features characterizing protozoa and slime molds.
2. List examples, habitats, reproductive methods, and unique features of the protozoans.
3. Become familiar with representative protozoans and slime molds.

Please visit connect.mheducation.com to review online resources tailored to this lab.

Protozoans (*proto* = first, *zoan* = animal) are among the most versatile of all eukaryotes on earth. Protozoa, however, like algae, is a descriptive term rather than a taxonomic group. Protozoans are eukaryotes with an animal-like, heterotrophic ecology, which means they are active consumers and not photosynthetic. (Remember that heterotrophs get their nutritional energy from organic molecules made by other organisms.) Typically, protozoans have food vacuoles to enclose food particles for digestion, and contractile vacuoles to expel excess water. Their single cells employ a variety of features for motility, and they live everywhere from a drop of pond water to the intestines of termites. Review table 25.1 and the supergroups of protists that include common protozoans.

AMOEBAS (SUPERGROUP AMOEBOZOA)

Amoebas occur throughout the world in marine, freshwater, and terrestrial environments. Their unifying characteristic is the presence of **pseudopods,** which are movable extensions of cytoplasm used for locomotion and gathering food. Amoebas are likely polyphyletic, lack flagella, and most reproduce asexually.

Amoeba

Amoeba is a genus among many organisms commonly called amoebas that has a structure and physiology typical of most amoeboid genera (fig. 26.1). *Amoeba* are **phagocytic,** meaning they engulf food particles and form a **food vacuole** surrounded by a membrane. They then secrete enzymes into the food vacuole for **intracellular digestion.** A **contractile vacuole** maintains the cell's water balance by accumulating and expelling excess water. Other common amoebas include *Difflugia,* which makes a protective case of sand grains called a **test.** Test is a general term referring to a secreted or partially secreted covering, much like a shell (fig. 26.2). Examine a prepared slide of *Difflugia.* Also examine the amoeba *Entamoeba histolytica,* a parasite that causes dysentery in humans.

Procedure 26.1 **Observe *Amoeba* movement and structure**

1. Review microscope use and procedures in Exercise 3, as well as the associated videos at the lab manual's website. Use a dissecting microscope (see Exercise 3) to examine a culture of living *Amoeba*. Locate individuals on the bottom of the culture.
2. Prepare a wet mount of living *Amoeba* by using an eyedropper to remove a few drops from the bottom of the culture of organisms.
3. Put the drops in a depression slide if one is available or use a standard slide.
4. Cover the preparation with a coverslip and examine it under low power (10×) (see Exercise 3). Soon the *Amoeba* should move by extending their pseudopods.
5. If nutrient broth is available, add a drop to the preparation and observe the *Amoeba*'s response.
6. Examine a prepared slide of stained *Amoeba,* and locate the structures shown in figure 26.1.
7. Examine any other live amoebas and prepared slides that are available. Draw the basic structure of these organisms.

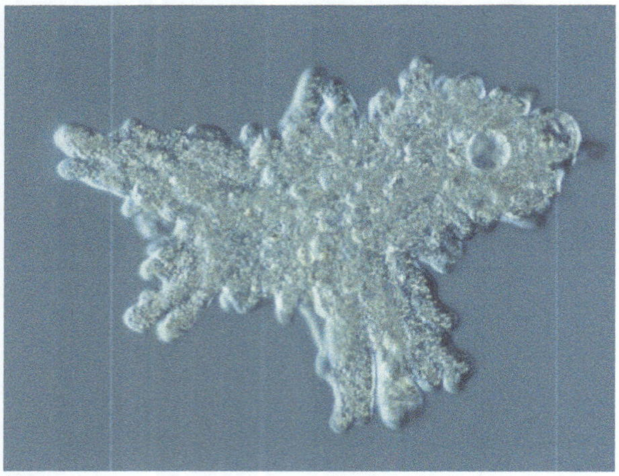
(a) M I (Spike) Walker/Alamy Stock Photo

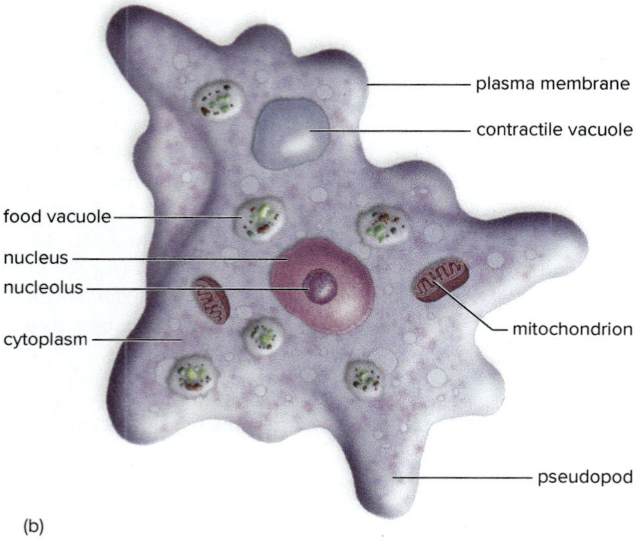

(b)

Figure 26.1 Amoebas. (*a*) Light micrograph of *Amoeba proteus* showing blunt pseudopodia (160×). (*b*) Anatomy of *Amoeba.*

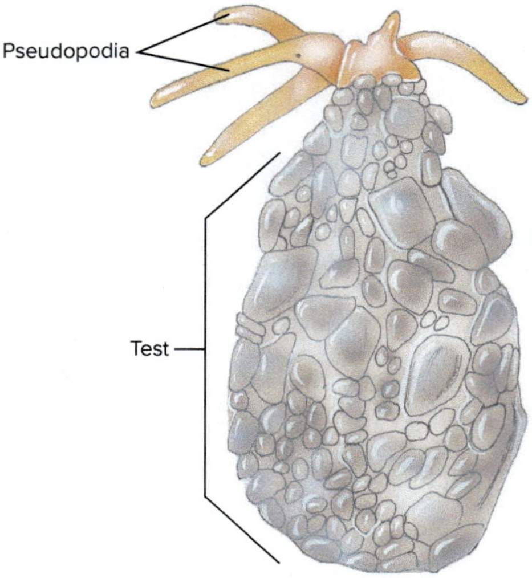

Figure 26.2 *Difflugia oblongata,* a common freshwater amoeba with a sand-grain case. The case consists of cemented mineral particles collected by the amoeba (150×).

Question 1

a. Can you detect moving cytoplasm in the extending pseudopods of *Amoeba?*

b. What do you suppose the living *Amoeba* is moving toward or away from?

c. How does the *Amoeba* respond to nutrient broth?

d. About how long would it take an *Amoeba* to move across the field of view on low power?

e. Why is a contractile vacuole of a protozoan often more difficult to see than a food vacuole?

f. Why would excess water tend to accumulate in *Amoeba?*

SLIME MOLDS (SUPERGROUP AMOEBOZOA)

Bizarre plasmodial slime molds such as *Physarum* stream along the damp forest floor in a mass of brightly colored protoplasm called a **plasmodium** in which individual cells are indistinguishable. Plasmodia are coenocytic (multinucleate) because their nuclei are not separated by cell walls, and the entire plasmodium resembles a moving mass of slime. The plasmodium of a slime mold should not be confused with the sporozoan genus with the same name.

Often plasmodia of slime molds live beneath detached bark on decomposing tree trunks. Slime mold plasmodia occasionally occur on lawns or on mulch beneath shrubs. The moving protoplasm of a plasmodium conspicuously pulsates back and forth as it engulfs and digests bacteria, yeasts, and other organic particles.

©BiologyImaging.com

Figure 26.3 Slime mold (*Physarum*). Slime molds share characteristics with fungi but also with protistans. Their life cycle includes a plasmodial stage, which is a thin, multinucleate mass of streaming protoplasm with amoeboid or flagellated cells.

Procedure 26.2 **Examine a culture of *Physarum***

1. Obtain a dissecting microscope and a petri plate containing a culture of *Physarum* growing on oatmeal flakes.
2. Examine the yellowish trails of the colony under the highest magnification of the microscope. Look for cytoplasmic movement.
3. Look for signs of the organism beginning to condense into darkly tipped sporangia. Sporangia are easily visible without a microscope.

Question 2

a. Is cytoplasmic movement of *Physarum* apparent?

b. Is the movement in a particular direction?

c. What is a possible function of cytoplasmic movement in *Physarum?*

The *Physarum* on demonstration is in the vegetative plasmodial stage and feeds on the oatmeal (fig. 26.3). However, if environmental conditions become less than optimal (e.g., if food or moisture decreases), the plasmodium may dry into a hard resistant structure called a **sclerotium** and remain dormant until conditions improve. Or, if light is available, the diploid plasmodium will move to the illuminated area and coalesce. The condensed structure will grow sporangia, and meiosis will produce spores for dispersal (fig. 26.4). Light is associated with an open environment that allows successful reproductive dispersal; dispersal under or within a tree trunk would be ineffective. Haploid spores produced by meiosis in the sporangia germinate as amoeboid or flagellated organisms. These haploid stages may later fuse as gametes and grow into a new plasmodium.

FORAMINIFERANS (SUPERGROUP RHIZARIA)

These marine organisms are called "shelled amoebas" because they surround themselves with a secreted test and have long, thin, rather stiff pseudopods protruding from their tests (fig. 26.5). The test is made of calcium carbonate and is perforated with pores. The marine fossil record is replete with old and well-preserved tests of forams and is often used by oil companies to locate oil-bearing strata.

Plasmodium, *Physarum*

Sporangia, *Hemitrichia*

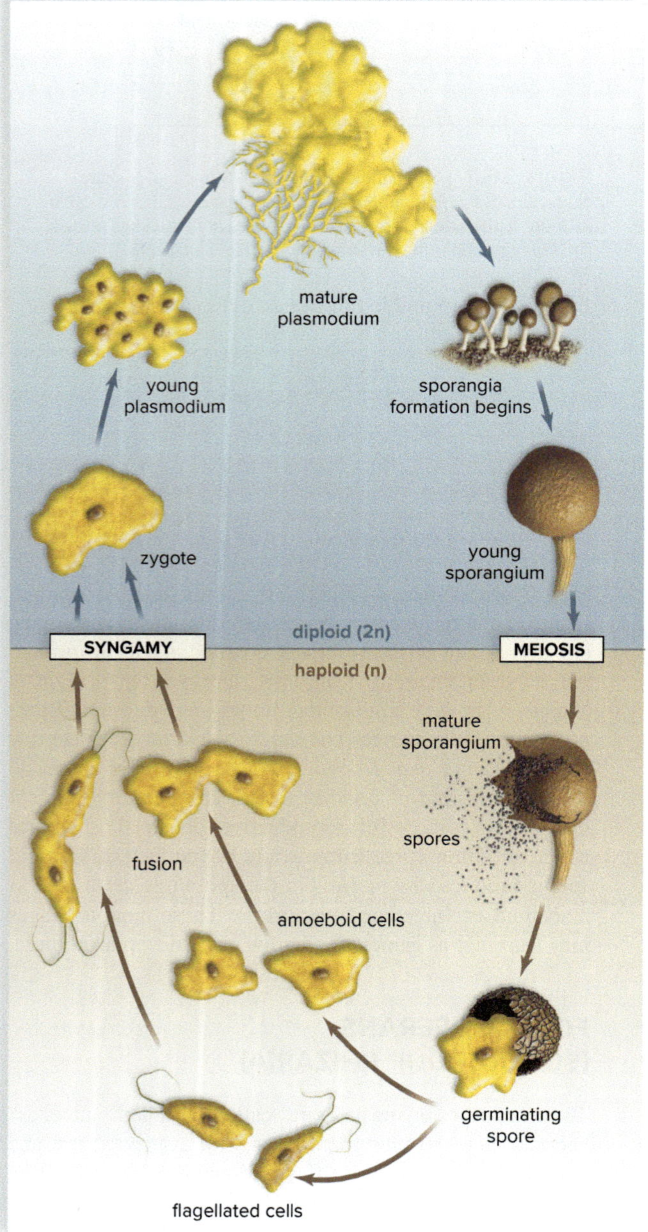

Figure 26.4 Life cycle of a plasmodial slime mold. When food or moisture is scarce, a diploid, mature plasmodium stops moving and forms sporangia. Haploid spores form by meiosis. The spores wait until conditions are favorable to germinate. Spores can give rise to flagellated or amoeboid gametes; the two forms convert from one to the other readily. Fusion of the gametes (syngamy) forms the diploid zygote, which gives rise to the mobile, feeding plasmodium by mitosis.

Procedure 26.3 **Examine foram tests**

1. Obtain a prepared slide of foram tests.
2. Search with low magnification the edges of the cover slips for the foraminiferan tests. They are relatively heavy and shift to the side easily.

Question 3

How could fossilized forams in different geological layers of rock or sediment indicate the probability of finding oil?

FLAGELLATES (SUPERGROUP EXCAVATA)

Flagellates have at least one **flagellum** and are likely the most primitive protozoans. Flagellates are parasitic as well as free-living heterotrophs.

Trypanosoma

Among flagellates, trypanosomes are pathogenic and cause African sleeping sickness and Chagas' disease. Charles Darwin may have died from Chagas' disease, for during his later years he suffered from general fatigue, irregular fever, and heart damage. All of these are symptoms of Chagas' disease transmitted by the bite of an assassin bug, which resembles a common stinkbug. Trypanosomes are common in the tropics and spread by infection from biting insects such as mosquitoes, sand flies, and tsetse flies (fig. 26.6*b*).

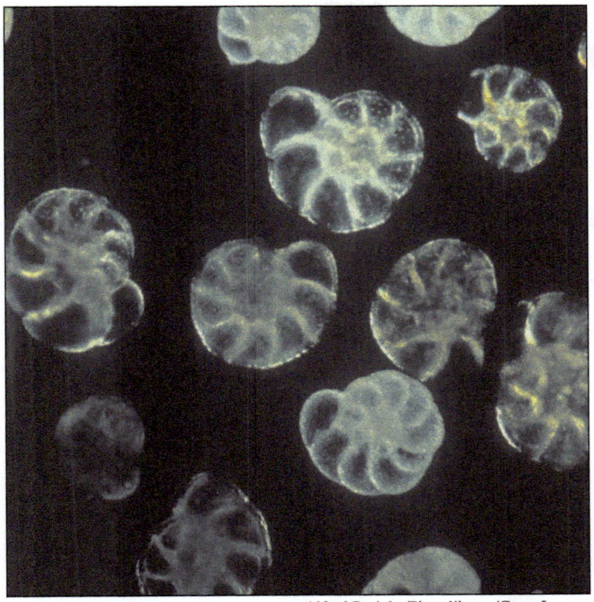

(a) Alfred Pasieka/Photolibrary/Getty Images

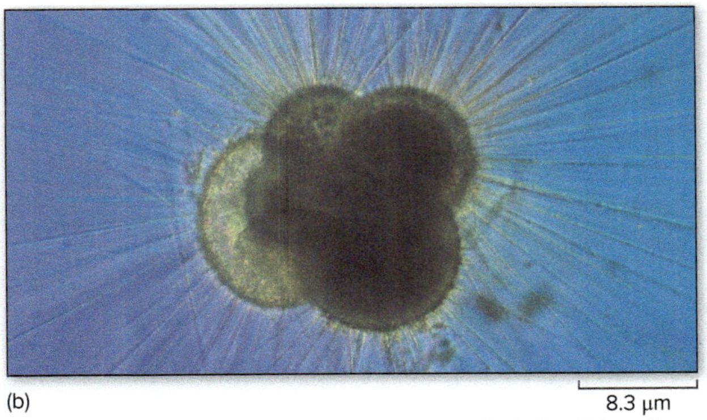

(b) Claude Carre/Science Source

Figure 26.5 Foraminifera. (*a*) The calcereous tests, or shells, of some representative foraminiferans (40×). (*b*) Podia, which are thin cytoplasmic projections, extend through pores in the calcareous test of this living foram (40×).

Procedure 26.4 Examine a prepared slide of *Trypanosoma* and compare its size to that of *Amoeba*

1. Obtain a prepared slide of *Trypanosoma* among blood cells.
2. Locate the organisms intermingled with the blood cells (fig. 26.6). The organisms are not inside the blood cells; they are in the surrounding plasma.
3. Try to distinguish the flagellum and undulating membrane of an individual. The **undulating membrane** is a thin, flat surface that can be undulated (i.e., waved) for locomotion. A rippling wave travels along the membrane and pushes the organism forward.
4. Trypanosomes are quite small. To estimate its size, first note the magnification of the objective you are using.
5. Refer to your notes from Exercise 3 (The Microscope) concerning the diameter of the field of view associated with the current magnification.
6. Estimate the portion of the diameter of the field of view occupied by a *Trypanosoma*. Calculate the size of a trypanosome.
7. Make a similar estimation of the size of *Amoeba* with a prepared slide at the same magnification.

Question 4

a. How large is a trypanosome relative to *Amoeba?*

b. What alga does a trypanosome superficially resemble?

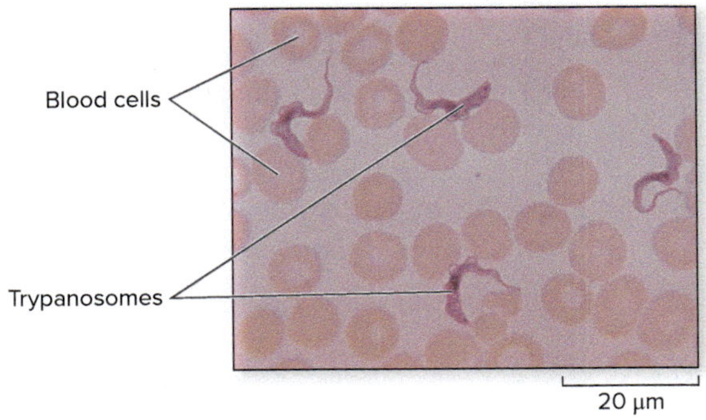

(a) Dr. Myron G. Schultz/Centers for Disease Control and Prevention

(b) Martin Dohrn/Science Photo Library/Getty Images

Figure 26.6 A parasite and its vector. (*a*) *Trypanosoma* among red blood cells. The nuclei (dark-staining bodies), anterior flagella, and undulating, changeable shape of the trypanosomes are visible in this photomicrograph (400×). (*b*) The tsetse fly, shown here sucking blood from a human arm, can carry and transmit trypanosomes.

African Sleeping Sickness

Human African trypanosomiasis, also known as sleeping sickness, is a vector-borne parasitic disease. Parasites of genus *Trypanosoma* are transmitted to humans by the bite of tsetse flies (*Glossina* spp.; fig. 26.6*b*), which acquire their infection from humans or other infected animals (World Health Organization Fact Sheet No. 259). The most vulnerable people are in rural populations dependent on agriculture, fishing, animal husbandry, or hunting. *Trypanosoma brucei gambiense (T.b.g)* accounts for 95% of reported cases of sleeping sickness (fig. 26.7). After continued control efforts, the number of cases reported in 2009 dropped below 10,000 for the first time in 50 years. This trend continued in 2010 with 7139 new cases reported. For unknown reasons, many regions of 36 sub-Saharan affected countries have tsetse flies, but no sleeping sickness. A person can be infected for months or even years without developing the major signs and symptoms of the disease, which include fever, headaches, joint pains, and itching. Advanced stages disrupt the sleep cycle and disrupt coordination.

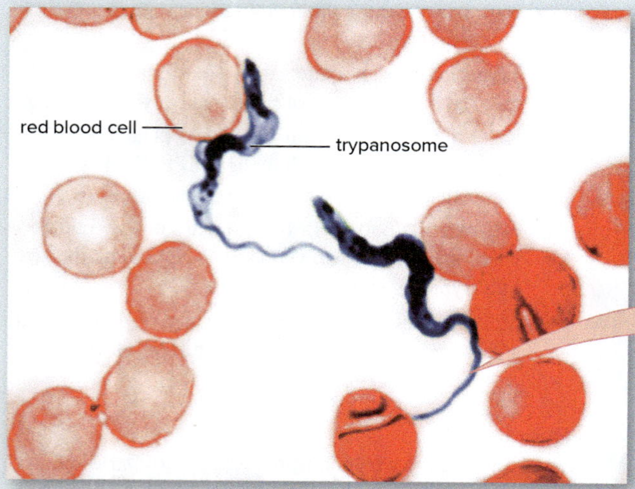

(a) Eric Grave/Science History Images/Alamy Stock Photo

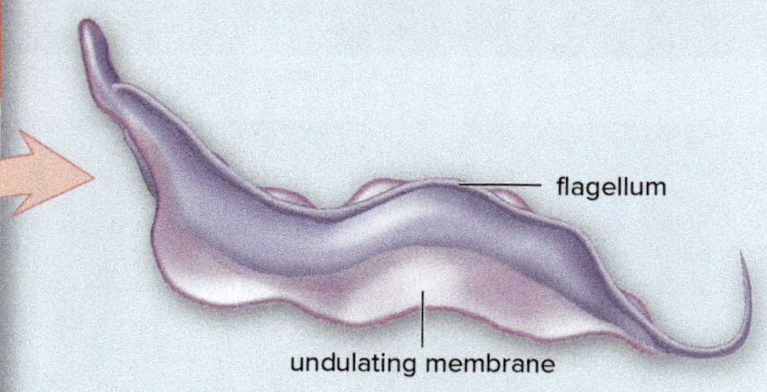

(b)

Figure 26.7
Trypanosoma brucei. (*a*) Micrograph of *Trypanosoma brucei,* a causal agent of African sleeping sickness, among red blood cells. (*b*) This drawing shows the general structure of *Trypanosoma brucei.*

CILIATES (SUPERGROUP CHROMALVEOLATA)

More than 8000 species of ciliates have been described, all having characteristically large numbers of **cilia.** Review in your textbook the difference between cilia and flagella. Many ciliates also have two types of nuclei: **micronuclei** and **macronuclei** (fig. 26.8). Macronuclei, which develop from micronuclei, control cellular function, and they divide when ciliates reproduce asexually by mitosis.

Paramecium

This free-living, freshwater genus is widely studied and easily observed. *Paramecium,* like most ciliates, undergoes a sexual process called **conjugation** (fig. 26.9*b*). During conjugation, individuals from two different strains align longitudinally and exchange nuclear material (fig. 26.9*b*). This exchange seems to stimulate metabolism of the individuals and is usually followed by frequent mitosis. Asexual reproduction is more common than conjugation and includes mitosis of the micronucleus and transverse fission (fig. 26.9*a*) of the macronucleus and cell body.

Procedure 26.5 **Examine conjugating and dividing *Paramecium***

1. Obtain a prepared slide of conjugating *Paramecium* and one of transversely dividing (binary fission) *Paramecium.*
2. Notice the alignment of the conjugating cells. Their nuclei are in close proximity.
3. Notice the plane of division of the transversely dividing cells.

Procedure 26.6 **Observe living *Paramecium***

1. Prepare a wet mount from a culture of living organisms.
2. Add a drop of methylcellulose to your wet mount to slow the *Paramecium* and make it easier to examine.

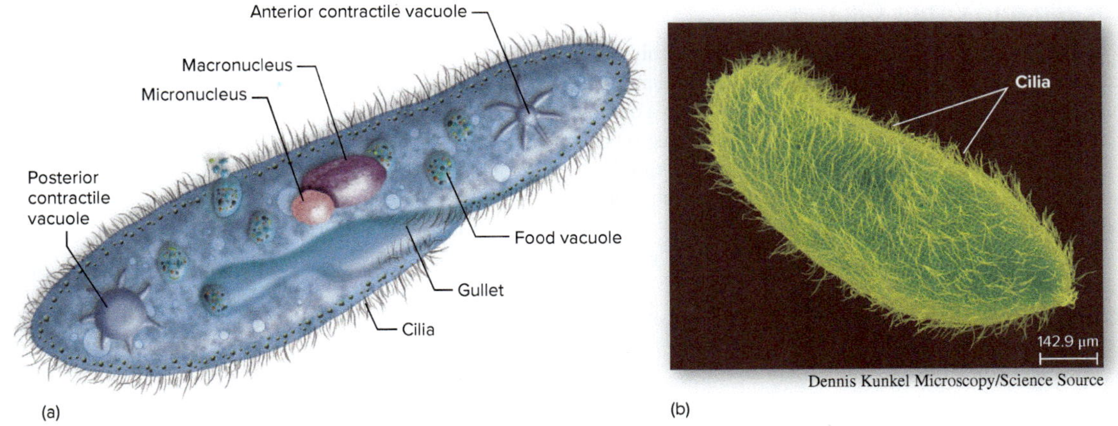

Figure 26.8 *Paramecium.* (*a*) The main features of this familiar ciliate include cilia, two nuclei, and numerous specialized organelles. (*b*) Scanning electron micrograph of *Paramecium* showing numerous cilia on the cell surface.

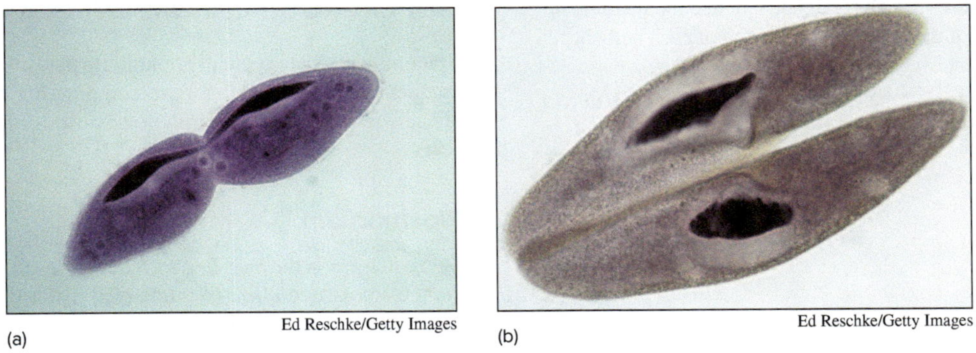

Figure 26.9 Reproduction among paramecia. (*a*) A mature *Paramecium* divides asexually by transverse fission (100×). (*b*) During conjugation, individuals exchange genetic material. Conjugation is a sexual process.

3. Describe aspects of their movement in comparison to *Amoeba.*

4. Identify as many structures as possible in figure 26.8.

Question 5

a. Are cilia visible on living or prepared *Paramecium?*

b. Does *Paramecium* rotate as it moves?

c. How does movement of *Paramecium* compare with that of *Amoeba?* With a flagellated alga?

Question 6

a. Why is the division of *Paramecium* cells called "transverse" fission?

b. Why is transverse fission not a sexual process?

c. What are the advantages and disadvantages of conjugation in *Paramecium?*

Vorticella and Other Ciliates

This freshwater ciliate is sessile (i.e., attached to a substrate) and has two notable features: (1) a contractile stalk that attaches the organism to the substrate and (2) a cell body with a corona of cilia. To feed, *Vorticella* extends its contractile stalk to push the cell body as far as possible from the substrate and from other individuals. Then it rapidly beats its cilia to capture food particles. This is a type of filter feeding (fig. 26.10).

Procedure 26.7 **Examine *Vorticella* and other ciliates**

1. Use a dissecting microscope to examine a living colony of *Vorticella,* which often grows on the glass and other hard substrates in stagnant aquaria.
2. Tap the sides of the dish and observe the contraction of each stalk.
3. When the stalks are extended, notice the whirling action of the cilia at the open end of each bell-shaped individual.
4. Examine a prepared and stained slide of *Vorticella* and draw its general shape.
5. Examine other cultures or prepared slides of ciliates available in the lab. Draw the general shape of these organisms and describe the movement of living specimens.

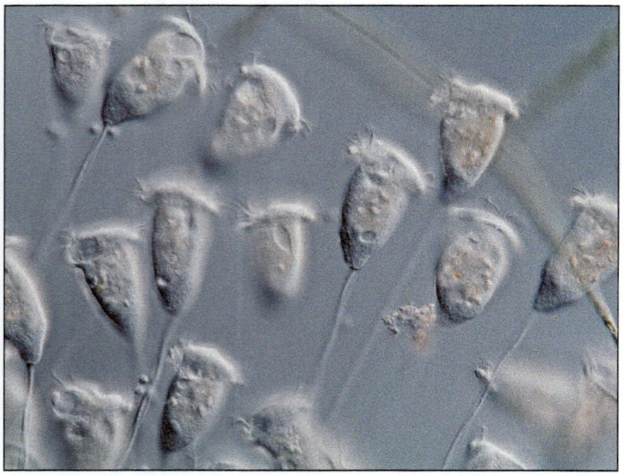

blickwinkel/Alamy Stock Photo

Figure 26.10 *Vorticella* are heterotrophic, feed largely on bacteria, and have retractable stalks (80×).

Question 7

a. What is the value or function of rapid contraction of the stalk of *Vorticella?*

b. What is the probable function of the moving cilia of *Vorticella?*

APICOMPLEXANS (SUPERGROUP CHROMALVEOLATA)

Apicomplexans are typically nonmotile parasites of animals. Frequently, these parasites have complex life histories and life stages with various morphologies occurring in multiple hosts.

Plasmodium

This pathogen is the best-known apicomplexan and has been the most common killer of humans in history. *Plasmodium* cause malaria, and mosquitoes of the genus *Anopheles* transmit *Plasmodium* from human to human (fig. 26.11). These malarial parasites infect and rupture red blood cells, causing cycles of fever and chills. Sexual reproductive stages occur in the mosquito host, and asexual stages occur in the human host.

Procedure 26.8 **Examine a blood smear from a victim of malaria**

1. Obtain a prepared slide with *Plasmodium.*
2. Locate the infected blood cells. The organisms are inside the infected blood cells, not in the surrounding plasma as are trypanosomes.
3. Locate and compare infected and uninfected blood cells. The infected cells will exhibit only faint signs of the circular ring stage of the parasite. Sketch what you see (fig. 26.12).
4. Review in your textbook the life cycle of *Plasmodium* (fig. 26.11).

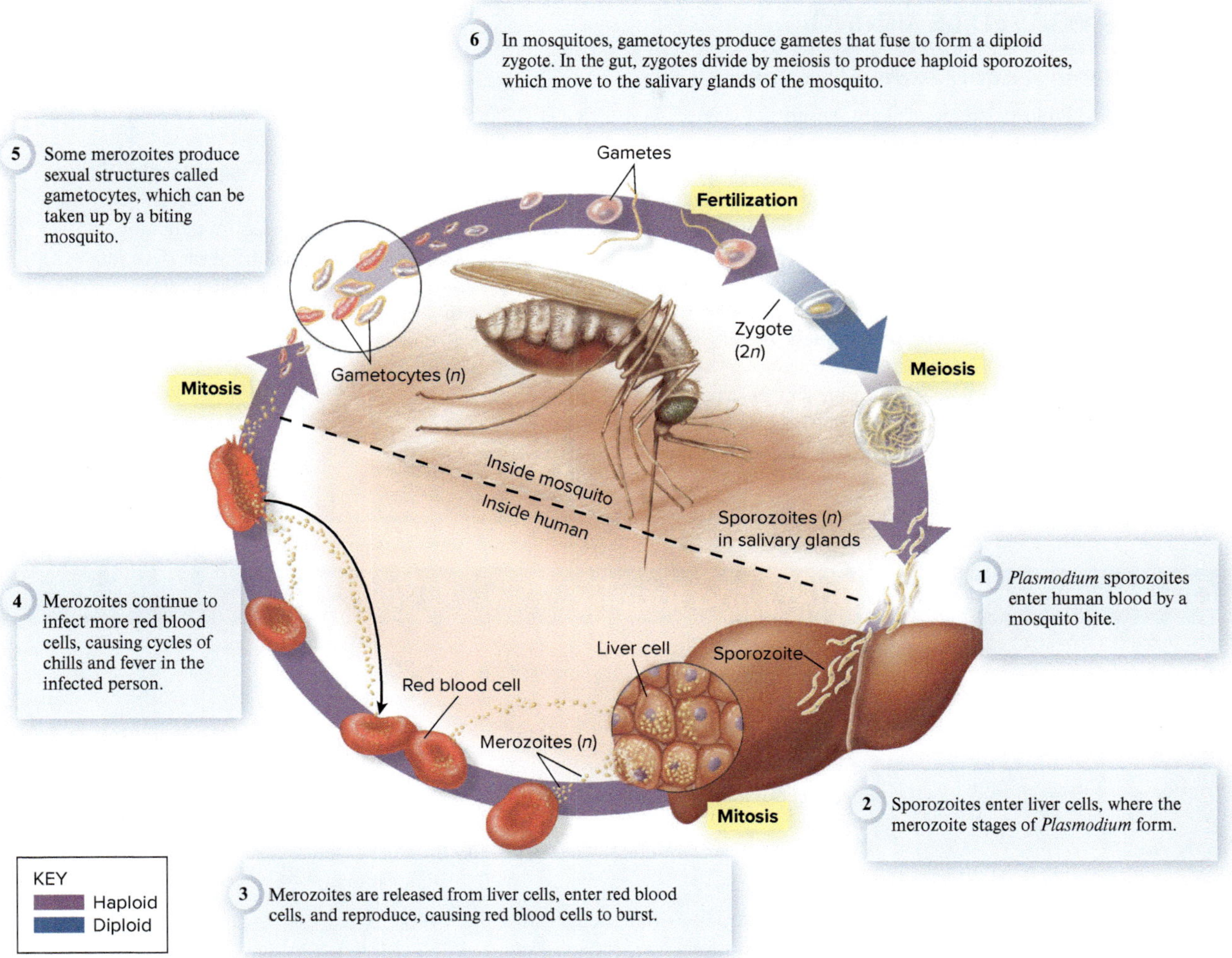

Figure 26.11 Diagram of the life cycle of *Plasmodium falciparum,* the agent of malaria. This life cycle requires two alternate hosts, humans and *Anopheles* mosquitoes.

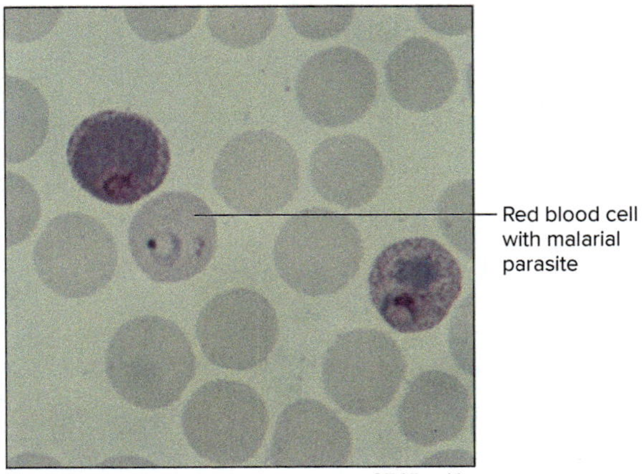

Figure 26.12 Malarial parasite, *Plasmodium*, within a red blood cell (500×).

INQUIRY-BASED LEARNING

Are protozoa sensitive to dissolved nutrients?

Observations: Protozoa are particularly sensitive to changes in their aquatic world, including changes in the amount and type of nutrients in the water. Nutrient broth elicits a response from protozoans seeking food. Nutrients include a variety of organic compounds, including fats, carbohydrates, and proteins.

Question: What kinds of nutrients elicit the strongest response from protozoa?

a. Establish a working lab group and obtain Inquiry-Based Learning Worksheet 26 from your instructor.

b. Discuss with your group ways to measure protozoan response to introduced solutes. Pose a well-defined question relevant to the preceding observation and question. Record your question on Worksheet 26.

c. Translate your question into a testable hypothesis and record it.

d. Outline on Worksheet 26 your experimental design and supplies needed to test your hypothesis. Ask your instructor to review your proposed investigation.

e. Conduct your procedures, record your data, answer your question, and make relevant comments.

f. Discuss with your instructor any revisions to your questions, hypotheses, or procedures. Repeat your work as needed.

Questions for Further Study and Inquiry

1. What requirements might make culturing parasitic zoomastigotes difficult in the lab?

2. Why do some scientists call conjugation "sexual reproduction" and others do not?

3. Is the cell the fundamental unit of life in plasmodial slime molds? Or is the "whole organism," regardless of cellular composition, the fundamental unit? Explain your answer.

4. What functions do cilia, flagella, and pseudopods have in common?

5. What factors may account for the ubiquitous occurrence and great structural diversity of unicellular organisms?

6. In what sense are protists "primitive" and in what sense are they "advanced"?

7. Why are unicellular organisms that reproduce by mitosis considered immortal?

Survey of the Kingdom Fungi

Molds, Sac Fungi, Mushrooms, and Lichens

EXERCISE 27

Learning Objectives

By the end of this exercise you should be able to:

1. Describe the characteristic features of the kingdom Fungi.
2. Discuss variation in structures and sequence of events of sexual and asexual reproduction for the major phyla of the kingdom Fungi.

Please visit connect.mheducation.com to review online resources tailored to this lab.

Fungi are as cute as a colorful mushroom and as ugly as the fuzzy slime on a spoiled pork chop. We slice some species to put on our salads, while others would kill us within minutes of eating them. Some species produce antibiotics vital to medicine, while others cause fatal disease. Fungi decompose dead organisms and recycle vital nutrients, while just as quickly they can attack our crops and our refrigerated foods. They do all this, yet they are among the organisms least understood by introductory biology students.

Fungi are basically filamentous strands of cells that secrete enzymes and feed on the organic material on which they are growing. That organic material may be humus in the soil where mushrooms grow or a stale loaf of bread where mold thrives. It may be the skin between your toes inhabited by athlete's foot fungus or a decaying animal on the forest floor being decomposed by fungi digesting the animal's dead tissue. Fungi not only cause disease; they are also important decomposers that recycle nutrients from dead organisms.

The basic structure of a fungus is the **hypha** (pl., *hyphae*)—a slender filament of cytoplasm and nuclei enclosed by a cell wall (fig. 27.1). A mass of these hyphae makes up an individual organism and is collectively called a **mycelium.** A mycelium can permeate soil, water, or living tissue; fungi certainly seem to grow everywhere. In all cases, the hyphae of a fungus secrete enzymes for **extracellular digestion** of the organic substrate. Then the mycelium and its hyphae absorb the digested nutrients. For this reason, fungi are called **absorptive heterotrophs.** Heterotrophs obtain their energy from organic molecules made by other organisms.

Fungi feed on many types of substrates. Most fungi obtain food from dead organic matter and are called **saprophytes.** Other fungi feed on living organisms and are **parasites.** Many of the parasitic fungi have modified hyphae called **haustoria,** which are thin extensions of the hyphae that penetrate living cells and absorb nutrients.

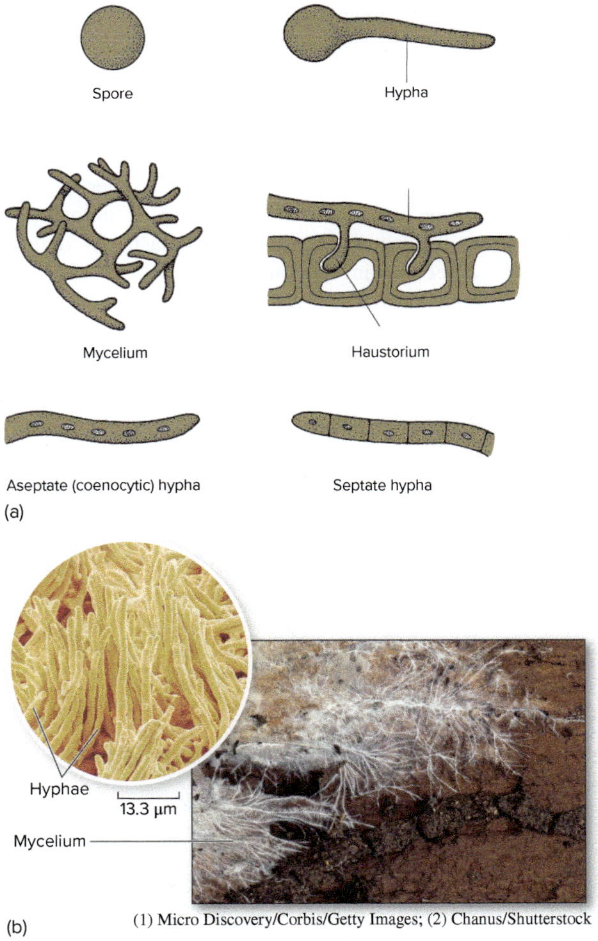

(1) Micro Discovery/Corbis/Getty Images; (2) Chanus/Shutterstock

Figure 27.1 Fungal mycelium. (*a*) Fundamental elements of fungal structure. (*b*) This mycelium, composed of hyphae, is growing through leaves on a forest floor in Maryland.

Hyphae of some species of fungi have crosswalls called **septa** that separate cytoplasm and nuclei into cells. Hyphae of other species have incomplete or no septa (i.e., are aseptate) and therefore are **coenocytic** (multinucleate). Notably, the cell walls of fungi are usually not cellulose, but instead are made of **chitin,** the same polysaccharides that comprise the exoskeleton of insects and crustaceans.

REPRODUCTION IN KINGDOM FUNGI

Reproduction in fungi can be sexual or asexual. Subtle variation in the patterns and morphology of sexual reproduction distinguish one phylum from another.

Asexual Reproduction

Asexual reproduction is usually fast and efficient. A single organism can do it. However, it does not include genetic recombination, because meiosis and fertilization are not part of asexual reproduction. Fungi commonly reproduce asexually by mitotic production of haploid vegetative cells called **spores** produced in **sporangia, conidiophores,** and other related structures. Spores are microscopic and surrounded by a covering well suited for the rigors of distribution into the environment. *Pilobolus,* an interesting fungus, points its sporangia toward the sun. This orientation of an organism to light is called **phototaxis.** *Pilobolus* ejects its entire sporangium as far as two meters to distribute its spores.

Budding and **fragmentation** are two other methods of asexual reproduction. Budding, which is mitosis followed by an uneven distribution of cytoplasm, is common in yeasts. After budding, an outgrowth of the original cell detaches and matures into a new organism. Fragmentation is the breaking of an organism into one or more pieces, each of which can develop into a new individual.

Sexual Reproduction

The sexual life history of fungi and other eukaryotes includes the familiar events of vegetative growth, genetic recombination, meiosis, and nuclear fusion (karyogamy). However, the timing of these events is unique in fungi. It's unique because the mature vegetative form of most fungi is haploid. Fungi reproduce sexually when hyphae of two genetically different individuals of the same species encounter each other. In many fungal life cycles, haploid cells (n) from two mating strains will fuse their cytoplasms, **plasmogamy,** and become dikaryotic ($n + n$) with two nuclei per cell. Later the nuclei will fuse, via a process called **karyogamy,** to become diploid ($2n$) zygotes. Karyogamy is equivalent to fertilization.

A generalized life cycle of fungi (fig. 27.2) illustrates four important features of fungi:

- Nuclei of a fungal mycelium are haploid during most of the life cycle.
- Gametes are produced by mitosis and differentiation of haploid cells rather than directly from meiosis of diploid cells.
- Meiosis quickly follows formation of the zygote, the only diploid stage.
- Haploid cells produced by meiosis are not gametes; rather, they are spores that grow into a mature haploid organism. Recall that asexual reproduction produces spores by mitosis. In both cases, haploid spores grow into mature mycelia.

None of these features of the sexual cycle are unique to fungi, but together they describe the typical fungal life cycle.

Classification of Fungi

The great diversity of fungi results from modifications of hyphae into varied and specialized reproductive structures often unique to a phylum, genus, or species. The five most diverse phyla of fungi are **Chytridiomycota, Zygomycota, Ascomycota, Basidiomycota,** and **Glomeromycota.**

As you examine members of these phyla, carefully note variations on the fundamental structure of vegetative mycelia

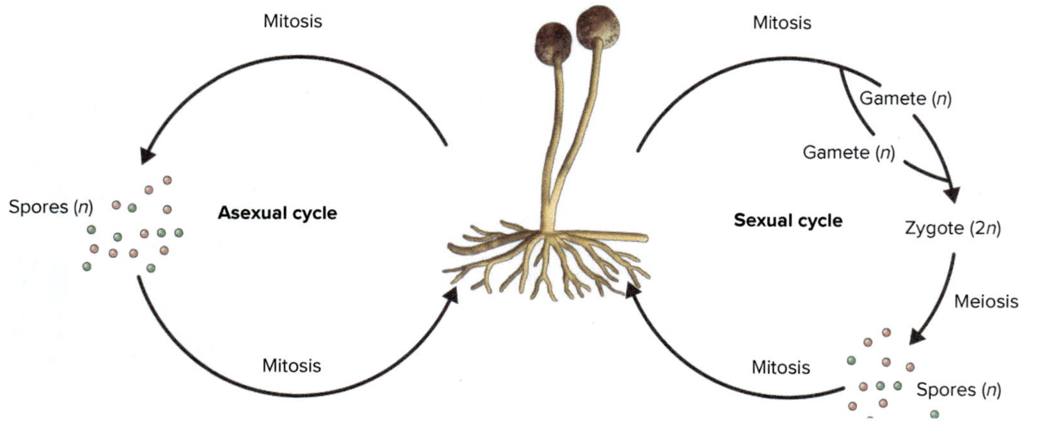

Figure 27.2 Generalized life cycle of a fungus.

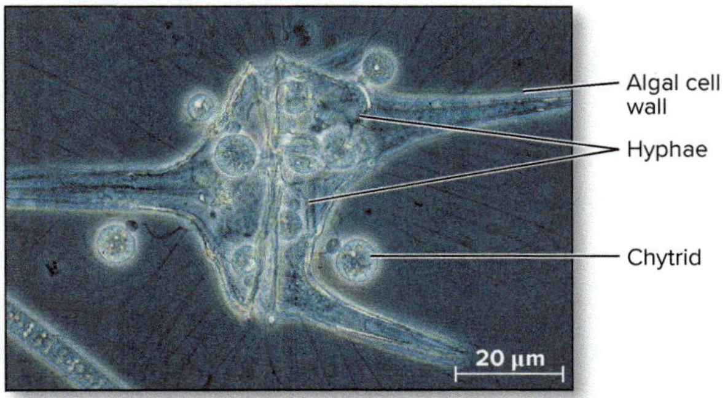

(a) Photographs by H. Cantor-Lund reproduced with permission of Freshwater Biological Association

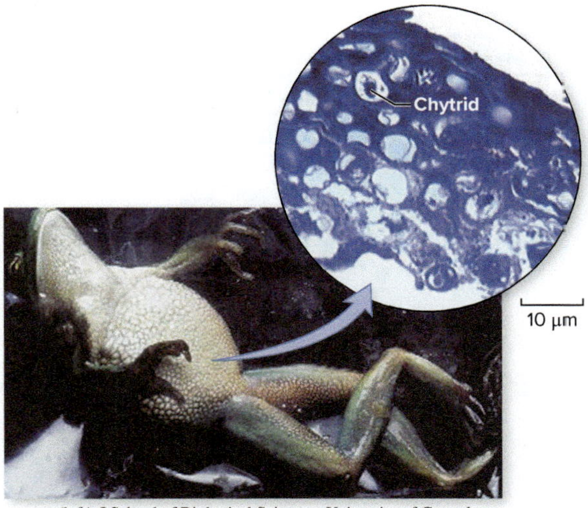

(b) (left) ©School of Biological Sciences, University of Canterbury, New Zealand; (inset) Courtesy of Dr. Peter Daszak

Figure 27.3 Chytridiomycota. (*a*) Chytrids growing on a freshwater alga. The colorless chytrids produce hyphae that penetrate the cellulose cell walls of the dinoflagellate *Ceratium hirundinella,* absorbing organic materials from the alga. Chytrids use these materials to produce spherical flagellate spores that swim away to attack other algal cells. (*b*) Frog killed by chytridiomycosis. Lesions formed by the chytrid can be seen on the abdomen of this frog.

and specialized structures associated with sexual and asexual reproduction. The names of phyla of fungi are derived from sexual reproductive structures rather than asexual structures (table 27.1). However, each phylum has various modifications for both sexual and asexual reproduction. Our research and understanding of less diverse phyla of fungi (that are not covered in this exercise) continue to develop.

PHYLUM CHYTRIDIOMYCOTA (CHYTRIDS)

Molecular genetic evidence indicates that chytrids may be the most ancient fungi. They are typically aquatic saprobes or parasites on plants, animals, and protists (fig. 27.3). They have flagella, which is unusual for fungi. They also have absorptive nutrition and chitinous cell walls and share proteins and nucleic acids common to other fungi. Their distinctive reproductive feature is motile spores with flagella. In some areas, infections by chytrids have significantly reduced the population of many amphibians (fig. 27.3*b*).

SAFETY FIRST Before coming to lab, you were asked to read this exercise so you would know what to do and be aware of safety issues. In the space below, briefly list the safety issues associated with today's procedures. If you have questions about these issues, contact your laboratory assistant before starting work.

Procedure 27.1 Examine a representative of phylum Chytridiomycota

1. Obtain a culture of *Allomyces* or *Chytridium,* two common genera of chytrids.
2. Prepare a wet mount and use your microscope to examine chytrid morphology. If necessary, review in Exercise 3 the proper use of a compound microscope.
3. Be prepared to compare chytrid morphology with the morphology common to the other major fungal groups.

Question 1

a. Are hyphae apparent?

b. Are the cells motile?

Table 27.1

Reproductive Feature of the Major Phyla of Fungi

Phylum	Key Reproductive Feature
Chytridiomycota (chytrids) 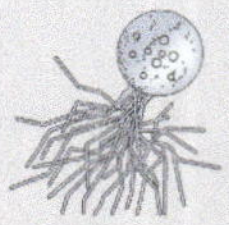	Motile spores with flagella 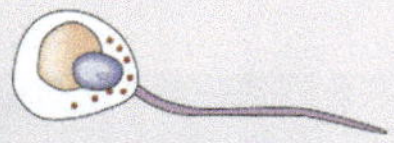
Zygomycota (zygote fungi) 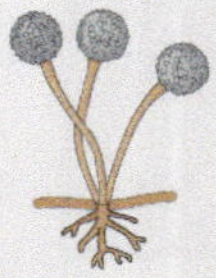	Resistant zygosporangium as sexual stage 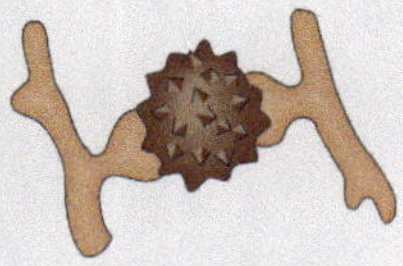
Ascomycota (sac fungi) 	Sexual spores borne internally in sacs called asci 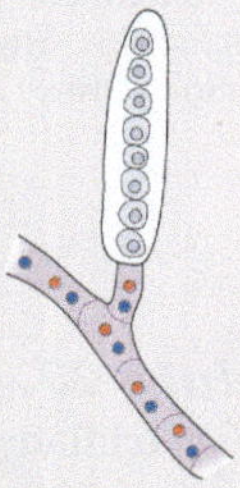
Basidiomycota (club fungi) 	Sexual spores borne externally on club-shaped structures called basidia 
Glomeromycota (plant symbionts) 	Reproduction typically asexual. Aseptate hyphae. 

PHYLUM ZYGOMYCOTA (BREAD MOLDS)

Zygomycetes (750 species), which include common bread molds, derive their name from resting sexual structures called **zygosporangia** that characterize the group. Most zygomycetes are saprophytic and their vegetative hyphae lack septa (i.e., they are aseptate).

Procedure 27.2 **Examine common bread mold**

1. Obtain from your instructor a petri dish containing a piece of moldy bread moistened and exposed to air for a few days.

2. Note the velvet texture and various colors of the molds.

3. Use a dissecting microscope to examine the mycelia and notice that they grow as a tangled mass of hyphae.

Question 2

a. How many species of mold are on the bread?

b. Is pigment distributed uniformly in each mycelium? If not, where is the pigment concentrated in each mold?

c. What is the adaptive significance of spores forming on ends of upright filaments rather than closer to the protective substrate?

A common genus of bread mold is *Rhizopus.* Its hyphae are modified into **rhizoids** (holdfasts), **stolons** (connecting hyphae), **sporangiophores** (supportive stem-like filaments), and **sporangia** (sites of asexual haploid spore formation), as shown in figure 27.4. Sporangiophores are upright hyphal filaments supporting sporangia (fig. 27.5). Within a sporangium, haploid nuclei become asexual spores and are separated by cell walls. These spores are released to the environment when the mature sporangium breaks open. *Rhizopus* appears dark from the thousands of black sporangia of a growing mycelium; hence, a common name of *Rhizopus* is black bread mold. This asexual formation of thousands of haploid spores is an efficient way to reproduce.

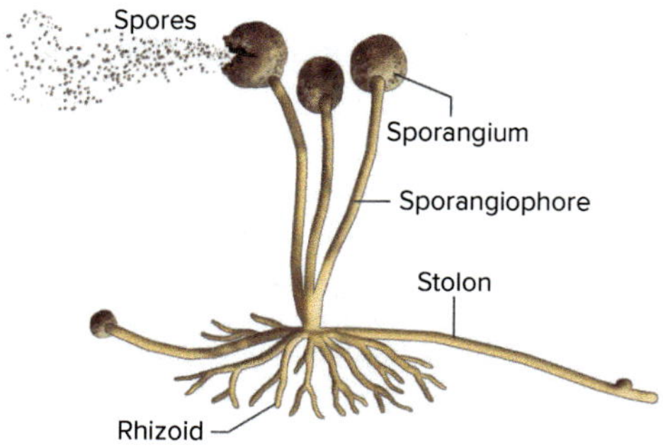

Figure 27.4 Vegetative and asexual reproductive structures of *Rhizopus,* a bread mold.

Dr. Jeremy Burgess/SPL/Science Source

Figure 27.5 A sporangium of *Rhizopus stolonifer* (500×) releases thousands of spores when it matures and breaks open.

Sexual reproduction is a more involved interaction requiring contact between two genetically different strains, i.e. a + strain and a – strain). Because *Rhizopus* is isogamous (meaning that gametes from both strains look alike), the strains are not called male and female, but instead are called + and –. In procedure 27.3 you will observe contact between + and – strains. On one side of the plate a + strain of *Rhizopus* was introduced, and on the other side a – strain was introduced. These strains grew toward each other and formed reproductive structures called **gametangia** where they came into contact.

Sexual Reproduction in *Rhizopus*

1. Sexual reproduction begins when hyphae of each strain touch each other (fig. 27.6).
2. Where the hyphae touch, **gametangia** form and appear as swellings. Within the gametangia many nuclei differentiate to serve as gametes from each strain.
3. The wall between the gametangia breaks down, and cells fuse (plasmogamy). Fusion of nuclei (karyogamy) follows and occurs in the zygosporangium.
4. A typically massive and elaborate **zygosporangium** differentiates around the zygotes. Except for the zygotes, all other nuclei are haploid.
5. Soon after the zygosporangium forms, the zygotes undergo meiosis. One or more of the resulting haploid spores soon germinate.
6. The hyphae of the germinating cells break out of the zygosporangium, produce a sporangiophore via mitosis, and form spores asexually.
7. The spores are released and produce a new generation of mycelia.

Procedure 27.3 Examine *Rhizopus*

1. Obtain a culture of asexually reproducing *Rhizopus* from your instructor. This culture is growing in a sealed petri plate containing nutrient-fortified agar. Do not remove the top of the dish because you may contaminate the culture and unnecessarily release spores into the room.
2. Your instructor has prepared a demonstration slide of *Rhizopus* stained with lactophenol cotton blue for you to examine. Examine this slide and sketch what you see.
3. Your instructor has also prepared some cultures of *Rhizopus* (or *Mucor,* or *Phycomyces*) for observing the structures for sexual reproduction. Obtain one of these plates. Locate the midline where the different strains have come in contact.
4. Locate sexual gametangia and zygosporangia along the midline of this living culture. In addition, dark, asexual sporangia will also be apparent throughout the culture.
5. Examine a prepared slide of *Rhizopus* with developed zygosporangia (figs. 27.6 and 27.7). Then observe the zygosporangia where the strains have touched. Be careful not to confuse zygosporangia with dark sporangia on top of sporangiophores.

Question 3

a. In what structure is the dark pigment of *Rhizopus* concentrated?

b. Is *Rhizopus* reproducing sexually as well as asexually in the same petri dish? How can you tell?

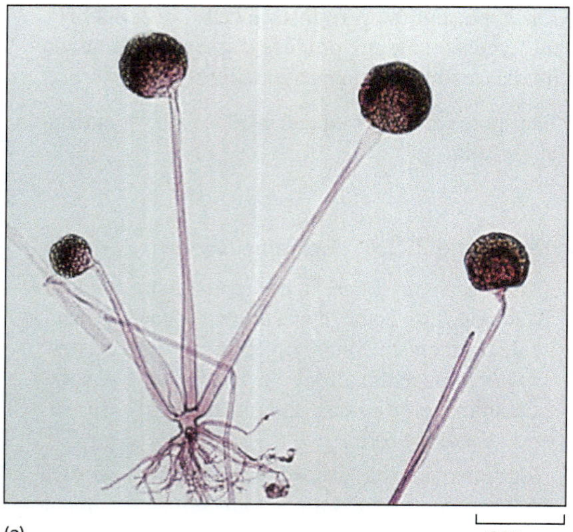

(a)

Carolina Biological Supply Company/Phototake/Medical Images

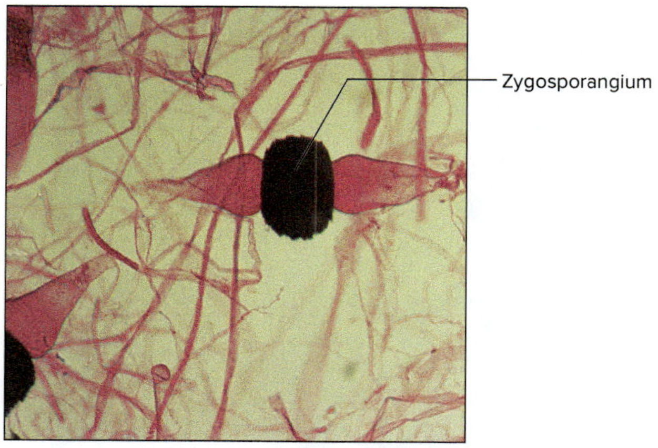

Richard H. Gross/Biological Photography

Figure 27.7 Sexual conjugation in *Rhizopus* (200×).

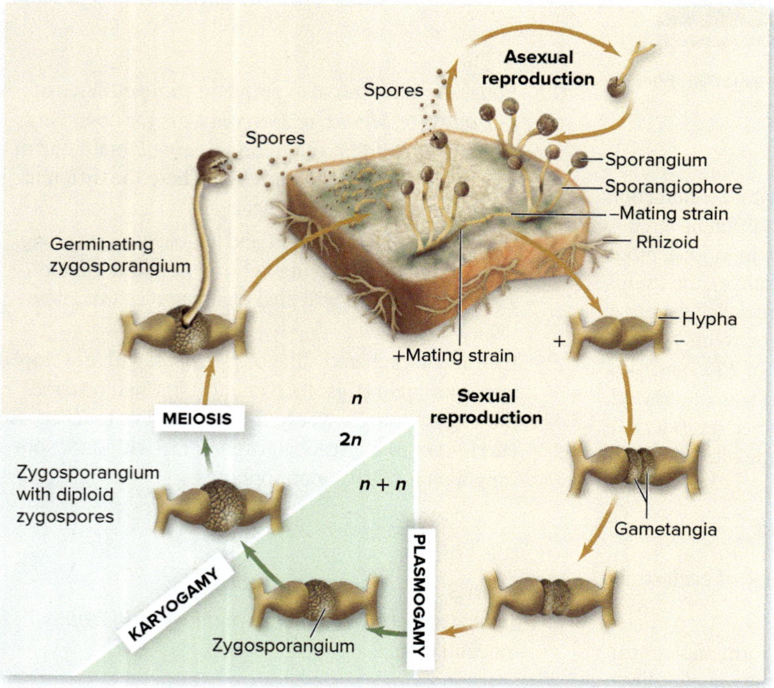

(b)

Figure 27.6 *Rhizopus* (*a*) and its life cycle (*b*). Hyphae grow and feed on the surface of the bread or other material and produce clumps of erect, sporangium-bearing stalks. If both + and – strains are present in a colony, they may grow together, and their nuclei may fuse to form diploid (2*n*) zygotes. These zygotes, which are the only diploid cells of the life cycle, form within a thick, dark structure called a zygosporangium. Meiosis occurs in the zygosporangium, and vegetative, haploid hyphae grow from the resulting haploid (*n*) cells.

PHYLUM ASCOMYCOTA (SAC FUNGI)

Phylum Ascomycota (30,000 species) includes yeasts, some molds, morels, and truffles (fig. 27.8). Its name is derived from a microscopic, sac-shaped, sexual reproductive structure called an **ascus.**

Ascomycetes reproduce asexually by forming spores called **conidia.** Modified hyphae called **conidiophores** partition nuclei in longitudinal chains of beadlike conidia (fig. 27.9). Each conidium contains one or more nuclei.

(a) ©BiologyImaging.com

(b) ©BiologyImaging.com

Figure 27.8 Representatives of Ascomycota. All visible structures of fleshy fungi, such as the ones shown here, arise from an extensive network of filaments (hyphae) that penetrate and interweave with the substrate on which they grow. (*a*) *Helvella esculenta* is a poisonous ascomycete known as "false morel." Many people have died after mistaking this mushroom for an edible species. (*b*) A cup fungus in the rain forest of the Amazon Basin.

Conidia form on the surface of conidiophores (in contrast to spores that form within sporangia in *Rhizopus*). When mature, conidia are released in large numbers and germinate to produce new organisms. *Aspergillus* and *Penicillium* are common examples of fungi that form conidia. For some ascomycetes, sexual reproduction has never been observed. However, most are likely ascomycetes that have lost their ability to sexually reproduce.

Many ascomycetes are economically important. For example, species of *Penicillium* are used to produce antibiotics. *Penicillium roquefortii,* which is abundant in caves near Roquefort-sur-Soulzon, France, gives a unique flavor to Roquefort cheese. *Aspergillus oryzae* is used to brew Japanese saki and to enrich food for livestock.

Procedure 27.4 Examine fungi with conidia

1. Obtain a culture plate of living *Aspergillus, Penicillium,* or *Neurospora*. Notice the soft texture of the colonies.
2. Use a dissecting microscope to examine the colonies' hyphae and their reproductive conidia. Note the rounded tufts of these reproductive cells.
3. Conidia are quite small. You will examine them more closely in procedure 27.5.

Procedure 27.5 Examine *Penicillium*

1. Examine a prepared slide of *Penicillium* and a pure, living culture provided by your instructor (fig. 27.10).
2. Notice the formation of conidia.

Question 4

What is the relative size of *Penicillium* hyphae compared with *Rhizopus* hyphae?

Yeasts are common unicellular ascomycetes and include about 40 genera. Most of their reproduction is asexual by cell fission or budding (i.e., the formation of a smaller cell from a larger one). However, sexual reproduction can occur. Occasionally, two sexually reproducing yeast cells will fuse to form one cell with two nuclei, a process

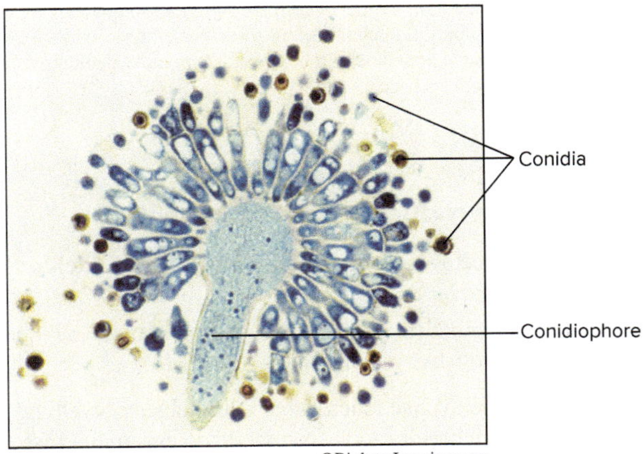

©BiologyImaging.com

Figure 27.9 Thin section through a conidiophore of *Aspergillis* sp.

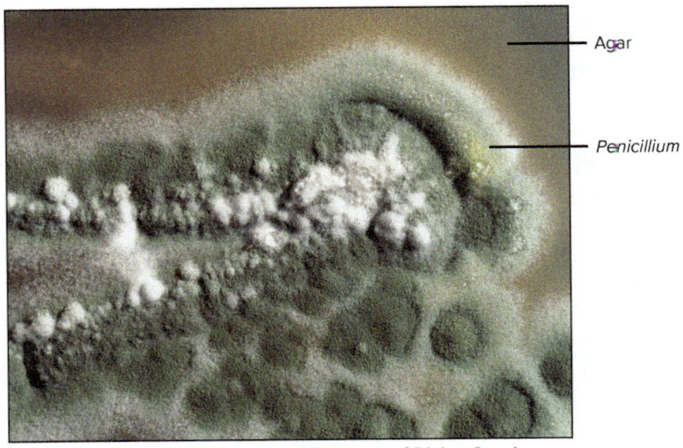

©BiologyImaging.com

Figure 27.10 *Penicillium*, an ascomycete, growing on agar.

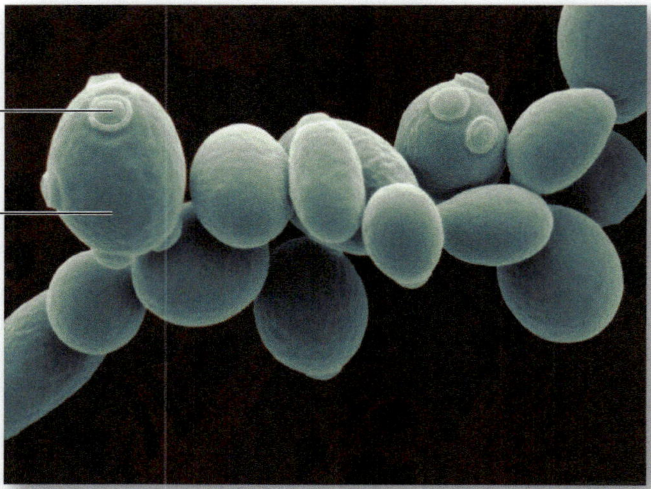

Mediscan/Alamy Stock Photo

Figure 27.11 Budding in the yeast *Saccharomyces cerevisiae* (500×).

called **plasmogamy.** This cell functions as an ascus in which fertilization is followed immediately by meiosis. The resulting haploid ascospores function directly as new yeast cells. Yeasts do not form conidia.

The yeast used to produce wine and beer is usually a strain of *Saccharomyces cerevisiae.* The yeasts and other fungi growing naturally on grapes used for making wine may impart a unique flavor to a wine more than does the specific variety of grapes.

Procedure 27.6 **Examine *Saccharomyces,* a yeast, and *Peziza,* a cup fungus**

1. Obtain and examine a stock culture of *Saccharomyces* (fig. 27.11).
2. Prepare a wet mount of the yeast from a culture dispensed by your instructor. Only a small amount of yeast is needed to make a good slide.
3. Review the description of sexual reproduction in cup fungi.
4. Examine a prepared slide of a cross section through the ascocarp of *Peziza* (fig. 27.12). Locate the asci.

Question 5

a. Do you see chains of yeast cells produced by budding?

b. How is the structure of yeast hyphae different from that of molds?

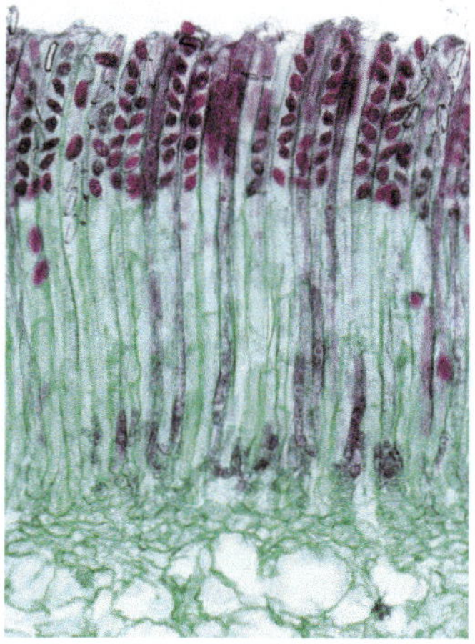
©BiologyImaging.com

Figure 27.12 Asci from the lining of a cup of *Peziza,* a cup fungus. The diploid ($2n$) zygote in each ascus divides meiotically to produce four haploid (n) nuclei. Each of these nuclei then divides mitotically. These meiotic and mitotic divisions produce a column of eight ascospores in each ascus (400×).

Sexual Reproduction in Ascomycota

Most ascomycetes such as *Peziza* follow a sexual life cycle including ascus production.

1. Sexual reproduction begins with contact of monokaryotic hyphae from two mating strains (fig. 27.13).
2. Where the hyphae touch, large multinucleate swellings appear (antheridia and ascogonia) and eventually fuse. Haploid nuclei of the two strains intermingle in the swelling (ascogonium).

(a) Robert Marien/Corbis/Getty Images

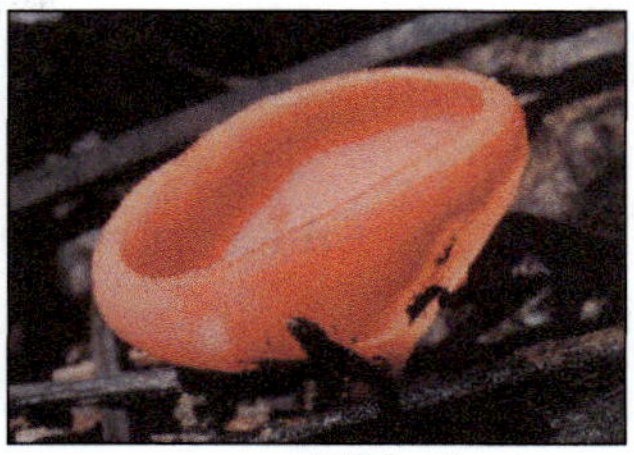

©BiologyImaging.com

(b)

Figure 27.13 The sexual life cycle of an ascomycete. Representative ascomycetes include (*a*) morels and (*b*) cup fungi, both of which undergo (*c*) a sexual life cycle including formation of characteristic asci.

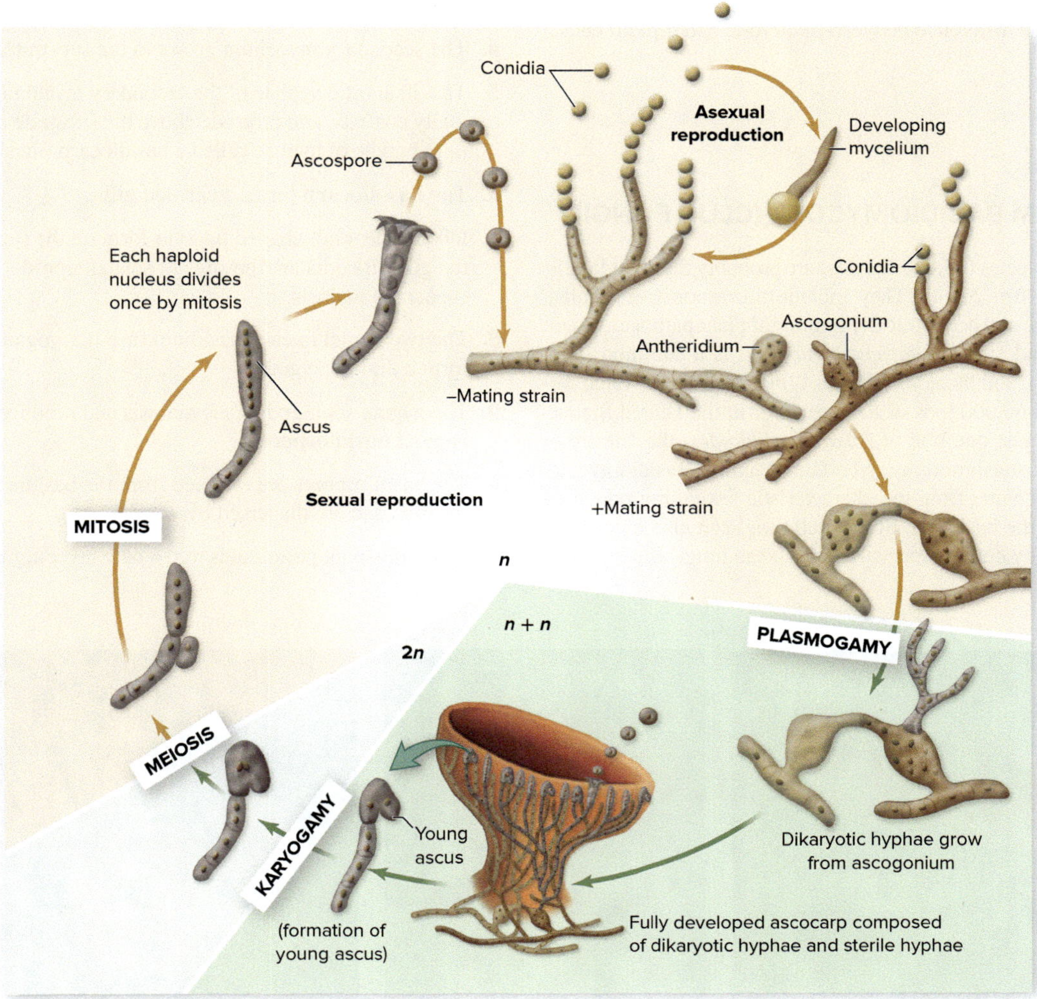

(c)

3. A **dikaryotic** mycelium grows from this swelling. Each dikaryotic cell has one nucleus from each parent; the nuclei do not fuse immediately.

4. Tightly bundled dikaryotic hyphae grow and mingle with **monokaryotic** hyphae from each parent to form a cup-shaped **ascocarp.**

5. Dikaryotic cells lining the inside of the ascocarp form sac-shaped **asci** (sing., ascus).
6. The nuclei fuse (karyogamy) in each ascus to form a zygote.
7. After fusion, meiosis produces four haploid **ascospores.**
8. Subsequent mitosis produces eight ascospores within each mature ascus.
9. The asci on the surface of the ascocarp rupture and release ascospores into the environment.
10. Each ascospore can produce a new mycelium.

Question 6
What is the difference between dikaryotic and diploid cells?

PHYLUM BASIDIOMYCOTA (CLUB FUNGI)

Basidiomycetes (25,000 species) are probably the most familiar fungi (fig. 27.14). They include mushrooms, puffballs, shelf fungi, and economically important plant pathogens such as rusts and smuts. *Agaricus campestris* is a common field mushroom, and its close relative *A. bisporus* is cultivated for more than 60,000 tons of food per year in the United States. However, just one bite of *Amanita phaloides,* the "destroying angel" mushroom, may be fatal. Phylum Basidiomycota derives its name from the characteristic sexual reproductive structure, the basidium. Some club fungi can also reproduce asexually by conidia formation, as do sac fungi. But here we will describe only sexual reproduction involving their familiar fruiting body, a mushroom.

Sexual Reproduction in Basidiomycota

1. Haploid hyphae from different mating strains permeate the substrate (fig. 27.15).
2. Septa form between the nuclei in the hyphae and form **monokaryotic primary mycelia.** A monokaryotic mycelium has one nucleus in each cell.
3. Cells of the primary mycelia of different mating strains touch, fuse (plasmogamy), and produce a **dikaryotic secondary mycelium.**
4. The secondary mycelium grows in the substrate.
5. The dikaryotic hyphae of the secondary mycelium eventually coalesce and protrude above the substrate as a tight bundle of hyphae called a basidiocarp (mushroom).
6. The **basidiocarp** forms a cap and gills.
7. Dikaryotic, club-shaped **basidia** form on the surface of the gills. Basidia are the sites of sexual reproduction, especially meiosis.
8. The two nuclei in each basidium fuse (karyogamy) to form a diploid zygote.
9. The zygote soon undergoes meiosis and produces haploid **basidiospores.**
10. The basidiospores are released from the basidia lining the gills and are dispersed by wind.
11. A basidiospore germinates into a new mycelium.

(a) ©BiologyImaging.com (b) siloto/Shutterstock (c) ©BiologyImaging.com

Figure 27.14 Representative basidiomycetes. (*a*) Fly amanita (*Amanita muscaria*). Many species of *Amanita* are poisonous. (*b*) A common stinkhorn fungus (*Phallus impudicus*). (*c*) Earthstar (*Geaster*).

(a) ©BiologyImaging.com

Figure 27.15 Mushrooms (*a*) and their life cycle (*b*). Coalescing mycelia from compatible strains produce dikaryotic hyphae, which coalesce and form the familiar fruiting bodies of mushrooms. On the gills of the mushroom cap the nuclei in dikaryotic cells fuse, undergo meiosis, and produce basidiospores that germinate into new mycelia. The basidium is the site of karyogamy, which is the fusion of nuclei to form a zygote.

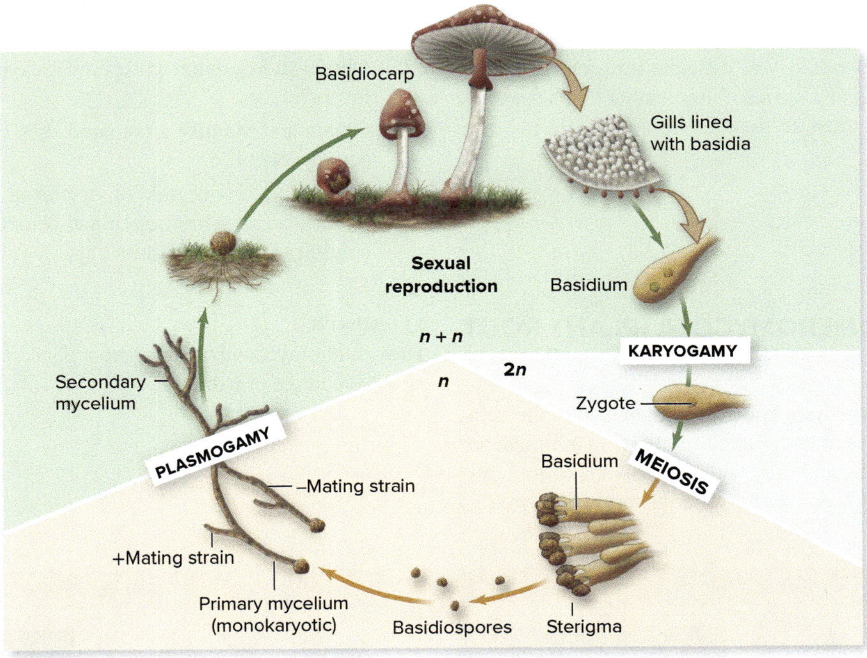

(b)

Procedure 27.7 Examine some common mushrooms and their relatives

1. Examine a specimen of an earthstar (*Geaster*). Earthstars are oddly structured basidiomycetes with an array of support structures shaped much like a star (fig. 27.14*c*).
2. Examine some mushrooms. Mushrooms are familiar examples of the aboveground portions of extensive mycelia permeating the soil. Note the mushroom's stalk and umbrellalike **cap** (also called the **pileus**).
3. Find the **gills** on the undersurface of the cap. Gills are lined with microscopic, club-shaped cells called **basidia** where sexual reproduction occurs. Phylum Basidiomycota is sometimes called the "club fungi" and derives its name from these characteristic basidia.

Procedure 27.8 Examine *Coprinus,* a common mushroom

1. Examine a prepared slide of gills from the cap of *Coprinus,* a common mushroom (fig. 27.16).
2. Note the dark basidiospores in rows along the surface of the gills. Interestingly, the gills form perpendicular to the ground and allow spores to free-fall and disperse. Research suggests that gravity influences the orientation of gills.

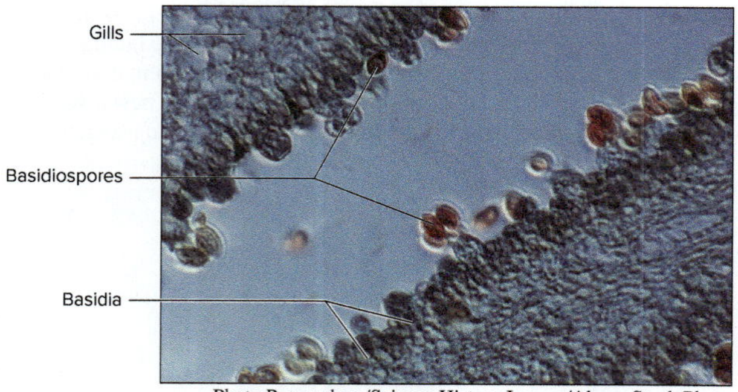

Photo Researchers/Science History Images/Alamy Stock Photo

Figure 27.16 Gills of the mushroom *Coprinus,* a basidiomycete (400×).

Question 7

How many spores would you estimate are present on the gills of a single cap of *Coprinus?* Remember that a prepared slide shows only a cross section.

PHYLUM GLOMEROMYCOTA (PLANT ROOT SYMBIONTS)

Glomeromycetes are best known for their symbiotic relationship with plant roots. Fungi in these symbioses are called **mycorrhizae,** and they act as extensions of root systems to greatly aid nutrient and water absorption as well as maintain beneficial soil consistency. Only in recent years have we realized their (1) immense biomass, (2) role in the global carbon budgets, and (3) value in facilitating nitrogen and nutrients absorption for at least 90% of all plant species. These plant species provide their symbionts with as much as 20% of the plant's fixed carbon as carbohydrates in return for the benefits of mycorrhizae.

Arbuscular mycorrhizae penetrate plant root cell walls and closely contact root cells' plasma membranes. All arbuscular mycorrhizae are glomeromycetes (fig. 27.17). **Ectomycorrhizae** are typically basidiomycetes and extend between host root cells but don't penetrate the cell wall.

Procedure 27.9 **Examine representatives of mycorrhizae.**

1. Obtain and examine a prepared slide of arbuscular mycorrhizae.
2. Obtain and examine a prepared slide of ectomycorrhizae.
3. Try to find in your field of view examples of the difference between arbuscular and ectomycorrhizal penetration of the root tissue.

Question 8

How might mycorrhizae promote a plant species occupation of a wide range of habitats?

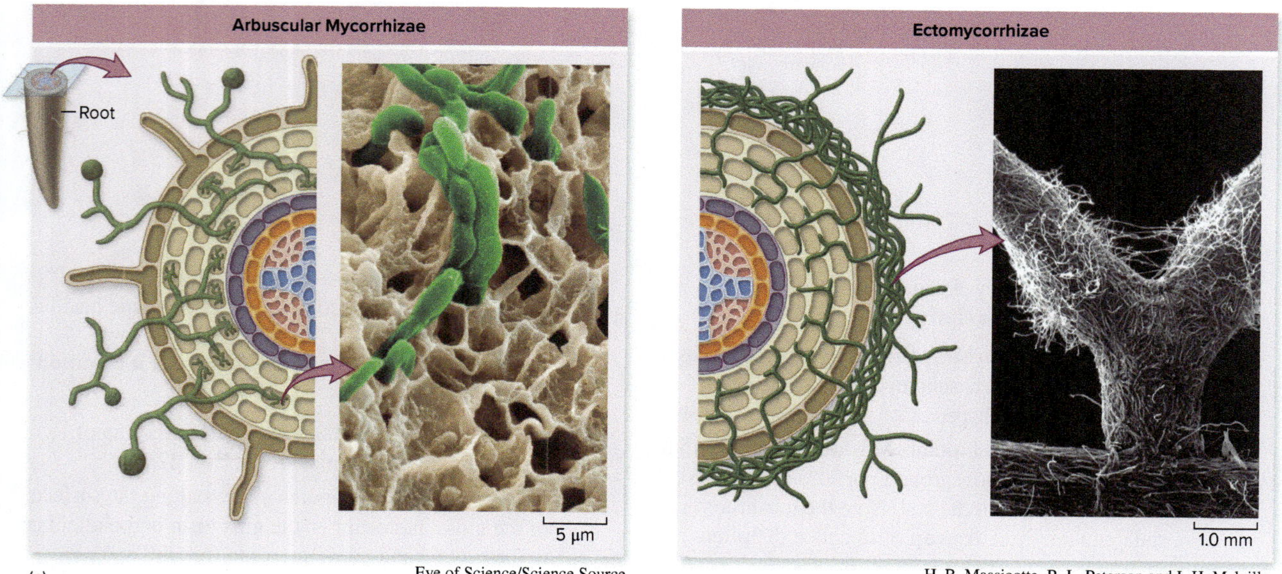

(a) Eye of Science/Science Source

(b) H. B. Massicotte, R. L. Peterson and L.H. Melville

Figure 27.17 Mycorrhizae. (*a*) In arbuscular mycorrhizae, fungal hyphae closely contact the cell's plasma membrane. (*b*) Ectomycorrhizae shown here on a tree root grow around and extend between the root cells but never penetrate the cell wall.

Question 9
To understand a forest's carbon budget, should we consider rising heat's effect on plant physiology? The effect of excess heat on fungal growth? The effect on a forest's photosynthetic rate? What might happen?

LICHENS

Lichens (25,000 species) are common, brightly colored organisms found on most hard substrates from the tropics to the artic. Trees, rocks, and firm soil provide all the support these slow-growing organisms need. A lichen includes an ascomycete (rarely other fungi) living symbiotically with a photosynthetic alga (a protist) or cyanobacterium. **Symbiosis** means living in a close and sometimes dependent association. About 26 genera of algae occur in different species of lichens. However, the green algae *Trebouxia* and *Trentepohlia* and the cyanobacterium *Nostoc* are in 90% of lichen species.

Lichens reproduce asexually by releasing fragments of tissue or specialized, stress-resistant packets of fungal and algal cells. Each of the two components (fungus and alga) may reproduce sexually by mechanisms characteristic of their phylum, and the new organisms may continue the lichen association.

The durable construction of fungi, linked with photosynthetic algae, enables lichens to proliferate in the harshest terrestrial habitats. Lichens have three basic growth forms: **crustose, foliose,** and **fruticose.** The thallus of crustose lichens grows close to the surface of a hard substrate such as rock or bark. The lichen is flat and two-dimensional. Foliose lichens adhere to their substrate, but some of the thallus peels and folds away from the substrate in small sheets. Fruticose lichens are three-dimensional and often grow away from the substrate with erect stalks. The tips of the stalks are often sites of ascus formation by the sexually reproducing ascomycete symbiont.

Lichens are extremely sensitive to air pollution. This is probably because they are adapted to efficiently absorb nutrients and minerals from the air. This makes lichens particularly susceptible to airborne toxins.

Procedure 27.10 Examine lichens

1. Examine a prepared slide of a cross section of a lichen thallus and note the intimate contact between the symbionts (fig. 27.18).
2. Examine the dried lichens on display and note the three basic growth forms: crustose, fruticose, and foliose (fig. 27.18).

Question 10
a. What are the advantages of having an alga and a fungus in a lichen? What could each organism contribute to the partnership?

b. Would you expect lichens to grow best in rural or urban environments? Why?

INQUIRY-BASED LEARNING

How effective are antimicrobial properties of common fungi?

Observation: Many fungi produce antibiotics that hinder or stop the growth of microbes. These antibiotics are adaptive because they reduce competition and protect the fungus from predators.

Question: Do common fungi such as bread mold produce antimicrobial compounds?

a. Establish a working lab group and obtain Inquiry-Based Learning Worksheet 27 from your instructor.
b. Discuss with your group well-defined questions relevant to the preceding observation and question. Choose and record your group's best question for investigation.
c. Translate your question into a testable hypothesis and record it.
d. Outline on Worksheet 27 your experimental design and supplies needed to test your hypothesis. Ask your instructor to review your proposed investigation.
e. Conduct your procedures, record your data, answer your question, and make relevant comments.
f. Discuss with your instructor any revisions to your questions, hypotheses, or procedures. Repeat your work as needed.

Figure 27.18 Lichen morphology. (*a*) A section of a compact crustose lichen shows the placement of the algal cells and the fungal hyphae, which encircle and penetrate the alga. (*b*) Thin sheets of crustose lichens adhere tightly to hard surfaces. (*c*) Fruticose lichens are shrublike. (*d*) Foliose lichens are leaflike. (*e*) This cross section of a lichen thallus shows the distribution of symbiotic algal cells and fungal hyphae. Algal cells are most dense at the upper surface to efficiently collect light, conduct photosynthesis, and provide carbohydrates to the hyphae.

Questions for Further Study and Inquiry

1. Mushrooms often sprout from soil in rows or circles commonly called "fairy rings." How would you explain the shapes of these formations?

2. What advantages does asexual reproduction have over sexual reproduction?

3. Does dominance of the haploid condition in a fungal life cycle offer an adaptive advantage? Why or why not?

4. Compare and contrast the structure of a fungal mycelium with the structure of a filamentous alga.

5. What is the advantage of maintaining a dikaryotic condition rather than immediate nuclear fusion?

6. In fungi, the only distinction between a spore and a gamete is function. Explain.

7. Describe three ways that fungi affect your life.

8. What are the major differences in the five phyla of fungi?

9. Draw and label a life cycle for one fungal phylum.

10. What products or activities by fungi benefit other organisms? What products or activities of fungi are harmful to other organisms?

DOING BIOLOGY YOURSELF

Wine-making is a multimillion-dollar industry, but the biology of wine-making is simple. Make your own special brand of wine by following the instructions in Exercise 12. Good luck!

WRITING TO LEARN BIOLOGY

Compare and contrast the fundamental life cycle of a fungus with that of plants and animals. When does meiosis occur in the sequence of events? Which stages are haploid and which are diploid?

Survey of the Plant Kingdom

Liverworts, Mosses, and Hornworts of Phyla Hepatophyta, Bryophyta, and Anthocerophyta

EXERCISE 28

Learning Objectives

By the end of this exercise you should be able to:

1. Describe the life histories and related reproductive structures of bryophytes.
2. Describe the distinguishing features of liverworts, mosses, and hornworts.
3. Describe some of the key adaptations that allow liverworts, mosses, and hornworts to live on land.
4. Describe the role of bryophytes in the environment.

Please visit connect.mheducation.com to review online resources tailored to this lab.

Plants evolved from a freshwater green algal species more than 550 million years ago and today comprise a remarkably diverse group of multicellular organisms. With few exceptions, plants are autotrophic, contain chlorophyll *a,* and have cell walls containing cellulose. Life cycles of all members of the plant kingdom are variations on alternation of generations. Be sure to review in your textbook this generalized life cycle and be familiar with the major stages presented in figure 28.1. In this and upcoming exercises, you'll learn how various modifications of this generalized life cycle characterize all of the phyla that comprise the plant kingdom.

You will study the following major groups of plants:

Phylum	*Example*
Hepatophyta	liverworts
Bryophyta	mosses
Anthocerophyta	hornworts
Pterophyta	ferns
Lycophyta	club mosses
Cycadophyta	cycads
Ginkgophyta	*Ginkgo*
Coniferophyta	conifers
Gnetophyta	gnetophytes
Anthophyta	flowering plants

These groups of plants are distinguished by morphology, life cycle, and the presence or absence of vascular tissues (fig. 28.2). Pay special attention to variations in each of these characteristics as you survey the plant kingdom in upcoming weeks.

Bryophytes include liverworts, mosses, and hornworts and are the most primitive group of terrestrial plants.

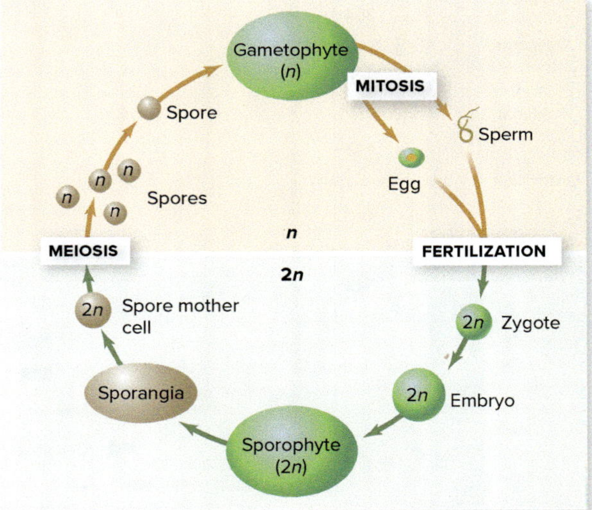

Figure 28.1 Alternation of generations. In a generalized plant life cycle, gametophytes, which are haploid (n), alternate with sporophytes, which are diploid ($2n$). Antheridia (male) and archegonia (female) are the sex organs produced by the gametophyte; they produce sperm and eggs, respectively. The sperm and egg fuse during fertilization to produce the first diploid cell of the sporophyte generation, the zygote. Meiosis occurs within sporangia, which are the spore-producing organs of the sporophyte. The resulting spores are haploid and are the first cells of the gametophyte generation.

Bryophytes are green, have rootlike structures called **rhizoids,** and may have stem and leaflike parts. Bryophytes do not generally possess specialized vascular tissues, which transport materials between roots and shoots. This lack of developed vascular tissues in bryophytes typically limits their distribution to moist habitats because their rhizoids neither penetrate the soil very far nor absorb many nutrients. Also, the lack

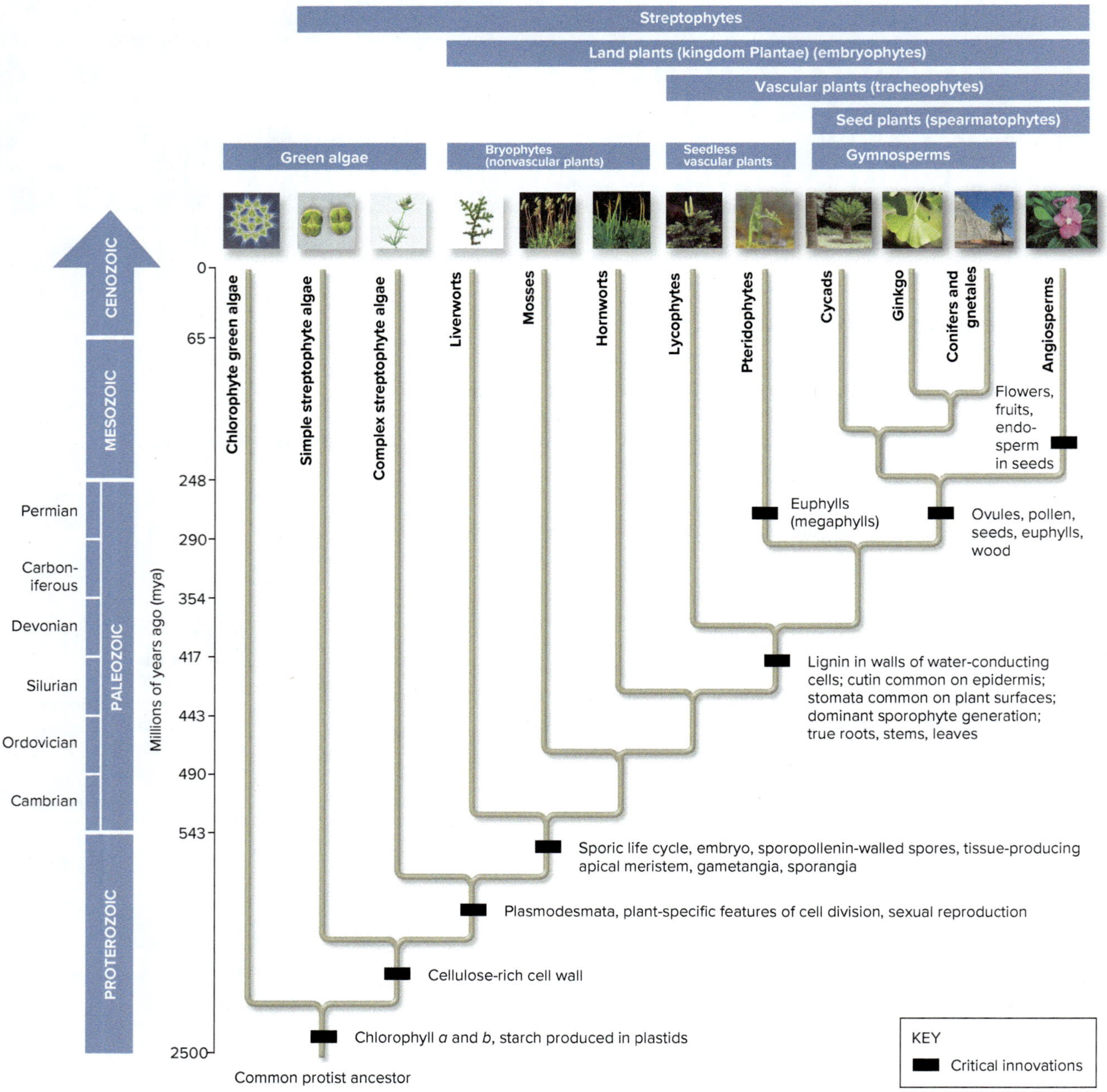

(a) Roland Birke/Phototake/Medical Images; (b) Pavel Skaloud; (c-f) Lee W. Wilcox; (g) Ed Reschke/Getty Images; (h) Lars Porta/Shutterstock; (i) Hoberman Collection/Alamy Stock Photo; (j) Peter Frank/Getty Images; (k) Kevin Schafer/Alamy Stock Photo; (l) itman__47/Shutterstock

Figure 28.2 The evolutionary history of plants. The evolution of plants has produced several important adaptations, including protected multicellular embryos that enabled plants to successfully invade a variety of terrestrial habitats. Land plants gradually evolved several adaptations that enabled them to thrive in terrestrial habitats. Streptophyte algae refers to those algal species with specific features in common with land plants.

of vascular tissues necessitates that their photosynthetic and nonphotosynthetic tissues be close together. Because vascular tissues, along with supporting tissues, are often absent, bryophytes are relatively small and inconspicuous. Despite their diminutive size, however, bryophytes occur throughout the world in habitats ranging from the tropics to Antarctica.

There are approximately 24,000 species of bryophytes, more than any other group of plants except the flowering plants. Bryophytes fix CO_2, degrade rocks to soil, stabilize soil, and reduce erosion. Humans have used bryophytes in a number of ways, including as a fuel, in the production of Scotch whiskey, and as packing materials. In many terrestrial ecosystems, bryophytes reduce the leaching of nitrogen and other nutrients from the soil.

The plant body of bryophytes is called a **thallus** (pl., *thalli*). Liverwort thalli are flattened dorsoventrally (from back and front plane, rather than from side to side plane) and are bilaterally symmetrical (i.e., have two equal halves). For comparison, moss thalli are erect and radially symmetrical (circular). Hornwort thalli are similar to those of liverworts.

The life cycle of bryophytes is characterized by a distinct alternation of generations that includes both gametophyte and sporophyte phases, *but the gametophyte is the dominant phase* (fig. 28.3) because it is larger and lives longer. Bryophytes have multicellular sex organs in which gamete-producing cells are enclosed in a jacket of sterile cells. **Antheridia** are male sex organs that produce swimming, biflagellate sperm. Bryophytes require free water—usually provided as rain or dew—for sexual reproduction because their sperm must swim to eggs. These sperm fertilize eggs produced in **archegonia,** the female sex organs. The fertilized egg is called a **zygote.** This zygote divides and matures in the **archegonium** to produce the **sporophyte,** which remains attached to and nutritionally dependent on the **gametophyte.** The mature sporophyte produces haploid spores (via meiosis), each of which can develop into a gametophyte.

Some bryophytes have a sperm-delivery system that improves their rate of sexual reproduction. For example, fertile moss plants produce scented compounds that attract tiny insects such as mites and springtails; sperm can attach to these insects as they wade through dew on these plants. When the insects are lured to another moss plant by its scent, the insects deliver the hitchhiking sperm and, in the process, increase the rate of fertilization.

Before continuing this exercise, review figure 28.1; be sure you understand the ploidy (i.e., haploid versus diploid) of the gametophyte and sporophyte generations of the life cycle.

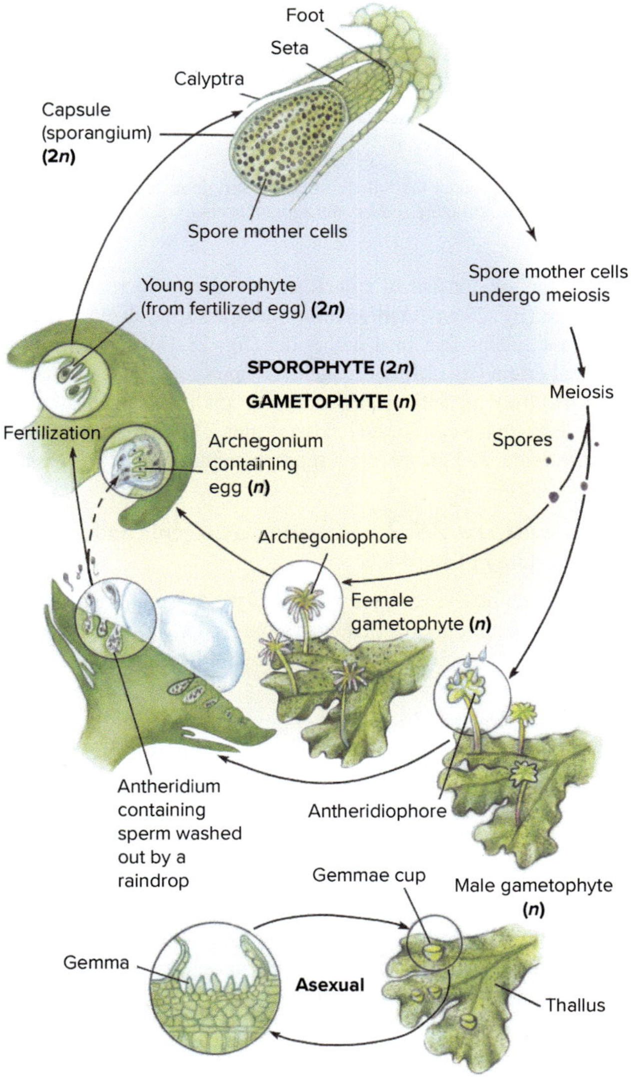

Figure 28.3 Life cycle of *Marchantia,* a liverwort. During sexual reproduction, spores produced in the capsule germinate to form independent male and female gametophytes. The archegoniophore produces archegonia, each of which contains an egg; the antheridiophore produces antheridia, each of which produces many sperm. After fertilization, the sporophyte develops within the archegonium and produces a capsule with spores. *Marchantia* reproduces asexually by fragmentation and gemmae.

PHYLUM HEPATOPHYTA: LIVERWORTS

Liverworts are the earliest land plants. Although many liverworts are "leafy," we will restrict our observations to a thallus-type liverwort, *Marchantia.* The gametophytic thallus of this liverwort grows as a large, flat, photosynthetic structure on the surface of the ground (fig. 28.4).

Liverwort Gametophyte

Procedure 28.1 **Examine the thallus of *Marchantia***

1. Observe some living *Marchantia* and note the Y-shaped (dichotomous) growth. Rhizoids extend downward from the lower (ventral) surface of the thallus.

Question 1

What are the functions of rhizoids?

2. View the upper (dorsal) surface of the thallus with a dissecting microscope and note the pores in the center of the diamond-shaped areas. Obtain a prepared slide of a thallus of *Marchantia* and locate a pore in cross section. These pores in the dorsal surface of the thallus overlie air chambers containing **chlorenchyma** (chloroplast-containing) cells.

(a) Ed Reschke/Stone/Getty Images

(b) Ed Reschke/Getty Images

Figure 28.4 *Marchantia.* The flat, leafy thallus of this liverwort grows close to the ground. (*a*) A thallus bearing upright male reproductive structures called antheridiophores (3×). (*b*) A thallus bearing upright female reproductive structures called archegoniophores (4×).

Question 2
What is the function of these pores?

3. Gently return the living materials to their containers.

Asexual Reproduction in Liverworts

Liverworts can reproduce asexually via fragmentation. In this process, the older, central portions of the thallus die, leaving the growing tips isolated to form individual plants.

Structures called **gemmae cups** occur on the dorsal (upper) surface of some thalli near the midrib (fig. 28.5). Gemmae cups represent another means of asexual reproduction by liverworts. Inside the gemmae cups are lens-shaped outgrowths called **gemmae** (sing., *gemma*), which are splashed out of the cup by falling drops of rain. If a gemma lands in an adequate environment, it can produce a new gametophyte plant.

Examine a prepared slide of gemmae cups. Also examine available live or preserved material. In the following space, diagram and label what you see, and compare it to figure 28.5.

Sexual Reproduction in Liverworts

Many species of *Marchantia* are **dioecious,** meaning that they have separate male and female plants. Gametes from each plant are produced in specialized sex organs borne on upright stalks (fig. 28.4). **Archegoniophores** are specialized stalks on female plants that bear archegonia. Each flask-shaped archegonium consists of a **neck** and a **venter,** which contains the **egg** (fig. 28.6*a*). **Antheridiophores** are specialized stalks on male plants that bear antheridia (fig. 28.4). Sperm form in antheridia (fig. 28.6*b*). Flagellated sperm are released and washed from the antheridia during wet conditions and eventually fertilize the egg located in the venter. The zygote remains in the venter and grows into a sporophyte plant.

Procedure 28.2 Examine archegonia and antheridia of liverworts

1. Examine living or prepared liverworts having mature archegoniophores that bear archegonia. Archegonia at various stages of development are located on the ventral surface.
2. Locate an egg in an archegonium. Notice a pattern of evolution in plants as well as in animals—eggs are larger and fewer in number, while sperm cells are smaller but greater in number.
3. Examine living or preserved liverworts with mature antheridiophores bearing antheridia.
4. Examine a prepared slide of cross sections of an antheridiophore. Antheridia are located just below the upper surface of the disk in a chamber that leads to the surface of the disk through a pore.

Question 3
How do the positions of the archegonium and antheridium relate to their reproductive function?

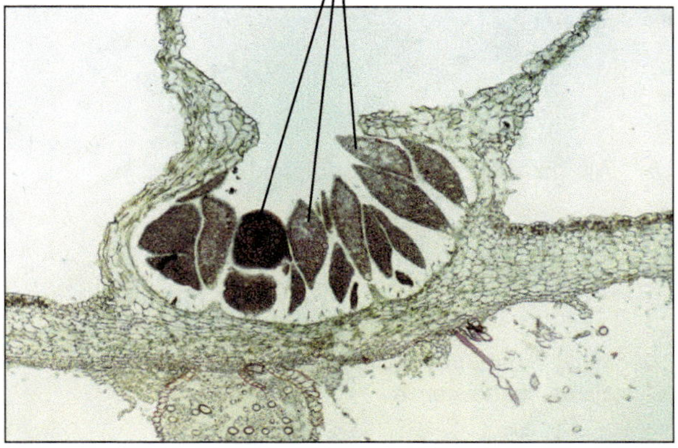

Figure 28.5 (*a*) Gemmae cups ("splash cups") containing gemmae on the gametophytes of a liverwort. Gemmae are splashed out of the cups by raindrops and can then grow into new gametophytes, each identical to the parent plant that produced it by mitosis (5×). (*b*) Longitudinal section of a gemmae cup (10×).

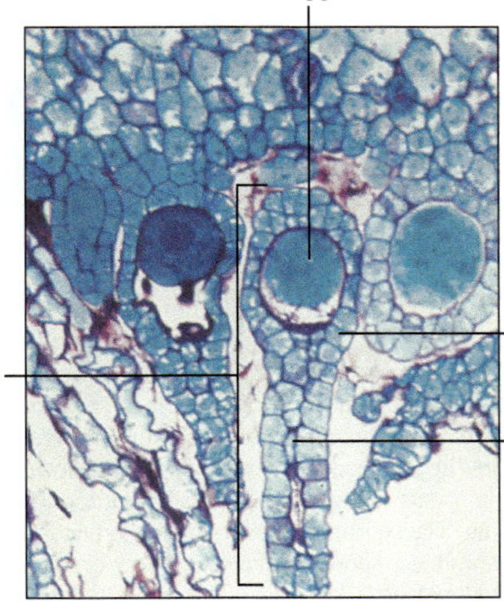

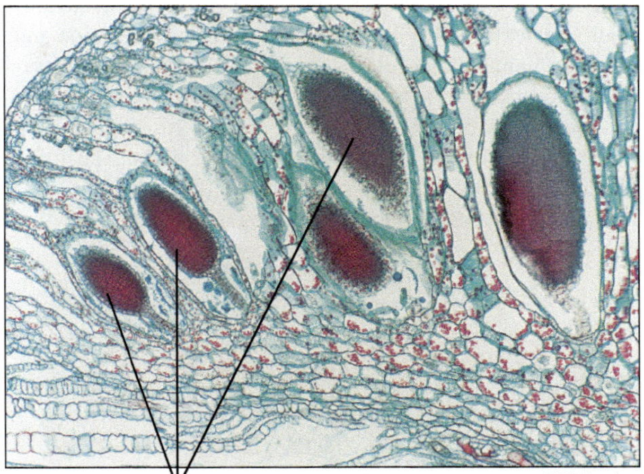

Figure 28.6 (*a*) Section through an archegonium of *Marchantia.* A single egg differentiates within the archegonium (320×). (*b*) Section through an antheridiophore, showing individual antheridia (90×).

5. Return materials and slides to their containers.

Liverwort Sporophyte

Procedure 28.3 **Examine sporophytes of liverworts**

1. Examine a prepared slide of a sporophyte of *Marchantia.* The nonphotosynthetic sporophyte is connected to the gametophyte by a structure called the **foot. Spores** are produced by meiosis in a **capsule** located on a **seta** (stalk) that extends downward from the foot (fig. 28.3).

2. Locate elongate cells called **elaters** among the spores. Elaters help disperse spores by twisting. In humid conditions the elaters coil, but when it is dry the elaters expand, pushing the spores apart and rupturing the spore case to release the spores.

3. Gamete release, fertilization, spore release, and germination are most efficient in individually specific environmental conditions.

Question 4

a. What is the function of the foot?

b. Are the spores haploid or diploid?

c. What is the functional significance of the response of elaters to moisture?

PHYLUM BRYOPHYTA: MOSSES

Mosses are often more visible than liverworts because of their greater numbers and more widespread distribution and because gametophyte plants of mosses are leafy and usually stand upright. Mosses also withstand desiccation better than do liverworts. Consequently, mosses often grow in a greater diversity of habitats than do liverworts. The moss gametophyte is radially symmetrical and is the most conspicuous phase of the moss life cycle (fig. 28.7).

Moss Gametophyte

Procedure 28.4 **Examine mosses**

1. Observe the living moss on display called *Polytrichum* (fig. 28.8). The "leafy" green portions of the moss are the gametophytes and are often only one cell thick (except at the midrib).
2. Make a wet mount of a single leaflet and examine it with low magnification.

Question 5

a. How many cells thick is the leaflet?

b. Is there a midrib vein?

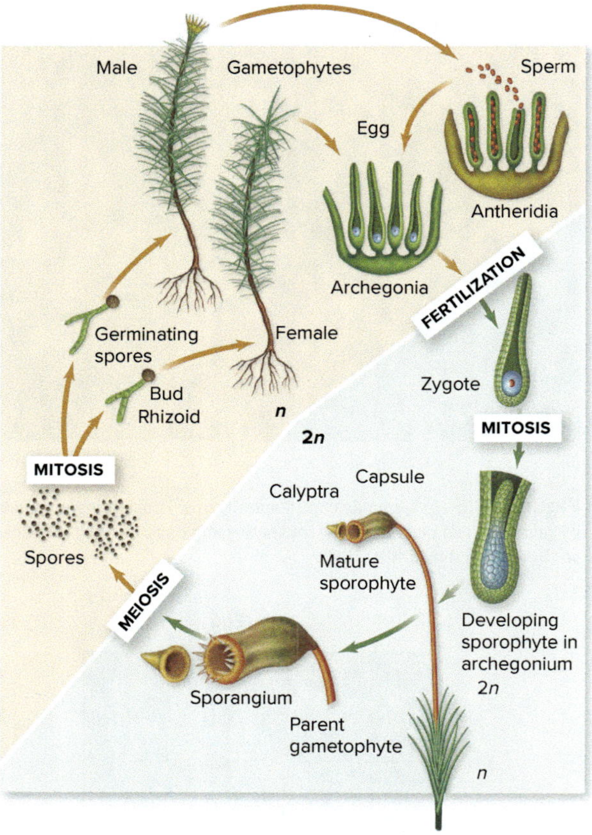

Figure 28.7 Moss life cycle. Haploid *(n)* sperm are released from antheridia on the male gametophytes. The sperm then swim through water to the archegonia and down their necks to fertilize the eggs. The resulting diploid (2*n*) zygote develops into a diploid sporophyte. The sporophyte grows out of the archegonium and differentiates into a slender seta with a swollen capsule at its apex. The capsule is covered with a cap, or calyptra, formed from the archegonium. The sporophyte grows on the gametophyte and eventually produces spores by meiosis. The spores germinate, giving rise to gametophytes. The gametophytes intially grow along the ground. Ultimately, buds produce leafy gametophytes.

c. Are pores visible on the leaf surface?

d. How does the symmetry of a moss gametophyte compare with that of a liverwort gametophyte?

e. In what ways are mosses morphologically similar to liverworts? In what ways are they different?

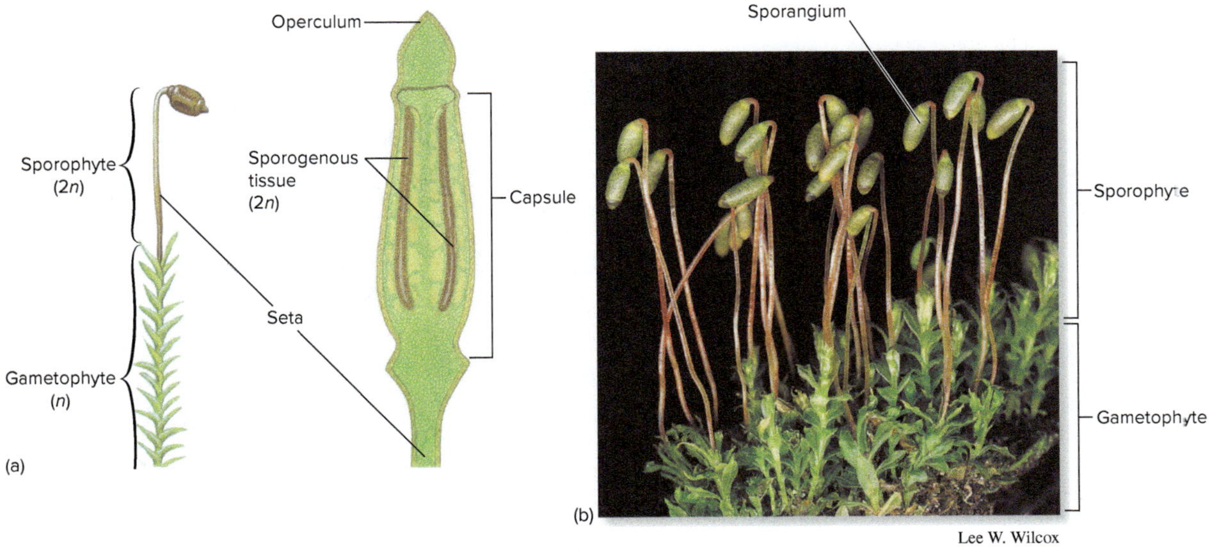

Figure 28.8 The structure of a moss. (*a*) Diagram of the parts of a mature mass sporophyte. (*b*) A hair cap moss, *Mnium.* The leaves belong to the gametophyte. Each of the stalks, with the capsule at its summit, is a sporophyte. Although moss sporophytes may be green and perform a limited amount of photosynthesis when they are immature, they are soon completely dependent, in a nutritional sense, on the gametophyte.

Moss gametophytes have specialized cells that aid in the absorption and retention of water. Mats of moss act, in effect, like sponges. The following procedure demonstrates the water-absorbing potential of mosses.

Procedure 28.5 **Water absorption by moss**

1. Weigh 3 g of *Sphagnum,* a peat moss, and 3 g of paper towel.
2. Add the moss and towel to separate beakers each containing 100 mL of water.
3. After several minutes, remove the materials from the beaker.
4. Measure the amount of water left in each beaker by pouring the water into a 100-mL graduated cylinder. Remember that 1 mL of water weighs 1 g.
5. Record your data.

Question 6

a. How many times its own weight did the moss absorb?

b. How does this compare with the paper towel?

c. Why is *Sphagnum* often used to ship items that must be kept moist?

6. Return all materials to their containers and clean your work area.

Asexual Reproduction in Mosses

Unlike liverworts, mosses lack structures such as gemmae for asexual reproduction. Mosses reproduce asexually by fragmentation.

Sexual Reproduction in Mosses

Most mosses, like liverworts, are dioecious (i.e., male and female reproductive structures are on separate individuals) Archegonia or antheridia are borne either on tips of the erect gametophyte stalks or as lateral branches on the stalks. The apex of stalks of the female plant (the plant that bears archegonia) appears as a cluster of leaves, with the archegonia buried inside.

Procedure 28.6 **Examine archegonia and antheridia of mosses**

1. Examine living or preserved mosses having mature archegonia.

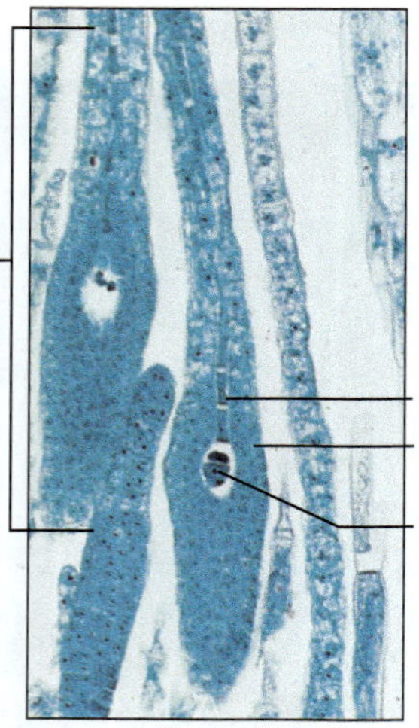

Courtesy G. S. Ellmore

Figure 28.9 A longitudinal section through the tip of a female gametophyte of a moss (45×).

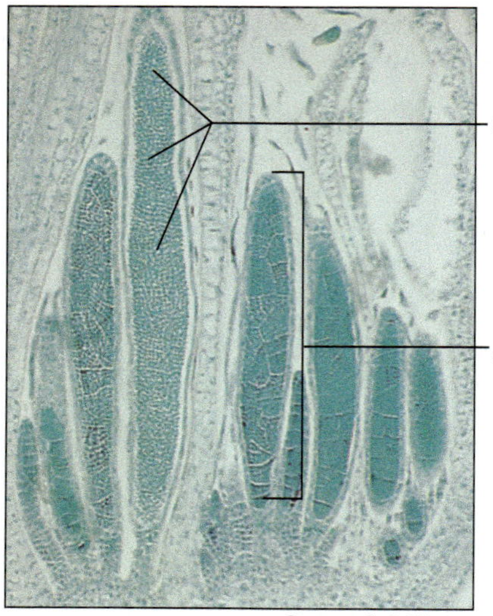

Courtesy G. S. Ellmore

Figure 28.10 A longitudinal section through the tip of a male gametophyte of a moss (45×).

2. Examine a prepared slide of moss archegonia (fig. 28.9). Note the canal that leads through the neck and terminates in the venter of the archegonium. When the archegonium matures, cells lining the neck disintegrate and form a canal leading to the egg. Sperm, following a chemical attractant released by the archegonium, swim through this canal to reach the egg.

Question 7
Where is the egg located in the archegonium?

3. Examine living or preserved mosses having mature antheridia. The male plant (i.e., the plant that bears antheridia) has a platelike structure on the tip with the "leaves" expanding outward to form a rosette. This structure is sometimes called a "moss flower" because of its appearance or a "splash cup" because of its function (i.e., the dispersal of sperm by falling raindrops).
4. Examine a prepared slide of moss antheridia, which appear as elongate, saclike structures (fig. 28.10).
5. Locate the outer sterile jacket and the inner mass of cells destined to become sperm.

Question 8
Are sperm haploid or diploid?

Moss Sporophyte

Moss sporophytes consist of **capsules** located atop stalks, called **setae,** that extend upward from the moss gametophyte (fig. 28.8*a*). A sporophyte is attached to the gametophyte by a structure called a **foot.**

Question 9
Is the sporophyte more prominent in mosses or liverworts?

The capsule atop the seta is covered by the hoodlike **calyptra,** which is a thin hood of tissue produced by the archegonium that covers the developing sporohyte as it matures. The calyptra falls off when the capsule matures. Inside the capsule are numerous haploid spores formed by meiosis.

Question 10
What is the adaptive significance of the seta of the sporophyte growing well above the mat of the gametophytes?

If enough living moss is available in the lab, remove the calyptra from a sporophyte capsule. On the tip of the capsule is a lidlike structure called the **operculum.** Remove the operculum and notice the hairlike teeth lining the opening of the lidlike structure. These teeth help control the release of spores from the capsule. In wet weather these teeth bend inward and prevent release of the spores. In dry weather they bend outward, facilitating distribution of spores by the wind.

Crush the capsule in water on a microscope slide and note the large number of spores released.

Question 11

a. What process produces spores?

b. Is the capsule haploid or diploid?

Moss spores germinate and form a photosynthetic **protonema,** which resembles a branching, filamentous alga. Leafy moss plants arise from "buds" located along the protonema.

Question 12

Can you think of any evolutionary implications of the similarity between a moss protonema and a filamentous green alga?

PHYLUM ANTHOCEROPHYTA: HORNWORTS

The hornworts are the smallest group of bryophytes; there are only about 100 species in six genera. Hornworts have several features that distinguish them from most other bryophytes. The sporophyte is shaped like a long, tapered horn that protrudes from a flattened thallus. Also, archegonia are not discrete organs. Rather, they are embedded in the thallus and are in contact with surrounding vegetative cells.

Figure 28.11 *Anthoceros,* a hornwort. The "horns" of a hornwort are sporophytes that grow up into the air,whereas the gametophytes grow flat along the ground. At maturity, the tips of hornwort sporophytes open and release spores.

The most familiar hornwort is *Anthoceros,* a temperate genus (fig. 28.11). Examine some living and preserved *Anthoceros.* Locate the gametophyte thallus and the rhizoids extending from its lower surface. Also locate the hornlike sporophytes extending from the upper surface. Spores are produced in the horn of the sporophyte. If prepared slides are available, locate spores in various stages of development within the sporophyte.

ORGANIZING WHAT YOU'VE LEARNED

Go to page 362, which follows Exercise 31. Complete the portions of the table that refer to the plants you studied in today's lab.

INQUIRY-BASED LEARNING

What are the roles of bryophytes in the environment?

Observation: Although they are usually overlooked, bryophytes often grow in places that other plants cannot grow and are important parts of many ecosystems. Find a place where bryophytes are growing on campus. Consider what these plants are doing in that ecosystem and why they are not in other ecosystems on campus.

Question: What conditions are best suited for the growth of bryophytes?

a. Establish a working lab group and obtain Inquiry-Based Learning Worksheet 28 from your instructor.

b. Discuss with your group well-defined questions relevant to the preceding observation and question. Choose and record your group's best question for investigation.

c. Translate your question into a testable hypothesis and record it.

d. Outline on Worksheet 28 your experimental design and supplies needed to test your hypothesis. Ask your instructor to review your proposed investigation.

e. Conduct your procedures, record your data, answer your question, and make relevant comments.

f. Discuss with your instructor any revisions to your questions, hypotheses, or procedures. Repeat your work as needed.

Questions for Further Study and Inquiry

1. Compare and contrast the complexity of bryophytes and algae regarding their morphology, habitat, asexual reproduction, and sexual reproduction.

2. What event begins the sporophyte phase of the life cycle? Where does this event occur in liverworts and mosses?

3. What event begins the gametophyte phase of the life cycle? Where does this event occur in liverworts and mosses?

4. What features distinguish a moss from a liverwort?

5. Diagram the life cycle of a liverwort, indicating which stages are sporophytic and which are gametophytic.

6. Diagram the life cycle of a moss, indicating which stages are sporophytic and which are gametophytic. Also indicate where meiosis and syngamy occur.

7. How did liverworts get their name?

8. What ecological roles do liverworts, mosses, and hornworts play in the environment?

9. Is the sporophyte of mosses ever independent of the gametophyte? Explain.

10. Why do you think that bryophytes are sometimes referred to as the amphibians of the plant kingdom?

11. What limits the height of mosses?

DOING BIOLOGY YOURSELF

Liverworts and mosses are surprisingly diverse. Make a simple collection of living liverworts and mosses from two or three sites. Bring the collection to class and compare yours with those of other students. Use reference books in your lab or library to identify the plants that you've collected.

WRITING TO LEARN BIOLOGY

Because water is required for the swimming sperm to reach the archegonium, would you say that this means that bryophytes are not truly land plants? Why or why not?

EXERCISE 29

Survey of the Plant Kingdom

Seedless Vascular Plants of Phyla Pterophyta and Lycophyta

Learning Objectives

By the end of this exercise you should be able to:

1. Discuss similarities and differences between ferns and other plants you have studied in the lab.
2. Describe the life cycles of ferns and their allies.
3. Describe the distinguishing features of true ferns, club mosses, whisk ferns, and horsetails.
4. Name some roles of ferns and fern allies.

Please visit connect.mheducation.com to review online resources tailored to this lab.

Seedless vascular plants include two phyla of nonflowering plants having a vascular system of fluid-conducting xylem and phloem: phylum Pterophyta (true ferns, whisk ferns, and horsetails) and phylum Lycophyta (club mosses) (table 29.1). The life cycle of seedless vascular plants includes both gametophyte and sporophyte phases, but in

Table 29.1

The Phyla of Extant Seedless Vascular Plants

Phylum	Examples		Key Characteristics	Approximate Number of Living Species
Pterophyta	Ferns		Primarily homosporous (a few heterosporous) vascular plants. Sperm motile. External water necessary for fertilization. Leaves are megaphylls that uncoil as they mature. Sporophytes and virtually all gametophytes photosynthetic. About 365 genera.	11,000
	Horsetails		Homosporous vascular plants. Sperm motile. External water necessary for fertilization. Stems ribbed, jointed, either photosynthetic or nonphotosynthetic. Leaves scalelike, in whorls, nonphotosynthetic at maturity. One genus (*Equisetum*).	15
	Whisk ferns		Homosporous vascular plants. Sperm motile. External water necessary for fertilization. No differentiation between root and shoot. No leaves; one of the two genera has scalelike enations and the other leaflike appendages.	6
Lycophyta	Club mosses		Homosporous or heterosporous vascular plants. Sperm motile. External water necessary for fertilization. Leaves are microphylls. About 12–13 genera.	1275

contrast to bryophytes (Exercise 28), in this group the sporophyte is the dominant phase (fig. 29.1). The sporophyte is the conspicuous adult stage in the life cycle of seedless vascular plants.

All ferns and fern allies possess **sporophylls** (*sporo* = spore forming, *-phyll* = leaf). Sporophylls are leaflike structures of the sporophyte generation that bear spores. They may be large **megaphylls** (*mega* = large, *-phyll* = leaf) with several to many veins (as in the megaphylls of true ferns), or they may be smaller **microphylls** (*micro* = small, *-phyll* = leaf) with one vein (as in whisk ferns, scouring rush, and club mosses). **Sporangia,** which form on sporophylls, are where spores are produced by meiosis (review fig. 28.1 and compare to fig. 29.1). Sporangia occur somewhere on all

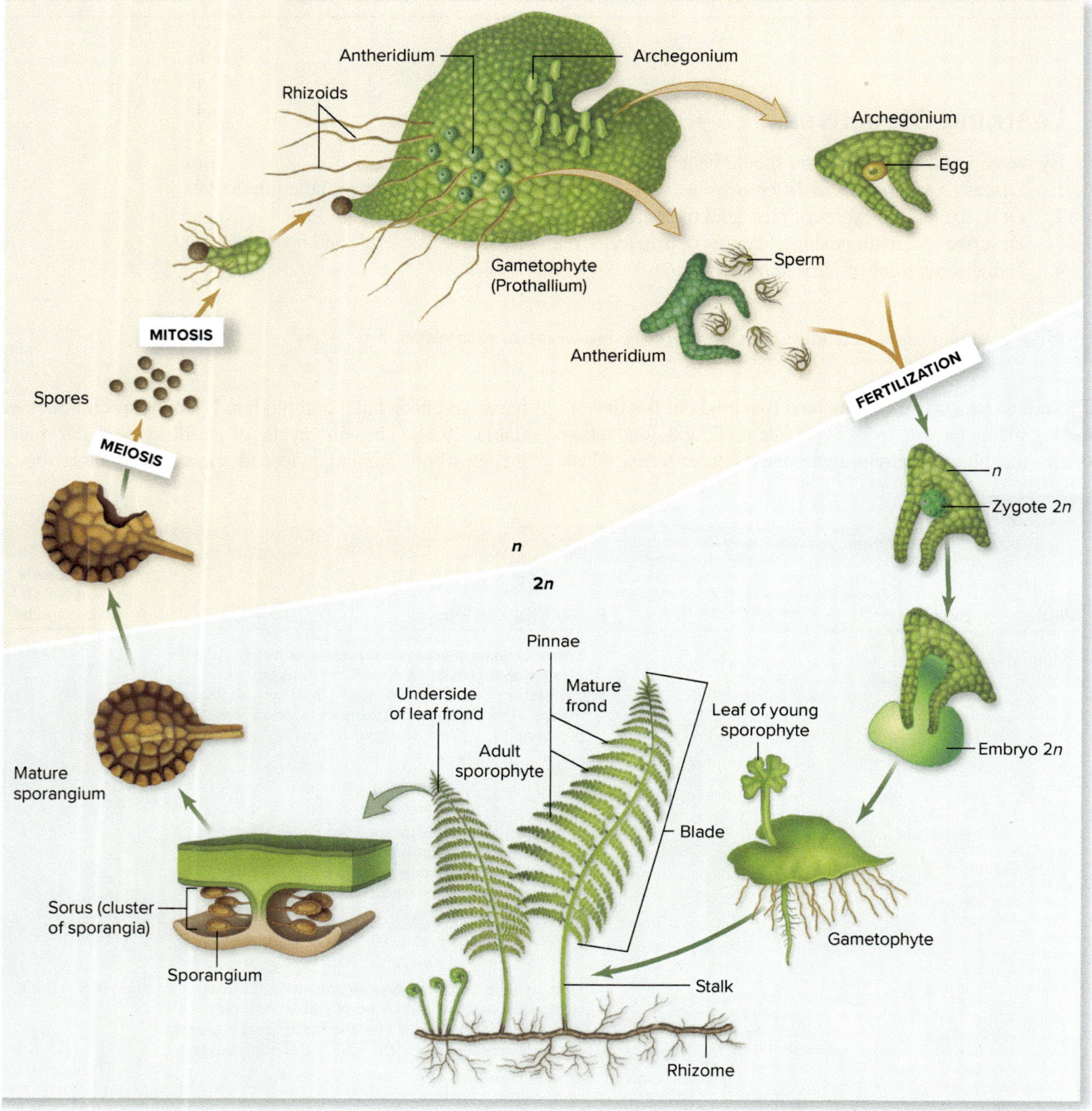

Figure 29.1 Fern life cycle. The haploid gametophyte (i.e., the prothallium) grows in moist places. Rhizoids are anchoring structures that project from the lower surface of a prothallium. Eggs and sperm develop in archegonia and antheridia, respectively, on gametophytes' lower surface. Sperm released from antheridia swim to archegonia and fertilize the single egg. The zygote—the first cell of the diploid sporophyte generation—starts to grow within the archegonium and eventually becomes much larger than the gametophyte. Most ferns have horizontal stems, called rhizomes, that grow below the ground. On the sporophyte's leaves (fronds) are clusters of sporangia within which meiosis occurs and spores are formed. When released, the spores germinate and become new gametophytes called prothallia.

plants. In ferns, the sporangia are on the backs of leaves; this is why the leaves are called sporophylls (fig. 29.2). Like bryophytes, ferns require water for fertilization.

Question 1

a. Like bryophytes, ferns require water for fertilization. Why? Explain your answer.

b. What are the primary functions (e.g., support, transport, photosynthesis, reproduction) of the major organs shown in fig. 29.1?

Figure 29.2 Fern sporangia. Most ferns have sporangia aggregated into clusters, called sori, on the undersides of their leaves. (*a*) In some ferns, such as the Japanese holly fern (*Crytomium falcatum*), each sorus is covered by a flap of leaf tissue called an indusium (also see fig. 29.3). (*b*) Other ferns bear uncovered sori, as shown here in the autumn fern (*Dryopteris erythrosora*). (*c*) In still other ferns, as in the maidenhair fern (*Adiantum tenerum*), sori are enfolded by the edge of the frond itself.

PHYLUM PTEROPHYTA (FERNS)

True ferns (phylum Pterophyta) inhabit almost all types of environments and possess characteristics of the more advanced seed plants as well as the less advanced bryophytes. Ferns have well-developed vascular tissue. Unlike bryophytes, ferns have an independent sporophyte (fig. 29 1) and stomata, which are pores that open and close on leaves and, in doing so, regulate gas exchange.

The diversity of ferns is striking; they range from majestic tree ferns to bizarre staghorn ferns. Tree ferns reach heights of up to 16 m. Along with other plants, these ferns once formed forests that were transformed into coal deposits. Today, humans use ferns as decorations and to grow rice. Review the fern life cycle shown in figure 29.1.

Question 2

a. Which parts of the life cycle are haploid?

b. Which are diploid?

c. What is the role of the gametophyte in ferns?

d. How is this different from that of the gametophyte in bryophytes?

e. In ferns, how does the size of the gametophyte compare with that of the sporophyte?

f. How does this compare with bryophytes, and why is this important?

Fern sporophytes grow indefinitely via underground stems called **rhizomes** (fig. 29.1). Examine the fern rhizomes on display. Also examine the different ferns available in the lab and note the different shapes of the leaflike fronds. Identify the **stalk, blade,** and **pinnae.**

Question 3

a. How many veins are present in each frond?

b. What tissues comprise a vein of vascular tissue?

c. What is the function of the stalk? The blade? The pinnae?

d. Rhizomes are involved in the asexual reproduction of ferns. How could this happen?

Groups of sporangia called **sori** (singular = **sorus**) form on the underside of fern fronds (fig. 29.2). Sporangia may be protected by a shield-shaped **indusium** (fig. 29.3), a specialized outgrowth of a frond. Meiosis in the sporangium produces haploid spores, the first stage of the gametophyte.

Procedure 29.1 Examine sori

1. Scrape a sorus into a drop of water on a microscope slide and use the low power on your microscope to observe the sorus.

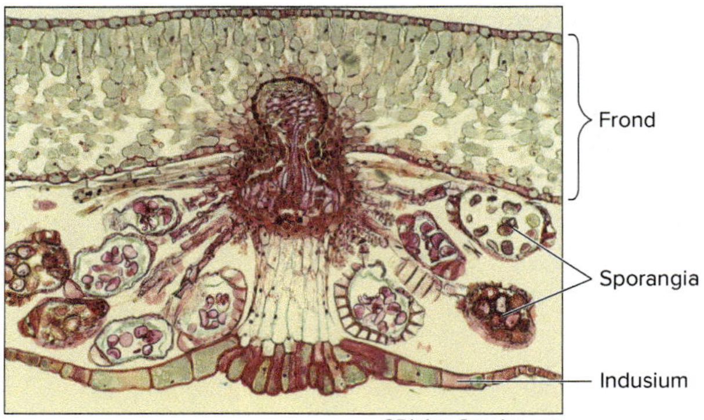

Figure 29.3 Fern sorus (20×). In most ferns, sori are on the undersides of leaves.

Question 4

Are any spores in the sporangium?

Be careful when handling acetone; it is a strong solvent. Wear safety glasses.

2. Place a few drops of acetone on the sporangium while you observe it with a dissecting microscope.
3. Watch the sporangium for a few minutes, adding acetone as needed.
4. Describe what you see.

Question 5

a. Did the application of acetone cause the spores of the fern to disperse?

b. How is the mechanism for spore dispersal in ferns similar to that of bryophytes?

5. Examine prepared slides of fern sori, referring to figure 29.3. Diagram and label each structure that you see, listing its function.

Fern Reproduction

When conditions are favorable, fern spores germinate and form a threadlike **protonema.** Subsequent cellular divisions produce an independent, heart-shaped **prothallium** ("valentine plant").

Question 6

a. Is the prothallium haploid or diploid?

b. Is the prothallium sporophyte or gametophyte?

Rhizoids and male and female reproductive structures occur on the underside of the prothallium. However, a prothallium rarely fertilizes itself because the antheridia and archegonia mature at different times. Globe-shaped antheridia form first, followed by archegonia. Archegonia are vase-shaped and are located near the cleft of the heart-shaped prothallium. After producing sperm, antheridia drop off, leaving sperm to swim to the archegonia of other prothallia.

Procedure 29.2 **Observe archegonia and antheridia**

1. Observe archegonia and antheridia (see fig. 29.1) on prepared slides.
2. Observe archegonia and antheridia on a living prothallium.

Question 7

a. What is the adaptive significance of having these structures on the lower surface of the prothallium rather than on the upper surface?

b. What is the adaptive significance of having sperm and egg produced at different times?

The zygote develops in the archegonium and is nutritionally dependent on the gametophyte for a short time (fig. 29.4). Soon thereafter, the sporophyte becomes leaflike and crushes the prothallium.

Figure 29.4 Young fern sporophyte growing out of its gametophyte parent (10X mag.). Shortly after this stage, the gametophyte dies and begins to decompose.

Fronds of the growing sporophyte break through the soil in a coiled position called a **fiddlehead** (fig. 29.5). The fiddlehead then unrolls to display the frond, a single leaf. Fiddleheads are considered a culinary delicacy in some parts of the world.

Most terrestrial ferns are **homosporous;** this means that they produce one kind of spore that develops into a single kind of gametophyte that produces both antheridia and archegonia (see fig. 29.1). Conversely, aquatic ferns such as

©BiologyImaging.com

Figure 29.5 A fiddlehead of a tropical tree fern.

Salvinia and *Azolla* are **heterosporous,** meaning that they produce two types of spores: **megaspores** and **microspores.** (You will learn more about heterospory in Exercise 30.) A megaspore forms a gametophyte with only archegonia, and a microspore forms a gametophyte with only antheridia. Examine the living *Salvinia* and *Azolla* in the lab (fig. 29.6).

Question 8
How do *Salvinia* and *Azolla* differ from other ferns you examined earlier?

(a) Source: Larry Allain/National Wetlands Research Center/USGS

(b) ©BiologyImaging.com

Figure 29.6 (*a*) *Salvinia* and (*b*) *Azolla* are ferns that grow in aquatic habitats.

WHISK FERNS

Whisk ferns are close relatives of ferns and include only two extant representatives: *Psilotum* (fig. 29.7) and *Tmesipteris*. *Psilotum* has a widespread distribution, whereas *Tmesipteris* is restricted to the South Pacific. *Psilotum* lacks leaves and roots and is homosporous.

Procedure 29.3 **Examine *Psilotum***

1. Examine a prepared slide of a *Psilotum* sporangium (fig. 29.8).
2. If available, examine living *Psilotum* plants in the lab.

Question 9

a. How would you describe the branching pattern of *Psilotum?*

b. Are any roots present?

c. Are any leaves present?

d. Where are the sporangia?

e. Where does photosynthesis occur in *Psilotum?*

Segundo Pérez/Alamy Stock Photo

Figure 29.7 Whisk ferns (*Psilotum* sp.) are so called because their branching pattern gives the impression of a whisk broom. The stems bear lobed sporangia (fig. 29.8).

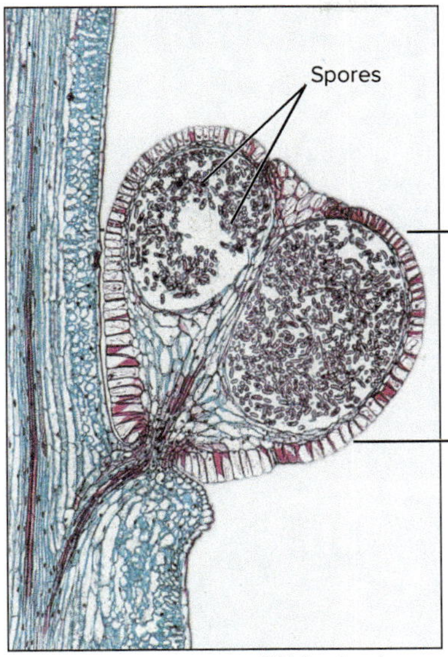

©BiologyImaging.com

Figure 29.8 *Psilotum* sporangium (18×).

HORSETAILS

Equisetum (also called scouring rush) is the only extant genus of horsetails, another small but distinctive group of ferns (fig. 29.9). *Equisetum* is an example of a plant whose vegetative structure identifies the plant better than does its reproductive structure: *Equisetum* is distinguished by its jointed and ribbed stem. Examine the living *Equisetum* plants available in lab.

Stuart Wilson/Science Source

Figure 29.9 Scouring rush, *Equisetum telmateia,* of the phylum Pterophyta. This species forms two types of erect stems, a green, photosynthetic type and a brownish type terminating in spore-producing cones. The spores produced by meiosis in the cones give rise to a single kind of tiny, green, nutritionally independent gametophyte.

Question 10

a. Where are the leaves?

b. What part of the plant is photosynthetic?

c. Which part of the life cycle of *Equisetum* is dominant, the sporophyte or gametophyte?

Feel the ribbed stem of an *Equisetum* plant. Its rough texture results from siliceous deposits in its epidermal cells. During frontier times, *Equisetum* was used to clean pots and pans, sand wooden floors, and scour plowshares, thus accounting for its common name of "scouring rush."

Strobili of *Equisetum* occur at the tips of reproductive stems. Within a strobilus, sporangia form atop umbrellalike modified branches called **sporangiophores.** Elaters in sporangia of *Equisetum* help disperse spores. Examine prepared slides of *Equisetum* strobili (fig. 29.10). Diagram, label, and state the function of each major structure composing the strobilus.

Question 11

How do elaters aid in the dispersal of spores?

PHYLUM LYCOPHYTA CLUB MOSSES

During the Devonian and Carboniferous periods (300–400 million years ago), club mosses, whisk ferns, and scouring rush were among the dominant plants on earth. Indeed, most of our coal deposits consist largely of these plants. However, the "giant" species of these phyla are now extinct, and modern representatives are relatively small compared to their giant ancestors.

Club mosses (phylum Lycophyta) possess true roots, stems, and leaves. Most asexual reproduction by club mosses occurs via rhizomes. If one is available, locate the rhizome on a *Lycopodium*, a club moss. Study the aboveground portion of the *Lycopodium* plant (fig. 29.11). *Lycopodium* is evergreen, as are some ferns, most gymnosperms, and some angiosperms.

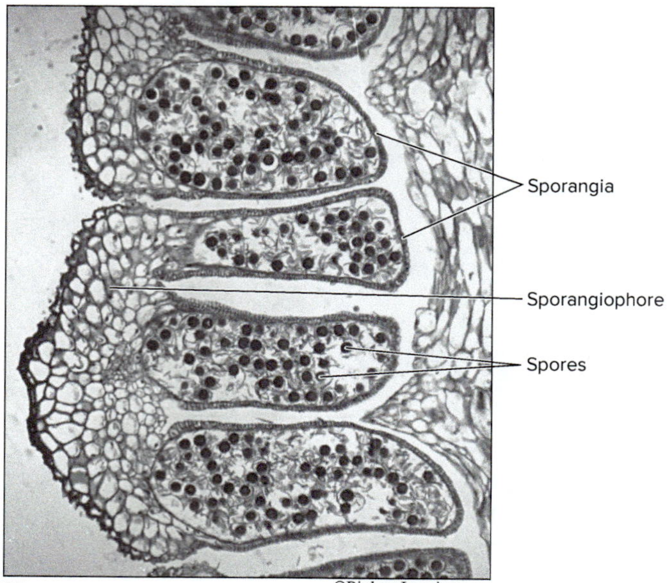

Figure 29.10 *Equisetum.* (*a*) Strobili occur at the tips of reproductive stems. (*b*) Cross-section of a strobilus showing spores within sporangia at the tips of sporangiophores (20×).

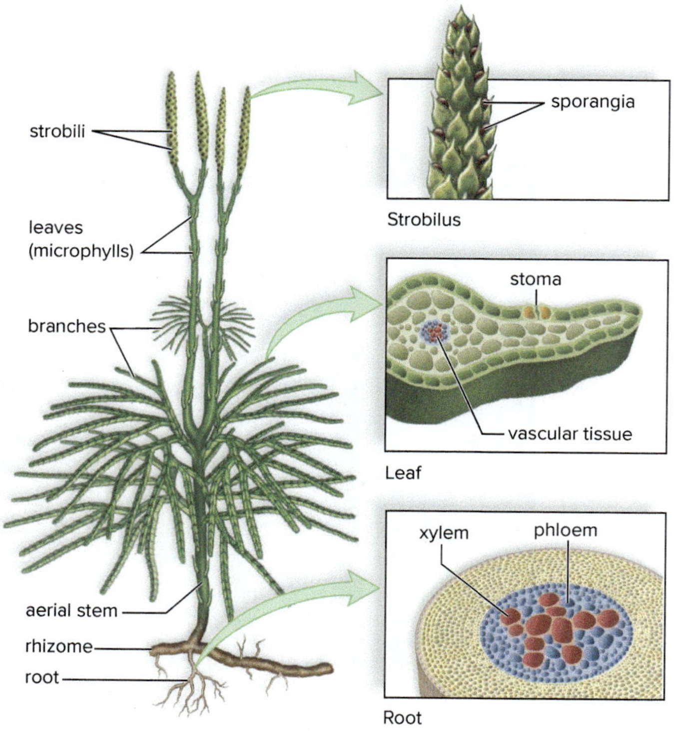

Figure 29.11 *Lycopodium,* a lycophyte. (*a*) The sporophyte develops from an underground stem called a rhizome. (*b*) The spore-producing structures occur in club-like structures called strobili, or cones, which are aggregates of closely packed sporangium-bearing branches or leaves.

Question 12

a. How could a rhizome be involved in asexual reproduction?

b. How is a rhizome different from a rhizoid?

c. Does the rhizome have leaves?

d. What are the shape and size of the leaves?

e. What is the significance of the form of the leaves?

f. Is a midvein visible?

g. What does the term "evergreen" mean?

h. Is being evergreen a good characteristic for classifying plants? Why or why not?

Sporangia of *Lycopodium* occur on small modified leaves called **sporophylls** clustered in **strobili** (cones) that form at the tips of branches. Species with these cones probably shared common ancestry with the familiar cone-bearing gymnosperms such as pine (*Pinus*) (see Exercise 30).

Procedure 29.4 **Examine club mosses**

1. Examine strobili on a living *Lycopodium* plant. Also examine prepared slides of strobili of *Lycopodium* (fig. 29.12).

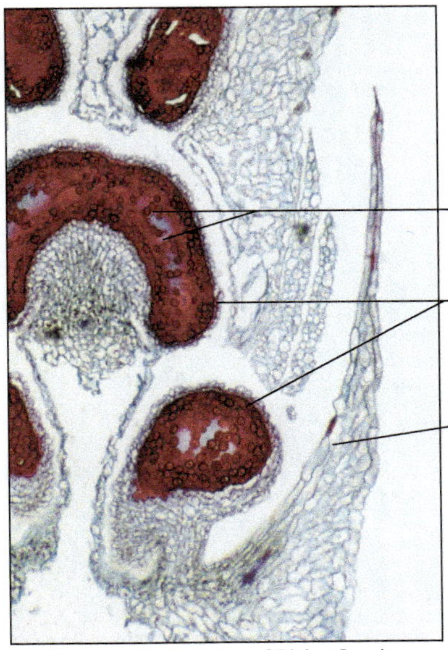

Figure 29.12 Sporophyll and sporangia of *Lycopodium*, a club moss (15×).

2. Diagram, label, and state the function of each major feature of the strobilus.

3. Examine prepared slides of spores of *Lycopodium* as well as those available on plants growing in the lab. Sketch what you see.

Question 13

a. How many sporangia occur on each sporophyll?

b. Can you see why spores of *Lycopodium* are sometimes called "vegetable sulfur"?

c. Why are the spores a good dry lubricant?

©BiologyImaging.com

Figure 29.13 *Selaginella,* a club moss.

d. Which is the dominant part of the *Lycopodium* life cycle, the sporophyte or gametophyte?

4. Examine living *Selaginella* plants (fig. 29.13). Many species of *Selaginella* produce two types of strobili, typically red and yellow. If strobili are present, examine spores derived from these cones.

5. Examine hydrated and dehydrated resurrection plants (*Selaginella lepidophylla*).

Question 14

a. Are spores of *Selaginella* similar in size?

b. What is this condition called?

c. What is the functional significance of the difference in the appearance of dehydrated and rehydrated *Selaginella?*

blickwinkel/Alamy Stock Photo

Figure 29.14 Quillworts (*Isoetes*) are so-named because of their narrow, quill-like leaves. Most quillworts live in aquatic environments.

d. Can you see why these plants are sometimes referred to as "resurrection plants"?

e. How does the formation of strobili in *Equisetum* compare with that in *Lycopodium* and *Selaginella*?

Isoetes, or quillwort, is an aquatic or semi-aquatic lycopod that usually grows in clear ponds and slow-moving streams. At the branching points along the stem of *Isoetes,* you'll see an unusual runnerlike organ. These proplike axes are called **rhizophores** and have structural features that are intermediate between stems and roots.

Procedure 29.5 Examine *Isoetes*

1. If available, examine living *Isoetes* (quillwort) plants (fig. 29.14).

2. Compare and contrast the following features of *Isoetes* with *Lycopodium* and *Selaginella*:

- Shape of aerial portion of plant
- Branching patterns
- Shape, size, and arrangement of leaves

ORGANIZING WHAT YOU'VE LEARNED

Go to page 362, which follows Exercise 31. Complete the portions of the table that refer to the phyla you studied in today's lab. Also, to review the structures and characteristics of seedless vascular plants, complete table 29.2 on the next page.

Table 29.2

Summary of the Structures Common to Seedless Vascular Plants

Plant Structure	Sporophyte or Gametophyte	Function
Prothallium		
Pinna		
Spore		
Frond		
Annulus		
Sporangium		
Antheridium		
Archegonium		
Microspore		
Megaspore		
Microphyll		
Megaphyll		

INQUIRY-BASED LEARNING

What happens during the "resurrection" of a "resurrection plant"?

Observation: The "resurrection" of a "resurrection plant" (*Selaginella*) involves the rapid absorption of water. This, in turn, triggers other changes in the plant that "revive" the plant.

Question: What changes accompany the "resurrection" of a "resurrection plant"?

a. Establish a working lab group and obtain Inquiry-Based Learning Worksheet 29 from your instructor.
b. Discuss with your group well-defined questions relevant to the preceding observation and question. Choose and record your group's best question for investigation.
c. Translate your question into a testable hypothesis and record it.
d. Outline on Worksheet 29 your experimental design and supplies needed to test your hypothesis. Ask your instructor to review your proposed investigation.
e. Conduct your procedures, record your data, answer your question, and make relevant comments.
f. Discuss with your instructor any revisions to your questions, hypotheses, or procedures. Repeat your work as needed.

Questions for Further Study and Inquiry

1. How are ferns and fern allies similar to and different from bryophytes? What are the implications of these similarities and differences?

2. What structures and features do ferns possess that bryophytes do not that may have contributed to their success in a broader range of environments?

3. What are the advantages of vascular tissues in land plants?

4. What are the distinguishing features of club mosses, whisk ferns, and horsetails? How are these plants different from ferns?

WRITING TO LEARN BIOLOGY

What problems did plants face as they moved onto land? What adaptations of mosses, liverworts, ferns, and other seedless plants are relevant to the transition?

EXERCISE 30

Survey of the Plant Kingdom

Gymnosperms of Phyla Cycadophyta, Ginkgophyta, Coniferophyta, and Gnetophyta

Learning Objectives

By the end of this exercise you should be able to:

1. Describe the distinguishing features of the gymnosperms.
2. Understand the life cycle of pine, a representative gymnosperm.
3. Understand the evolutionary significance of pollen and seeds.
4. Identify the parts and understand the function of a cone.
5. Identify the parts and understand the function of a seed.

Please visit connect.mheducation.com to review online resources tailored to this lab.

Gymnosperms were the first group of plants to (1) protect their developing embryos in nutrient-containing seeds and (2) make the evolutionary transition from swimming sperm to pollen-enclosed sperm. These adaptations helped make the gymnosperms a remarkably successful group of plants.

Gymnosperms produce exposed seeds borne on scale-like structures called **cones** (strobili). Like ferns, gymnosperms have a well-developed alternation of generations. Unlike most ferns, however, gymnosperms are **heterosporous;** that is, they produce two types of spores (fig. 30.1). **Microspores** occur in male cones and form male gametophytes. **Megaspores**

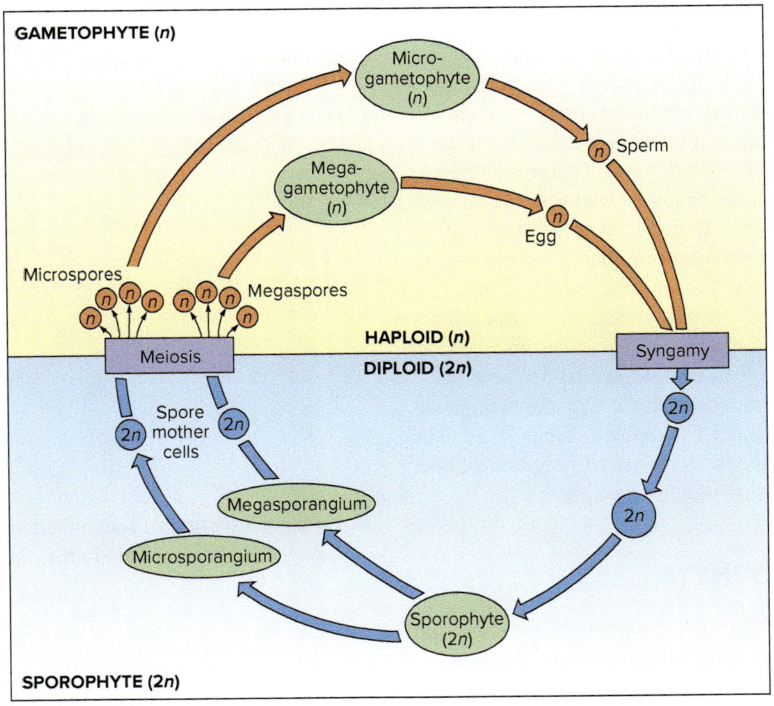

Figure 30.1 Diagram of the life cycle of a heterosporous vascular plant.

Table 30.1

The Five Phyla of Extant Gymnosperms

Phylum	Examples		Key Characteristics	Approximate Number of Living Species
Cycadophyta	Cycads		Heterosporous vascular seed plants. Sperm flagellated and motile but confined within a pollen tube that grows to the vicinity of the egg. Palmlike plants with pinnate leaves. Secondary growth slow compared with that of the conifers. Ten genera.	320
Ginkgophyta	*Ginkgo*		Heterosporous vascular seed plants. Sperm flagellated and motile but conducted to the vicinity of the egg by a pollen tube. Deciduous tree with fan-shaped leaves that have evenly forking veins. Seeds resemble a small plum with fleshy, ill-scented outer covering. One species: *Ginkgo biloba*	1
Coniferophyta	Conifers (including pines, spruces, firs, yews redwoods, and others)		Heterosporous seed plants. Sperm not motile; conducted to egg by a pollen tube. Leaves mostly needlelike or scalelike. Vascular. Trees, shrubs. About 50 genera.	630
Gnetophyta	Gnetophytes		Heterosporous vascular seed plants. Sperm not motile; conducted to egg by a pollen tube. The only gymnosperms with vessels. Trees, shrubs, vines. Three very diverse genera (*Ephedra, Gnetum, Welwitschia*).	65

occur in female cones and form female gametophytes. Gametophytes of gymnosperms are microscopic and completely dependent on the large, free-living sporophyte.

In gymnosperms, pollen grains are the male gametophyte. Pollination is the transfer of pollen from male cones (where the pollen is produced) to female cones, which house the eggs. In gymnosperms, pollen is carried from male cones to female cones by wind. Gymnosperms were the first plants to evolve that did not need free water to transfer sperm to egg, and were therefore able to colonize terrestrial habitats. Gymnosperms were the dominant land plants during the Age of Dinosaurs (i.e., during the Mesozoic Era, 225–65 million years ago). Today, we use gymnosperms for products ranging from lumber (e.g., pine, fir, spruce, cedar) to soaps, varnish, nail polish, gum, food, and perfume.

Gymnosperms include four phyla: Cycadophyta, Ginkgophyta, Coniferophyta, and Gnetophyta (table 30.1). The last of these, the Gnetophyta, is discussed only briefly here because the phylum consists of a few rare genera.

PHYLUM CYCADOPHYTA

The Cycadophyta (cycads) once flourished, but today the phylum consists of only about 10 genera and 320 species. Cycads resemble palms because they have unbranched trunks and large, closely packed leaves that are evergreen and tough (fig. 30.2). Unlike conifers, cycads have pollen and seed cones on separate plants. The cones, which are surrounded by leaves at the top of the stem, can be more than a meter long and weigh as much as 35 kg.

Sperm of cycads are flagellated. Examine a branch of a cycad such as *Zamia* bearing developing seeds. The seeds are fleshy and exposed to the environment.

Question 1

a. Based on your knowledge of reproduction in bryophytes (Exercise 28) and ferns (Exercise 29), is the presence of flagellated sperm in cycads surprising? Why or why not?

b. Is the possession of flagellated sperm a primitive or advanced characteristic in the plant kingdom? Why?

(a) ©BiologyImaging.com

(b) ©BiologyImaging.com

Figure 30.2 Cycads. (*a*) A male cycad (*Cycas revoluta*) with a strobilus. (*b*) A female cycad with strobili. *Zamia* sp. is the only genus of cycad native to the United States. The starchy roots and stems (mostly underground) of cycads were used by Native Americans for food.

David Stephenson/Lexington Herald-Leader/Tribune News Service/Getty Images

Figure 30.3 Maidenhair tree, *Ginkgo biloba,* is the only living representative of the phylum Ginkgophyta, a group of plants abundant 200 million years ago. Among living seed plants, only the cycads and *Ginkgo* have swimming sperm. This photograph shows *Ginkgo* leaves and fleshy seeds.

PHYLUM GINKGOPHYTA

The Ginkgophyta consists of one species, *Ginkgo biloba* (maidenhair tree), a large dioecious tree that does not bear cones. *Ginkgo* are hardy plants in urban environments and tolerate insects, fungi, and pollutants. Males are usually planted because females produce fleshy, smelly, and messy fruit that superficially resemble cherries. Leaves of *Ginkgo* have a unique shape (fig. 30.3). *Ginkgo* has not been found in the wild and would probably be extinct but for its cultivation in ancient Chinese gardens.

Question 2
What does dioecious mean?

PHYLUM CONIFEROPHYTA

The Coniferophyta are a large group of cone-bearing plants that includes the 5000-year-old bristlecone pines, the earth's oldest living plants. The cones they bear are reproductive structures of the sporophyte generation that consist of several scalelike sporophylls arranged about a central axis (fig. 30.4). Sporophylls, also present in ferns and their allies, are modified leaves specialized for reproduction. Sporophylls bear spores. In conifers, sporophylls of male cones are called **microsporophylls** (see fig. 30.8). On the surface of each microsporophyll is a layer of cells called a **microsporangium** that produces spores. Male cones live only a few weeks; after these cones release their pollen, they fall off the tree. Sporophylls of female cones are **megasporophylls;** each of these "scales" of the female cone

(a) ©BiologyImaging.com

(b) ©BiologyImaging.com

Figure 30.4 Pinecones. (*a*) First-year ovulate (seed) cones open for pollination. (*b*) Second-year ovulate pinecones at time of fertilization.

bears two spore-producing **megasporangia** on its upper surface. Microsporangia and megasporangia are patches of cells near the central axis of the sporophylls composing the cones on the respective sporophylls. Male cones are small and similar in all conifers (fig. 30.5). However, female cones are variable; they may be small (1–3 cm in *Podocarpus*) and fleshy (a *Juniperus* "berry") or large and woody (6–40 cm in *Pinus*).

In this exercise you'll study pine (*Pinus*), a representative and widely distributed conifer. Pine has considerable economic value because it is used to produce lumber, wood pulp, pine tar, resin, and turpentine. Other examples of conifers include spruce, cedar, and fir.

The Pine Sporophyte

The life cycle of *Pinus* is typical of conifers (fig. 30.6). You are already familiar with the sporophyte of pine—it is the tree.

Procedure 30.1 Examine pine twigs and leaves

1. Examine pine twigs having leaves (needles) and a terminal bud. Notice that the leaves are borne on short branches only a few millimeters long. The length and number of leaves distinguish many of the species of *Pinus*.
2. Examine a prepared slide of a cross section of a pine leaf; locate the structures labeled in figure 30.7.

Question 3

a. How are needles (i.e., leaves) arranged?

©BiologyImaging.com

Figure 30.5 Pollen-bearing cones of *Pinus*. These cones are usually found on the lower branches of pine trees. The sporophylls of the cones have opened to release pollen.

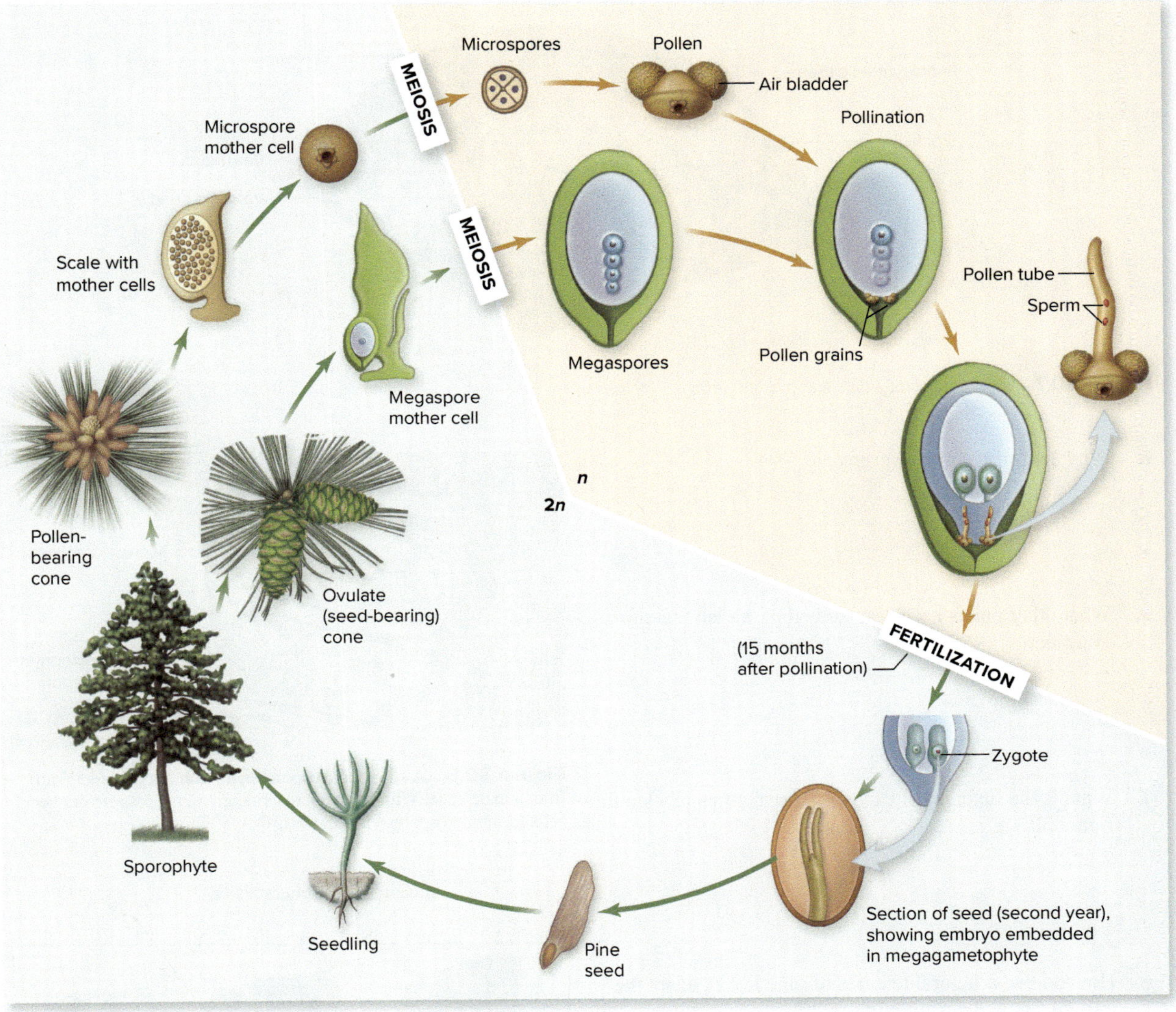

Figure 30.6 Pine life cycle. In seed plants, the gametophyte generation is greatly reduced. A germinating pollen grain is the mature microgametophyte of a pine. Pine microsporangia are borne in pairs on the scales of the delicate pollen-bearing cones. Megagametophytes, in contrast, develop within the ovule. The familiar seed-bearing cones of pines are much heavier than the pollen-bearing cones. Two ovules, and ultimately two seeds, are borne on the upper surface of each scale of a cone. In the spring, when the seed-bearing cones are small and young, their scales are slightly separated. Drops of sticky fluid, to which the airborne pollen grains adhere, form between these scales. Pollination occurs more than a year before the ovule produces a mature female gametophyte. These pollen grains germinate, and slender pollen tubes grow slowly toward the egg. When a pollen tube grows to the vicinity of the megagametophyte, sperm are released, fertilizing the egg and producing a zygote there. The development of the zygote into an embryo occurs within the ovule, which matures into a seed. Seeds mature up to six months after fertilization. Eventually, the seed falls from the cone and germinates, the embryo resuming growth and becoming a new pine tree.

b. How many leaves are in a bundle?

c. Pine "needles" are modified leaves. How are pine needles different from the leaves of broad-leaved trees such as maples and oaks?

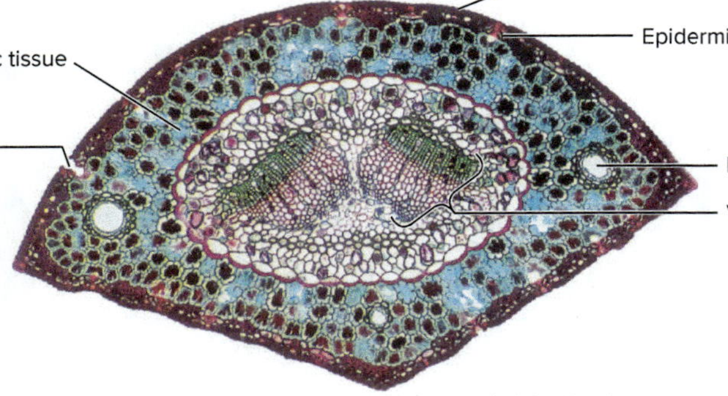

Figure 30.7 Cross section of a pine leaf (needle) (40×).

d. Why are pines called evergreens?

e. What other plants have you studied in the lab that are evergreen?

f. What is the function of each of the structures labeled in figure 30.7?

g. How do the structural features of pine leaves adapt the tree for life in dry environments?

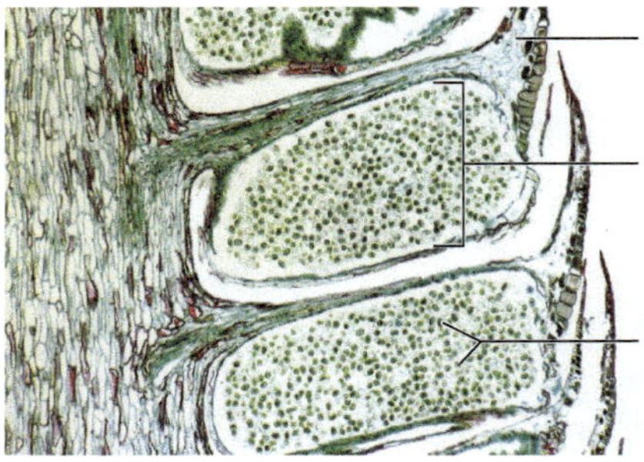

Figure 30.8 A single microsporophyll (with microsporangium) from a male cone. Within the microsporangium, microspore mother cells form microspores via meiosis (10×).

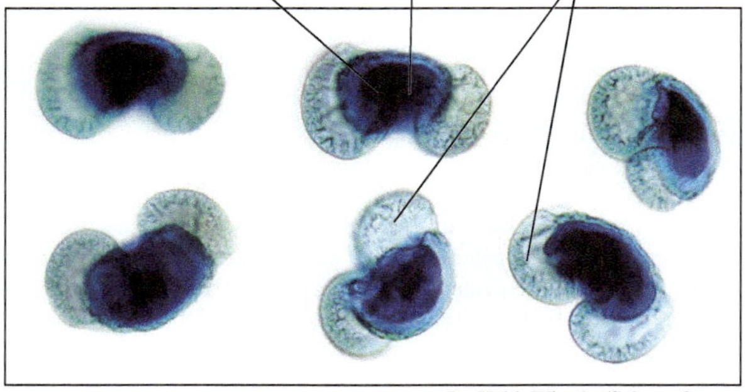

Figure 30.9 Pollen grains of *Pinus,* each with enclosed male gametophyte. Each gametophyte includes a small generative cell and a larger tube cell. When the pollen grain germinates, the pollen tube will emerge between the two bladder-shaped wings (20×).

Male cones usually form on the lower branches of pine trees, and female cones usually form on the upper branches. Examine pine branches with staminate (male) and ovulate (female) cones. Examine a prepared slide of a young staminate cone and note the pine pollen in various stages of development. Each scale (microsporophyll) of the male cone bears a microsporangium, which in turn produces diploid **microspore mother cells** (fig. 30.8). Microspore mother cells undergo meiosis to produce **microspores** that develop via mitotic divisions into microgametophytes called pollen grains (fig. 30.9). Each **pollen grain** consists of four nuclei and a pair of bladderlike wings. The gametophytes of gymnosperms are reduced to only a few cells.

Procedure 30.2 **Examine staminate pinecones and pollen**

1. If male cones are available, prepare a wet mount of some pine pollen; notice their characteristic shape.

2. Remove a scale from a mature staminate cone and tease open the microsporangium.
3. Prepare a wet mount of the microsporangium and its contents, and examine the contents with your microscope.
4. In the space below, draw the shape of a pine pollen grain. If male cones are not available, examine a prepared slide of pine pollen grains; sketch the shape of the pollen.

Question 4

a. Are all of the cones the same size?

b. Recall that in ferns the antheridia and archegonia on a prothallium mature at different times to avoid self-fertilization. How might the different locations of male and female cones on a pine tree help to do the same?

c. What is the probable function of the wings of a pine pollen grain?

Each ovuliferous scale of the female cone bears two megasporangia, each of which produces a diploid **megaspore mother cell.** Each megaspore mother cell undergoes meiosis to produce a **megaspore** that develops via mitotic divisions into a **megagametophyte.** The tissue of the megasporangium immediately surrounding the megagametophyte is the **nucellus** (nutritive tissue) surrounded by **integuments** (which will form the seed coat). A megagametophyte and its surrounding tissues constitute an ovule and contain at least one archegonium with an egg cell.

Procedure 30.3 **Examine ovulate pinecones and ovules**

1. Examine young living or preserved ovulate cones. These cones will develop and enlarge considerably before they are mature.
2. Examine a prepared slide of a young ovulate cone ready for pollination.
3. Examine a prepared slide of an ovulate cone that has been sectioned through an ovule (fig. 30.10). An ovule develops into a seed.
4. Examine a mature ovulate cone and notice its spirally arranged **ovuliferous scales.** These scales are analogous to microsporophylls of staminate cones. At the base of each scale you'll find two naked seeds. Notice that the seeds are exposed to the environment.

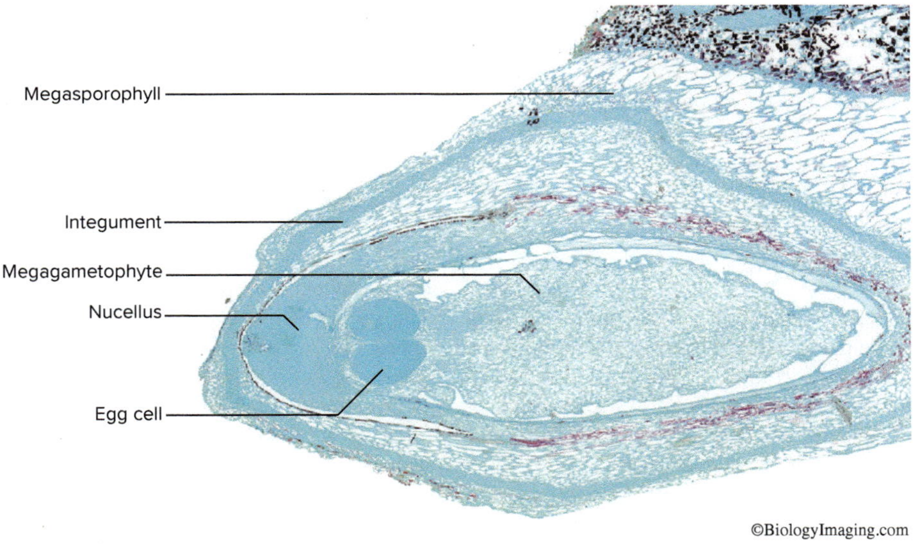

Figure 30.10 Pine ovule section before fertilization. The egg will be fertilized later to form a new sporophyte inside the seed. The nucellus is a nutritive tissue, and integuments form the seed coat (15×). When this specimen was prepared, the soft megagametophyte shrank away from the firmer exterior layers.

Question 5

a. On which surface of the scale are the seeds located?

b. How large is a staminate cone compared to a newly pollinated ovulate cone? Compared to a mature ovulate cone?

Pollination and Seed Formation in Pine

Pollination is the transfer of pollen to a receptive surface. Pollen grains released in late spring are carried by wind to ovulate cones, in which the grains sift through the ovuliferous scales and stick in a drop of resin at the micropylar end of an ovule. Pollination occurs more than a year before the ovule produces a mature female gametophyte. The pollen grain then germinates and grows a tube into the archegonium, where it releases its two nonmotile sperm nuclei. One of these sperms disintegrates; the other fuses with the egg to form a zygote. The zygote, while in the ovule, develops into the embryo of a new sporophyte. The ovule is now a seed and consists of an embryo, a seed coat (integuments of the megasporangia), and a food supply (tissue of the megagametophyte). Seeds mature up to six months after fertilization.

Procedure 30.4 **Examine a pine seed**

1. Examine a prepared slide of a pine seed. Locate the embryo, seed coat, and food supply. Seeds are released when the cone dries and the scales separate. This usually occurs 13–15 months after pollination.
2. Examine some mature pine seeds, noting the winglike extensions of the seed coat.

Seed

The evolution of seeds is one of the most significant events in the history of the plant kingdom. Indeed, the evolution of seeds is one of the factors responsible for the dominance of seed plants in today's environment because a seed permits a small but multicellular sporophyte to remain dormant until conditions are favorable for continued growth. While dormant, the young sporophyte is protected by a seed coat and is surrounded by a food supply.

Refer to the pine life cycle shown in figure 30.6 to answer Question 6. If plant material and time are available, examine other gymnosperms in the lab.

Question 6

a. Where are spores located?

b. What is the male gametophyte?

c. What is the female gametophyte?

d. What is an ovule?

e. What is an integument?

f. How are other gymnosperms similar to pine?

g. How are they different?

h. What evolutionary advantages might arise from not needing free water for fertilization?

PHYLUM GNETOPHYTA

The gnetophytes (71 species in three genera) include some of the most distinctive (if not bizarre) of all seed plants (fig. 30.11). They have many similarities with angiosperms, such as flowerlike compound strobili, vessels in the secondary xylem, loss of archegonia, and double fertilization. The slow-growing *Welwitschia* plants are extraordinary in appearance and live only in the deserts of southwestern Africa. Most of their moisture is derived from fog that rolls in from the ocean at night. *Welwitschia* stems are short and broad (1 m). Mature, 100-year-old plants have only two wide, straplike leaves, which fray as they age.

Ephedra, another gnetophyte, grows in North America and Southeast Asia. It is a shrub with small, scalelike leaves and conelike reproductive structures. The stimulant ephedrine, which is used to prevent low blood-pressure, is extracted from *Ephedra*.

(a) ©Robert & Linda Mitchell

(b) LOETSCHER CHLAUS/Alamy Stock Photo

(c) ©BiologyImaging.com

Figure 30.11 Gnetophytes. (*a*) Leaves and immature seeds of *Gnetum*. *Gnetum* grows as a shrub or woody vine in tropical or subtropical forests. (*b*) *Welwitschia* plants in the Namib Desert of southwest Africa. (*c*) *Ephedra* is the only living genus of Gnetophyta found in the United States. Most photosynthesis in *Ephedra* occurs in its photosynthetic stems, not its small, scaleline leaves. Dietary supplements containing ephedrine, which is extracted from *Ephedra*, are illegal, except those used in traditional Chinese remedies.

ORGANIZING WHAT YOU'VE LEARNED

Go to page 362, which follows Exercise 31. Complete the portions of the table that refer to the phyla you studied in today's lab.

INQUIRY-BASED LEARNING

How much pollen is released by a male pinecone?

Observation: Male pinecones produce and release large amounts of pollen. In some people this pollen contributes to hay fever and other sinus-related problems.

Question: How much pollen is released by a male pinecone?

a. Establish a working lab group and obtain Inquiry-Based Learning Worksheet 30 from your instructor.

b. Discuss with your group well-defined questions relevant to the preceding observation and question. Choose and record your group's best question for investigation.

c. Translate your question into a testable hypothesis and record it.

d. Outline on Worksheet 30 your experimental design and supplies needed to test your hypothesis. Ask your instructor to review your proposed investigation.

e. Conduct your procedures, record your data, answer your question, and make relevant comments.

f. Discuss with your instructor any revisions to your questions, hypotheses, or procedures. Repeat your work as needed.

Questions for Further Study and Inquiry

1. What is the difference between pollination and fertilization?

2. What does the term *gymnosperm* mean? Is this term an appropriate description? Explain your answer.

3. How is alternation of generations different in ferns and pines?

4. How are the environmental agents for uniting sperm and egg different in pines and bryophytes?

5. Compare and contrast alternation of generations in mosses, ferns, and pines.

6. What is the evolutionary significance of pollen and seeds?

7. What are the distinguishing features of each group of plants that you studied in this exercise?

WRITING TO LEARN BIOLOGY

Describe the ecological advantages that the production of seeds has over the reproductive process in ferns.

EXERCISE 31

Survey of the Plant Kingdom

Angiosperms

Learning Objectives

By the end of this exercise you should be able to:

1. Relate the life cycle of angiosperms to the other phyla of the plant kingdom.
2. Discuss the events associated with development of microspores, megaspores, gametophytes, gametes, seeds, and fruit.
3. List and discuss the reasons why angiosperms are considered to be the most advanced land plants.
4. Understand the structure and function of flowers.

Please visit connect.mheducation.com to review online resources tailored to this lab.

Flowering plants (phylum Anthophyta—also referred to as angiosperms)—are the most abundant, diverse, and widespread of all land plants. They owe their success to several factors, including their structural diversity, efficient vascular systems, and mutualisms with fungi and insects (fig. 31.1). Angiosperms' flowers attract insects and other pollinators, and thereby make reproduction more efficient. Fruits, which contain seeds and nutritive tissue, help disperse angiosperms to new environments.

BiologyImaging.com

Figure 31.1 This bee is pollinating a rock rose in central Texas. The evolution of flowers and the success of flowering plants correlate with the development of mutualisms with insects and other pollinators.

PHYLUM ANTHOPHYTA

Angiosperms range in size from 1 mm (*Wolffia*) to over 100 m tall (*Eucalyptus*). As in gymnosperms, the sporophyte of angiosperms is large and heterosporous. The microgametophyte (pollen) and megagametophyte (embryo sac) are completely dependent on the sporophyte. Angiosperms are important to humans because our world economy is overwhelmingly based on them. We eat and use **vegetative structures** (roots, stems, leaves) as well as **reproductive structures** (flowers, fruits, seeds) of angiosperms.

There are more than 250,000 living species of angiosperms. Until the late 1990s, most botanists grouped angiosperms into two groups, depending on the number of **cotyledons** (seed leaves) in the embryo: species having one cotyledon were called **monocots,** and those with two cotyledons were called **dicots.** However, genetic analyses revealed that species traditionally called dicots evolved from more than one common ancestor. As a result, the vast majority of species once classified as dicots are now known as **eudicots.** (The other former dicots comprise several small lineages that will not be considered in this exercise.) These studies also showed that all monocots evolved from a common ancestor; as a result, their classification as monocots remains unchanged (fig. 31.2). The "typical" features of these groups of plants are described in table 31.1.

Having studied gymnosperms, you probably concluded that vegetative adaptations of angiosperms and gymnosperms are similar. This is true. However, in today's exercise you will study two unique adaptations of angiosperm reproduction: flowers and fruit.

Figure 31.2 In monocots such as this lily (*Lilium* sp.), flower parts usually occur in multiples of three. A lily has six stamens, three petals, three petal-like sepals, and a three-chambered ovary.

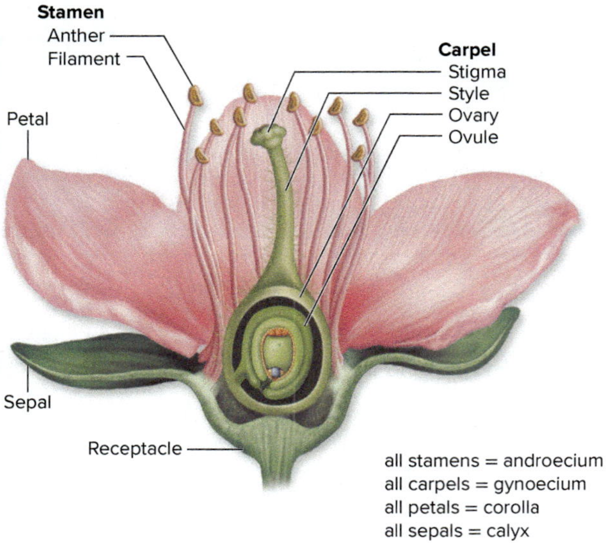

Figure 31.3 The parts of a flower. This is a generalized flower with four primary parts: sepals, petals, stamens, and carpels. The carpels and stamens are the fertile parts of a flower. The roles of carpels and stamens in the life cycle of angiosperms are shown in figure 31.4.

STRUCTURE AND FUNCTION OF FLOWERS

Luckily for florists, angiosperms include a seemingly infinite variety of flowers, ranging from the microscopic flowers of *Lemna* (duckweed) to the rare, monstrous blossoms of *Rafflesia,* which can be up to 1 m across. In today's exercise, we'll consider only the "typical" flower depicted in figure 31.3.

Table 31.1

Characteristics and Examples of the Two Major Classes of Angiosperms

Monocots (~70,000 species)
1. One cotyledon per embryo
2. Flower parts in sets of three
3. Parallel venation in leaves
4. Multiple rings of vascular bundles in stem
5. Lack a true vascular cambium (lateral meristem)
6. Pollen grains have one pore.
Examples: Orchids, barley, corn, lily, wheat, daffodils, onion, bamboo, banana
Eudicots (~170,000 species)
1. Two cotyledons per embryo
2. Flower parts in sets of four or five
3. Reticulate (i.e., netted) venation in leaves
4. One ring of vascular bundles or cylinder of vascular tissue in stem
5. Have a true vascular cambium (lateral meristem)
6. Pollen grains have three pores.
Examples: Zucchini, roses, oaks poppy, peas, beans, mint, daisies, tomato

Examine the flower model(s) available in the lab and the living flowers on display. Identify the following parts:

- **Peduncle**—flower stalk.
- **Receptacle**—the part of the flower stalk that bears the floral organs; located at the base of the flower; usually not large or noticeable.
- **Sepals**—the lowermost or outermost whorls of structures, which are usually leaflike and protect the developing flower; the sepals collectively constitute the **calyx.**
- **Petals**—whorls of structures located inside and usually above the sepals; may be large and pigmented (in insect-pollinated flowers) or inconspicuous (in wind-pollinated plants); the petals collectively constitute the **corolla.**
- **Androecium**—the male portion of the plant that rises above and inside the petals; consists of **stamens,** each of which consists of a **filament** atop which is located an **anther;** inside the anthers are **pollen grains,** which are the microgametophytes and contain the male gametes.
- **Gynoecium**—the female portion of the plant that rises above and inside the androecium; consists of one or more **carpels,** each made up of an **ovary, style,** and **stigma;** the ovary contains **ovules** that contain the megagametophyte; the

megagametophyte is called the embryo sac and contains female gametes.

The sepals and petals are usually the most conspicuous parts of a flower, and a variety of flower types are distinguished by their sepal and petal characteristics. In radially symmetrical flowers such as tulips, the different whorls of the flower consist of similarly shaped parts that radiate from the center of the flower and are equidistant from each other. In other flowers such as snapdragons, one or more parts of at least one whorl are different from other parts of the same whorl. These flowers are usually bilaterally symmetrical.

Procedure 31.1 **Examine flowers and their parts**

1. Obtain a flower provided in the laboratory.
2. Remove the parts of the flower, beginning at the lower, outside whorls of sepals and petals.
3. Locate the petals and sepals and determine their arrangement and point of attachment.

Question 1
How would you describe this flower?

4. Remove the petals and sepals.
5. Locate and remove a stamen and place it on a slide. Examine the stamen with low magnification.
6. Examine a prepared slide of pollen grains. Use high magnification to locate the generative and tube nucleus of a pollen grain. The **generative nucleus** is usually small, spindle-shaped, and off center (you cannot see it easily). The **tube nucleus** is larger and centered. You'll learn more about pollen nuclei later in this exercise.
7. Locate the gynoecium of the flower, and make longitudinal and transverse sections. The gynoecium consists of one or more fused carpels, each with an interior cavity called a locule containing ovules.

Question 2
How many carpels (locules) are apparent?

8. Examine the other types of flowers available in the lab. Repeat procedure 31.1 to guide your examination. Preserved specimens of *Lilium* (lily) and *Ranunculus* (buttercup) may also be provided for you to study and dissect.

Alternation of Generations in Flowering Plants

The life cycle of flowering plants involves the alternation of a multicellular haploid stage with a multicellular diploid stage, as is typical for all plants (fig. 31.4, also see fig. 30.1). The diploid **sporophyte** produces haploid spores by meiosis. Each haploid spore develops into the **gametophyte** by mitosis and cellular differentiation. In angiosperms the sporophyte is the large, mature organism with flowers that you easily recognize. The gametophyte within the flower is reduced to a pollen grain (that contains a sperm nucleus) or an embryo sac (that produces an egg) within an ovule. Be sure to review your textbook's description of alternation of generations.

Production of spores in the sporophyte by meiosis is part of a larger process called **sporogenesis.** Flowering plants produce two types of spores: **microspores** and **megaspores.** Production of gametes by the gametophytes is called **gametogenesis.**

Microsporogenesis, Production of Pollen, and Microgametogenesis

Many of the terms associated with flowers are similar to those you've learned in earlier exercises, except that with angiosperms those terms have a prefix. For example, sporogenesis (i.e., the production of spores) in angiosperms is called microsporogenesis. **Microsporogenesis** is the production of microspores within **microsporangia** of a flower's anthers via meiosis of **microspore mother cells** (microsporocytes) (fig. 31.5). These microspores grow and mature into microgametophytes, which are also known as pollen grains (fig. 31.6). The haploid nuclei in a mature pollen grain include a **tube nucleus** (or vegetative nucleus) and a **generative nucleus.** During pollination, pollen grains are transferred to the stigma, where they germinate and grow a tube through the style to the ovary. The tube nucleus controls the growth of the pollen tube, and the generative nucleus replicates to produce two **sperm nuclei.**

Pollen of some plants cause allergies in many people. However, studies of pollen are important to science beyond the treatment of allergies. For example, geologists examine pollen brought up in sediment cores during oil drilling. Dark brown to black pollen often indicate temperatures too high for oil deposits and suggest that a well will likely produce natural gas. Orange pollen suggest the less intense heat associated with high-quality oil production. In addition, examination of fossil pollen tells us about the diversity of ancient flora and climatic change through the ages, and helps us locate ancient seas and their shorelines where pollen is known to accumulate.

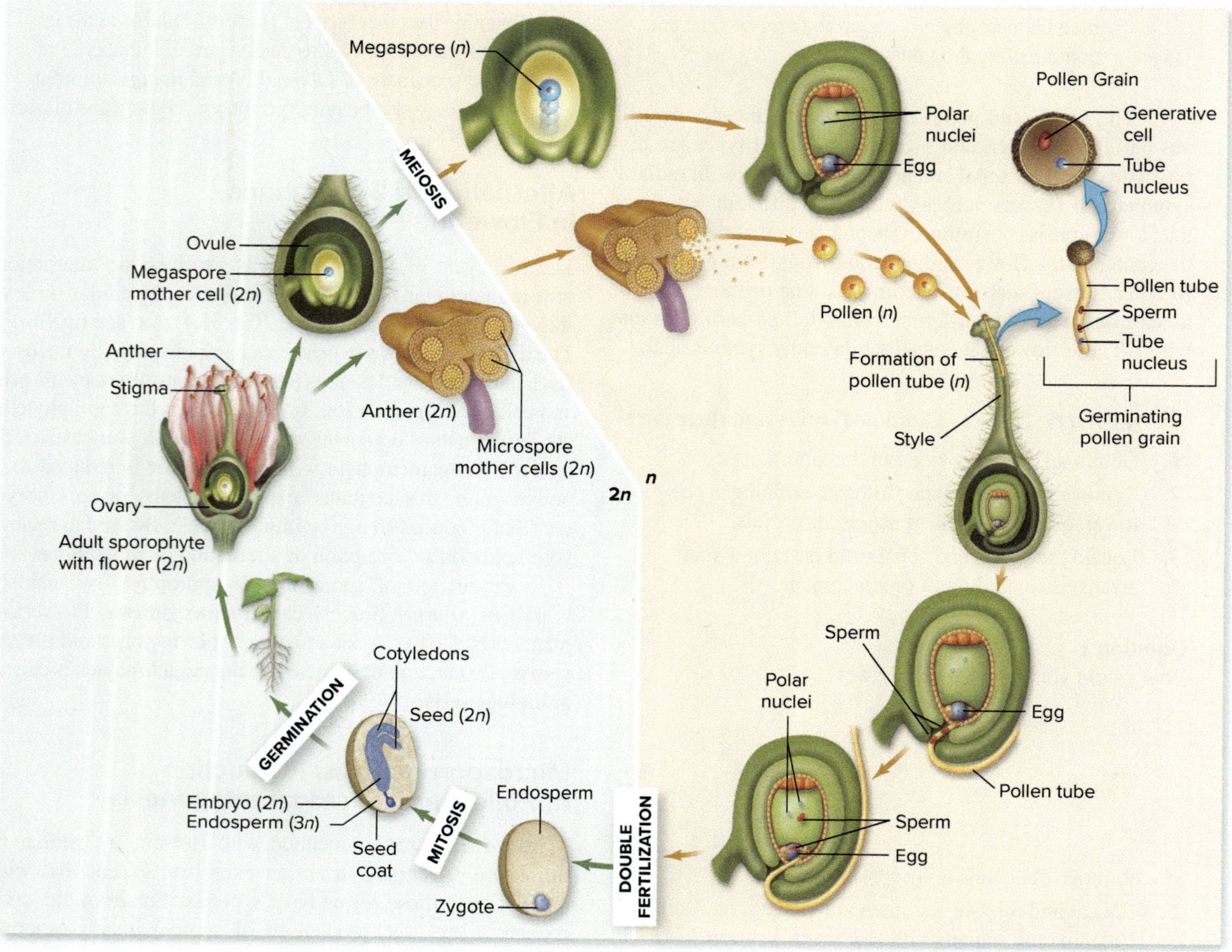

Figure 31.4 Angiosperm life cycle. Eggs form within the embryo sac inside the ovules, which, in turn, are enclosed in the carpels. The pollen grains, meanwhile, form within the sporangia of the anthers and are shed. Fertilization is a double process. A sperm and an egg fuse to produce a zygote; at the same time, another sperm fuses with the polar nuclei to produce the endosperm. The endosperm is the tissue, unique to angiosperms, that nourishes the embryo and young plant.

Procedure 31.2 Examine stages of microsporogenesis and microgametogenesis in *Lilium*

1. Examine fresh or preserved specimens of dehiscent (split-open) and predehiscent anthers (fig. 31.7).

Question 3

a. What differences can you detect between the structures of dehiscent and predehiscent anthers?

b. Which stage is the most mature?

2. Examine a prepared cross section of a young anther. Note the immature sporangia tissue that will form microsporocytes.
3. Examine a prepared cross section of a lily anther showing microsporocytes in early prophase I (fig. 31.8*a*).
4. Examine a prepared slide showing stages of meiosis II (fig. 31.8*b*).
5. Examine a prepared slide showing pollen tetrads of microspores produced by meiosis (fig. 31.8*c*).
6. Examine a prepared slide showing mature pollen with two or more nuclei (fig. 31.8*d*).
7. Examine living or prepared pollen from various plants if available. Note any differences among pollen grains and differences between pollen of monocots and eudicots.

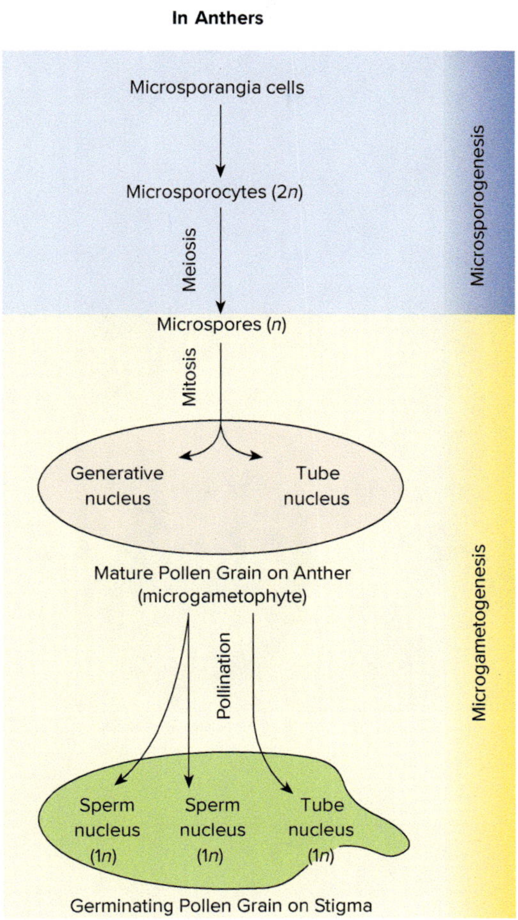

Figure 31.5 Microsporogenesis and microgametogenesis occur in the anthers of flowers. Mature pollen grains (see fig. 31.6) are microgametophytes. Micrographs of microsporogenesis and microgametogenesis are shown in figure 31.8.

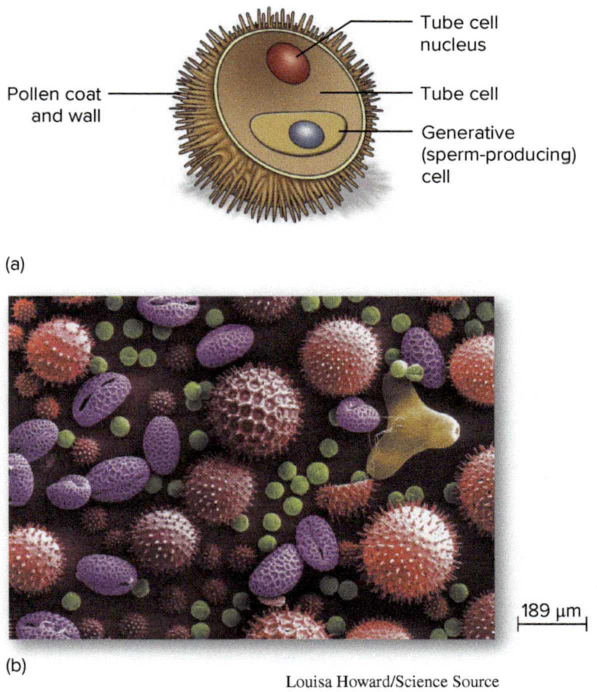

Louisa Howard/Science Source

Figure 31.6 Pollen grains. (*a*) A cut pollen grain with two nuclei. (*b*) Scanning electron micrograph of pollen grains of different species (1200×).

Question 4

a. How many pollen grains germinated?

b. Can you see vegetative and generative nuclei in the pollen tubes?

Procedure 31.3 Examine germinated pollen grains

Examine a prepared slide of germinated pollen grains.

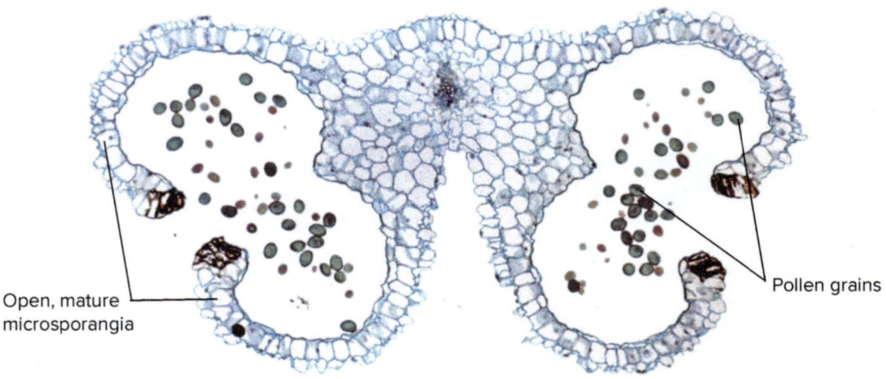

BiologyImaging.com

Figure 31.7 *Lilium* anther (40×). Mature pollen grains are the product of microsporogenesis and microgametogenesis within the microsporangia of flower anthers. A higher-magnification view of developing pollen grains is shown in figure 31.8.

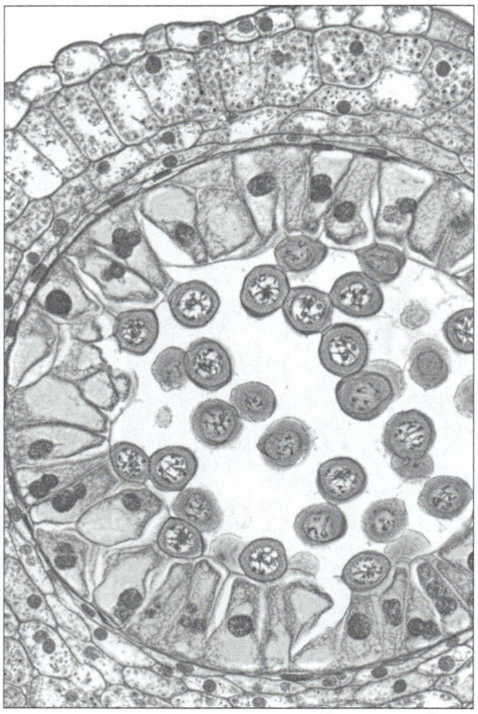

(a) M. I. Walker/Science Source

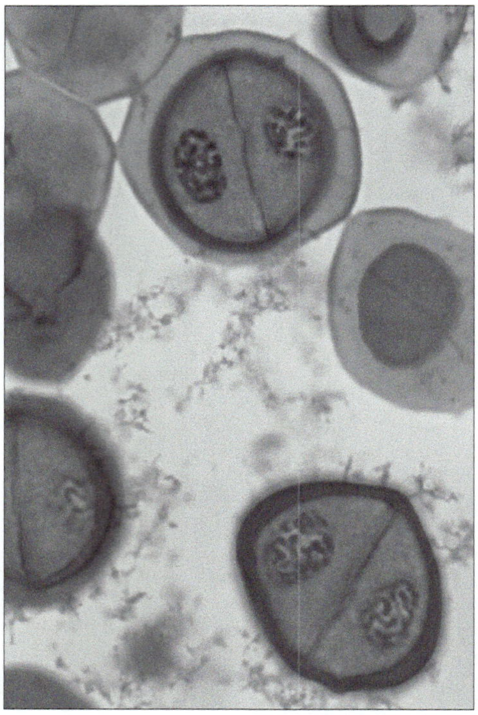

(b) Biophoto Associates/Science Source

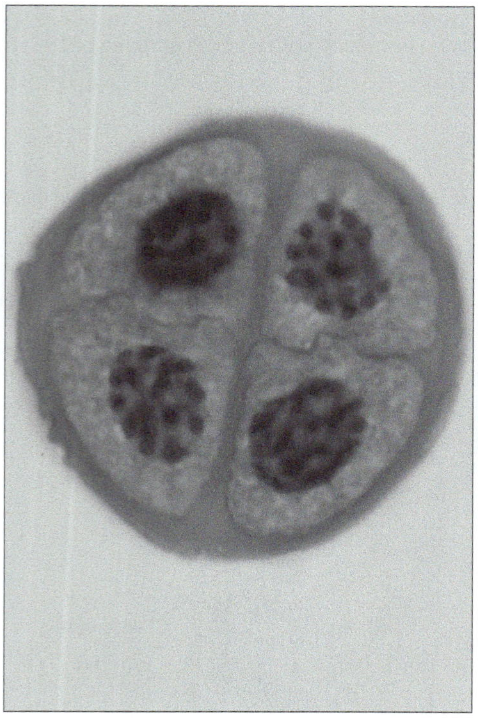

(c) ©Ed Reschke/Getty Images

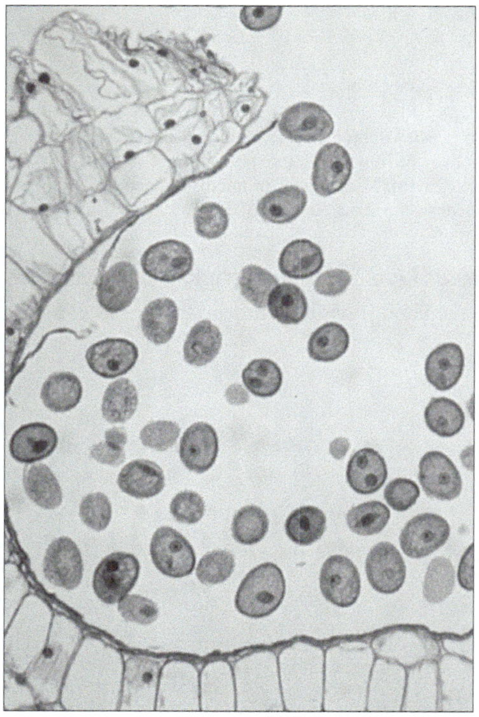

(d) Noble Proctor/Science Source

Figure 31.8 Stages of microsporogenesis and microgametogenesis in *Lilium*. (*a*) Cells in early prophase I (350×). (*b*) Cells in second meiotic division (500×). (*c*) Tetrads of microspores from meiosis (680×). (*d*) Mature pollen (230×). For an overview of these stages, see figure 31.5.

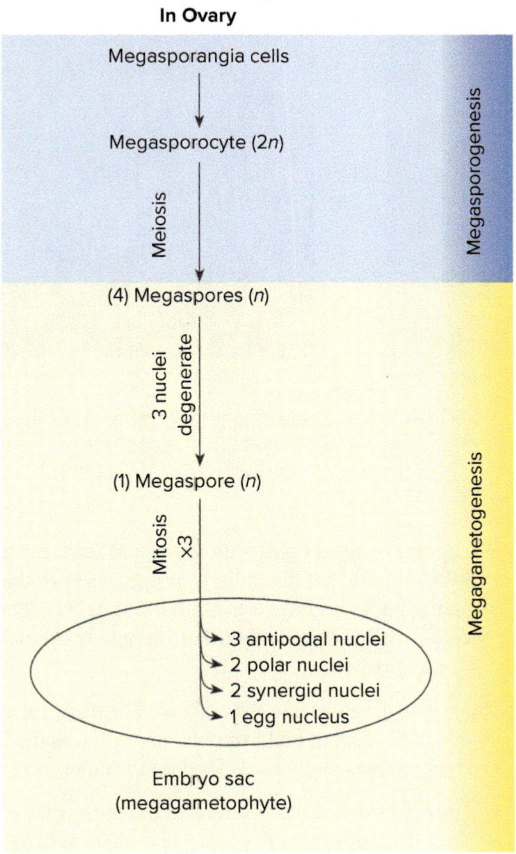

Figure 31.9 Megasporogenesis and megagametogenesis occur in the ovaries of flowers. The embryo sac is the mature megagametophyte. The locations of these processes are shown in figures 31.10 and 31.11.

Megasporogenesis, Production of Ovules, and Megagametogenesis

Megasporogenesis is the production of megaspores (fig. 31.9); it occurs in the sporangia of the flower ovary by meiosis of **megaspore mother cells** (megasporocytes). These megaspores undergo **megagametogenesis;** that is, they develop into **megagametophytes.** The megagametophyte and its surrounding tissues are called an **ovule.** Ovules usually have two coverings called **integuments.** The entire haploid structure is called the **embryo sac** and consists of only six to ten nuclei, one of which is an egg. A seven- or eight-celled embryo sac is most common (fig. 31.10).

Procedure 31.4 Examine stages of megasporogenesis and megagametogenesis in *Lilium*

1. Examine a cross section of a *Lilium* ovary and locate the six megasporangia. Within each of these megasporangia a megasporocyte will form and develop. The stages for development of a megasporocyte are shown in figures 31.9, 31.10, and 31.11.
2. Examine a prepared slide of a cross section of a *Lilium* ovary showing a diploid megasporocyte within the sporangium before meiosis (fig. 31.11*a*).
3. Examine a prepared slide showing the four-nucleate embryo sac after meiosis (fig. 31.11*b*). Meiosis produced four haploid megaspores in the embryo sac. In most angiosperms, three of the four nuclei degenerate and the single remaining nucleus passes through two mitotic divisions before the next stage. *Lilium* is atypical because all four products of meiosis contribute nuclei to the embryo sac, but only one develops into the egg cell.
4. Examine a prepared slide showing the eight-nucleate embryo sac (fig. 31.11*c*). Because of the large size of the embryo sac, it is seldom possible to observe all eight nuclei in the same section.
5. At the end of the megagametophyte toward the micropyle (the small opening where the pollen tube enters the ovule), locate the eight nuclei including the egg and two synergid nuclei associated with fertilization

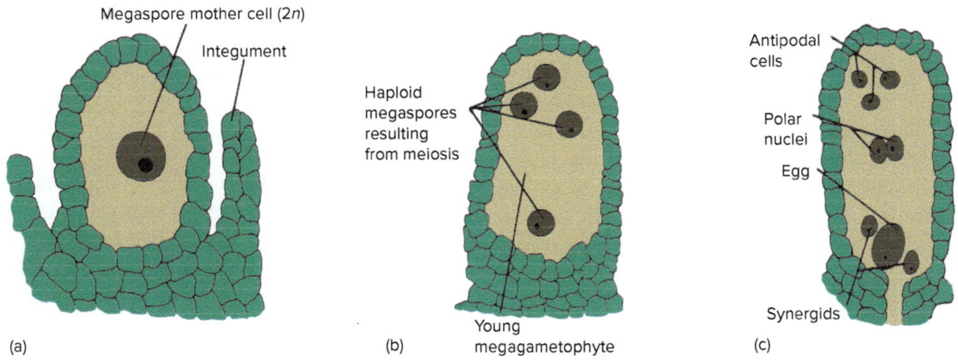

Figure 31.10 Megasporogenesis and megagametogenesis in *Lilium.* (*a*) The megaspore mother cell (megasporocyte) is diploid and undergoes meiosis. (*b*) Four haploid megaspores result from meiosis. (*c*) The mature megagametophyte contains eight nuclei, one of which is the egg.

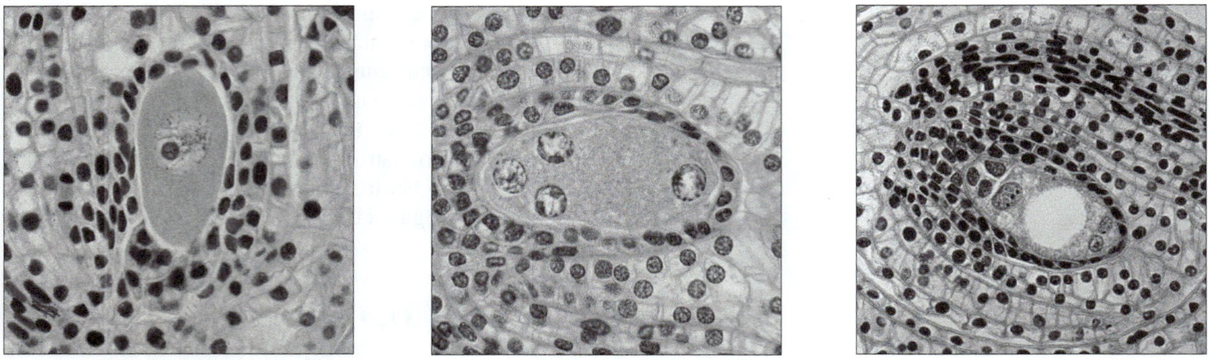

Figure 31.11 Stages of megasporogenesis and megagametogenesis in *Lilium* (200×). (*a*) Diploid megasporocyte in lily ovary. (*b*) Embryo sac with four nuclei produced by meiosis. (*c*) Embryo sac (mature megagametophyte) with eight nuclei (only six are visible in this image). For an overview of these stages, see figure 31.9.

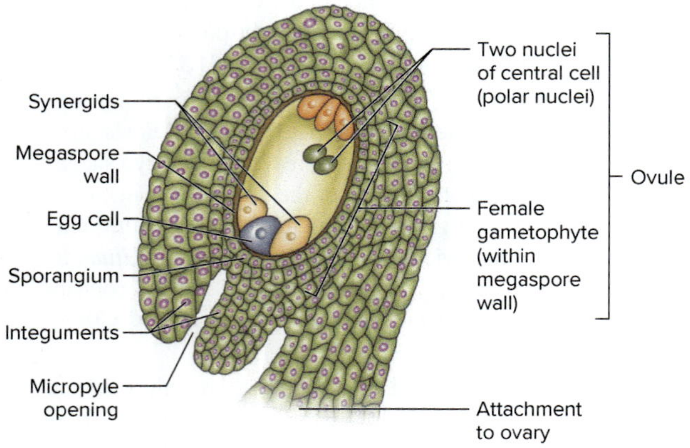

Figure 31.12 Mature female gametophyte within an ovule.

(figures 31.11c and 31.12). At the opposite end locate three antipodal cells that usually do not participate in reproduction. In the center are two polar nuclei that migrated from each pole of the megagametophyte.

POLLINATION AND FERTILIZATION

Biologists have long been interested in plant pollination, for obvious economic reasons. Angiosperm reproduction depends on pollination. However, you may have a more personal interest in pollen if you suffer from hay fever (fig. 31.6).

After the development of a microgametophyte (pollen grain) with sperm and a megagametophyte (ovule) with an egg, sexual reproduction in angiosperms occurs as follows:

1. Pollination occurs when pollen is transported to the surface of the flower's stigma (fig. 31.13).
2. The pollen grain germinates, and a pollen tube grows through the stigma and style to the ovary. Its growth is governed by the style and the tube nucleus of the pollen grain.
3. One sperm nucleus fuses with the egg to form the diploid (2*n*) zygote, and the other sperm fuses with the two polar nuclei to form a triploid (3*n*) nucleus. This process is called **double fertilization** and is characteristic of angiosperms.
4. The zygote develops into the embryo. The triple fusion of the sperm nucleus and two polar nuclei forms the triploid endosperm that provides food for the embryo.
5. The integuments of the ovule form the seed coat, and the fruit develops from the ovary and other parts of the flower.

SEED AND EMBRYO DEVELOPMENT

A **seed** is a mature ovule that includes a seed coat, a food supply, and an embryo. Seeds range in size from tiny (e.g., a broomrape seed weighs only a few micrograms) to massive (e.g., an avocado seed can weigh more than 50 g). Embryology of the mature zygote occurs within the seed and before germination. This embryology and its controlling factors are complex, but various stages of development are easily observed.

The stages of embryo development in the seed of *Capsella* (a eudicot) are shown in figure 31.14. The developing embryo grows, absorbs endosperm, and stores those nutrients in "seed leaves" called cotyledons. Development includes the following stages:

- **Proembryo stage.** During development, the **zygote** divides to form a mass of cells called the **embryo.** Initially the embryo consists of a **basal cell, suspensor,** and a two-celled proembryo. The suspensor is the column of cells that pushes the embryo into the **endosperm.** The endosperm is extensive but is being digested.
- **Globular stage.** Cell division of the proembryo soon leads to the globular stage that is radially symmetrical and has little internal cellular organization.

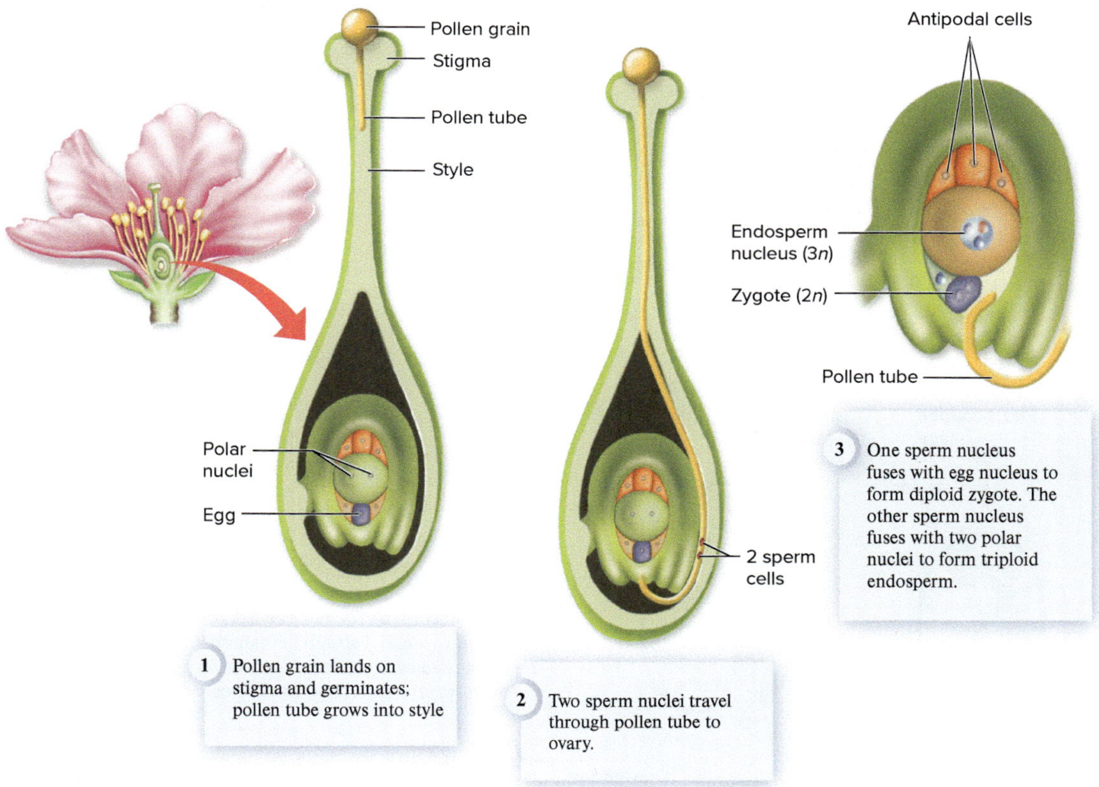

Figure 31.13 Pollen germination and double fertilization. Pollen sticks to a stigma on a flower of the same species. A pollen tube grows toward the ovule and transports two sperm nuclei. One sperm nucleus fertilizes the egg to form a zygote, and the other fertilizes the polar nuclei to yield the endosperm.

- **Heart-shaped stage.** Differential division of the globular stage produces bilateral symmetry and two **cotyledons** forming the heart-shaped embryo. The enlarging cotyledons store digested food from the endosperm. Tissue differentiation begins, and root and shoot meristems soon appear.
- **Torpedo stage.** The cotyledons and root axis soon elongate to produce an elongate torpedo-stage embryo. Procambial tissue appears and will later develop into vascular tissue.
- **Mature embryo.** The mature embryo has large, bent cotyledons on each side of the **stem apical meristem.** The **radicle,** later to form the root, is differentiated toward the suspensor. The radicle has a **root apical meristem** and **root cap.** The **hypocotyl** is the region between the apical meristem and the radicle. The endosperm is depleted, and food is stored in the cotyledons. The **epicotyl** is the region between attachment of the cotyledons and the stem apical meristem; it has not elongated in the mature embryo.

Procedure 31.5 Examine development of a *Capsella* embryo

1. Obtain and examine prepared slides showing the various stages of embryo development.

2. The cross section most likely passes through an entire fruit of *Capsella* and shows a number of sectioned, developing seeds each in a slightly different stage. Some seeds were sectioned off-center and the stage may not be obvious. Nevertheless, each slide should include an example of at least one stage.

3. Locate as many of the following stages of seed development as possible on your slide: globular, heart, torpedo, and mature embryos. These stages are continuous, so each slide may have multiple stages and intermediates between stages.

4. You may need to examine several slides. Find examples of as many stages as you can.

5. Compare your observations with figure 31.14.

Question 5

a. Why is the endosperm being digested?

b. Is *Capsella* a monocot or a eudicot? How can you tell?

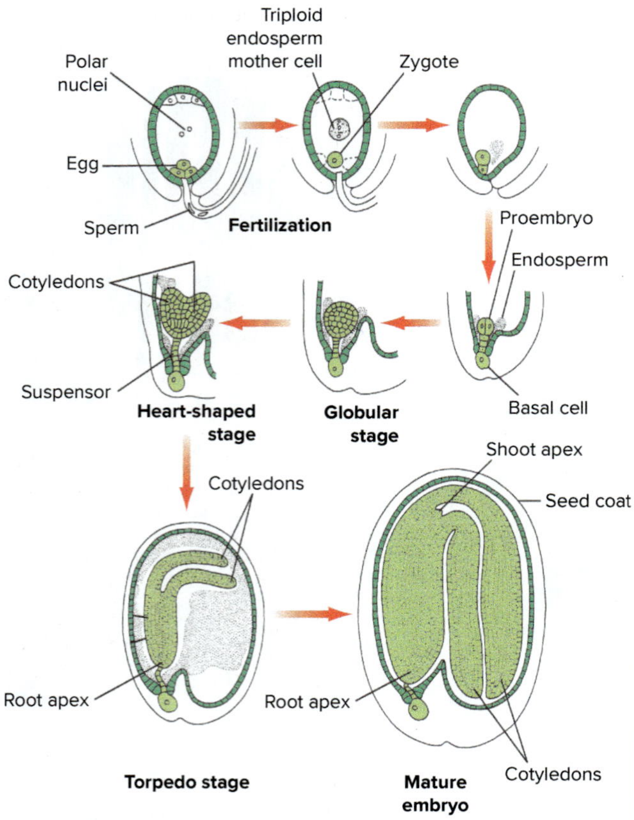

Figure 31.14 Stages of development in a *Capsella* embryo. This development transforms a zygote into a mature embryo.

SEED STRUCTURE

Do not eat any seeds or fruits used in this laboratory.

Procedure 31.6 Observe parts of a bean seed

1. Obtain some beans that have been soaked in water for 24 h.
2. Peel off the seed coat and separate the two cotyledons. Between the cotyledons you'll see the young root and shoot.
3. Examine the opened seed and compare its structure with that shown in figure 31.15. Look for these features:
   - **Micropyle**—a small opening on the surface of the seed through which the pollen tube grew.
   - **Hilum**—an adjacent, elliptical area at which the ovule was attached to the ovary.
   - **Cotyledon**—food for the embryo.
   - **Embryo with young root and shoot**—develops into the new sporophyte.

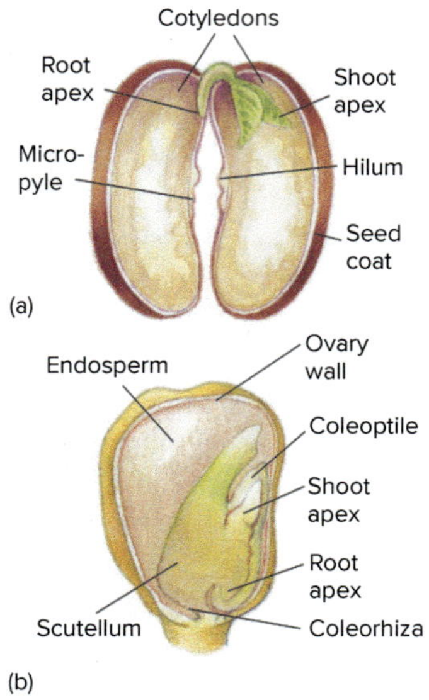

Figure 31.15 Seed structure of a garden bean (eudicot) and corn (monocot). (*a*) The two cotyledons in each seed of garden bean (*Phaseolus vulgaris*) absorb the endosperm before germination. (*b*) Corn (*Zea mays*) has seeds in kernels (grains); the single cotyledon is an endosperm-absorbing structure called a scutellum.

4. Add a drop of iodine to the cut surface and observe the staining pattern. (Recall from Exercise 6 that iodine stains starch black.) Show this pattern on figure 31.15.
5. Repeat steps 2–4 with soaked peas.

Question 6

How are seeds of peas and beans similar? How are they different?

Procedure 31.7 Examine a corn grain with embryo

1. Examine a prepared slide of a corn grain (fig. 31.16). Identify the following features:
   - **Endosperm**
   - **Scutellum** (cotyledon; this structure helps absorb the endosperm)
   - **Coleoptile** (sheath enclosing shoot apical meristem and leaf primordia of grass embryos)
   - **Root**
   - **Root cap**
   - **Coleorhiza** (sheath enclosing embryonic root of grass embryo)
   - **Shoot apical meristem**

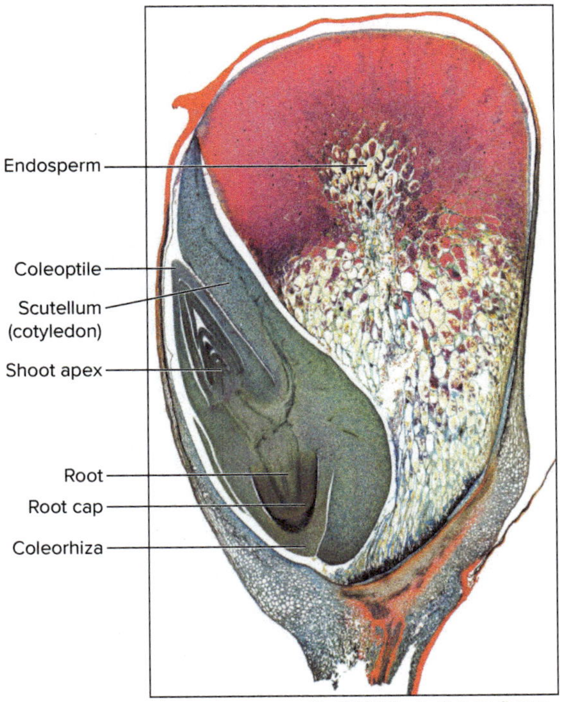

Figure 31.16 The structure of a corn grain, longitudinal section. The embryo is nourished by the starchy endosperm (7×).

2. Use a razor blade to longitudinally split a water-soaked corn grain.

3. Add a drop of iodine to the cut surface and observe the staining pattern. Indicate this pattern on figure 31.16.

Question 7

a. What does this staining pattern tell you about the content of the endosperm and embryo?

b. Do mature seeds of monocots or eudicots store most of their food in cotyledons? How can you tell?

FRUIT

A **fruit** is a mature, ripened ovary plus any associated tissues. Therefore, a fruit contains seeds. Are you surprised that tomatoes, beans, and okra are fruit? A mature fruit is often larger than the ovary at the time of pollination and fertilization, which indicates that a great deal of development occurs while the seeds are maturing.

Most fruits are either dry or fleshy. Dry fruits crack or split at maturity and release their seeds. Sometimes the dry wall surrounds the seed until it germinates. The fruit wall is usually tough and hard and is sometimes referred to as stony. The seeds of fleshy fruits remain in the tissue until germination.

A typical fruit has an outer wall called a **pericarp** composed of an **exocarp, mesocarp,** and **endocarp.** Within the pericarp are seeds, various partitions, and placental tissues. The fruit often includes the receptacle of the flower. The structure of a fruit depends on the number of carpels in each flower and whether or not the carpels are fused.

Procedure 31.8 **Observe the diversity of fruits**

1. Use the following descriptions to study and classify fruits available in the lab. This list is not complete but describes a few common types of fruit.

Procedure 31.9 **Observe the structure of a bean, sunflower, corn grain, apple, and tomato**

1. Examine the pod of a string bean. This single carpel has two seams that can open to release seeds. Remove the seeds; then locate and remove the seed coat. Split the seed and locate the embryo.

Question 8

a. Does the pod appear to be a single carpel with one cavity containing seeds?

b. Is the micropyle near the attachment of the seed to the pod?

2. Crack the outer coat of a sunflower fruit and remove the seed. Locate the embryo and determine whether sunflower is a monocot or eudicot.

Question 9

a. Before today would you have referred to the uncracked sunflower achene as a fruit? Why or why not?

b. Is sunflower a monocot or eudicot?

Dichotomous Key to Major Types of Fruit

I. Fleshy fruits

A. Simple fruits develop from a single ovary.

1. Flesh mostly of ovary tissue

a. Endocarp hard and stony; ovary superior and single-seeded (cherry, olive, peach, plum, coconut): **drupe**

b. Endocarp fleshy or slimy; ovary usually many-seeded (tomato, grape, green pepper): **berry**

2. Flesh mostly of receptacle tissue (apple, pear, quince): **pome**

B. Complex fruits develop from more than one ovary.

1. Fruit from many carpels on a single flower (raspberry, blackberry): **aggregate fruit**

2. Fruit from carpels of many flowers fused together (pineapple, mulberry): **multiple fruit**

II. Dry fruits

A. Fruits that split open at maturity (usually more than one seed)

1. Split occurs along two seams in the ovary. Seeds borne on one of the halves of the split ovary (pea and bean pods, peanuts): **legume**

2. Seeds released through pores or multiple seams (poppies, irises, lilies): **capsule**

B. Fruits that do not split open at maturity (usually one seed)

1. Pericarp hard and thick, with a cup at its base (acorn, chestnut, hickory): **nut**

2. Pericarp thin and winged (maple, ash, elm): **samara**

3. Pericarp thin and not winged

(sunflower, buttercup): **achene**

(cereal grains): **caryopsis**

3. Examine an ear of corn from which the husks have been removed without disturbing the "silk." The strands of silk are styles of the gynoecium.

Question 10

a. If a corn grain is actually a fruit, where is the pericarp?

b. Does a corn grain have a cotyledon as well as an endosperm?

4. Examine an apple, which is an example of a pome. Section the apple transversely and count the number of carpels. Then section an apple longitudinally and compare its structure with figure 31.18. Most of the outer flesh of the apple is derived from tissue other than the ovary. Locate the outer limit of the pericarp and the limit of the endocarp.

Question 11

a. How many carpels are fused to form an apple?

b. How might the fleshy pericarp aid in seed dispersal?

5. Examine a tomato and section it transversely. Compare its structure with that shown in figure 31.18. The jelly-like material is the placenta giving rise to the ovules.

Question 12

How many carpels are fused to form a tomato?

Our Uses of Angiosperms

We have countless uses for angiosperms, including wood (e.g., oak, maple), food (e.g., wheat, rice, corn, apples, tomatoes), textiles (e.g., jute), shade, decorations, insect repellants, spices, and landscaping. Modern agriculture is based almost entirely on angiosperms.

We also used angiosperms as sources of numerous drugs. For example, Taxol (a drug used to treat cancer) and morphine (a painkiller) are derived from flowering plants, as are illicit drugs such as cocaine, opium, and heroin. So, too, is caffeine, a drug ingested by more than 100 million Americans every day.

ORGANIZING WHAT YOU'VE LEARNED

Go to page 362, which follows this exercise. Complete the portions of the table that refer to the phyla you studied in today's lab.

That Morning Cup of Coffee

Coffee, which is grown in more than 50 countries worldwide, is the world's second-most traded commodity (petroleum is first). Coffee is a $60 billion industry. Brazil is the world's leading producer of coffee; it produces 30% of the world's crop. Worldwide, nearly 7 million tons of green coffee beans are produced per year. About 80% of this coffee comes from *Coffea arabica,* and the remaining 20% from *Coffea canephora.* It takes three to five years for a coffee plant to reach maturity and produce its scented white flowers and cherry-like fruit (figure 31.17). The red/purple coffee fruits are called "cherries." Except for peaberry coffee, whose fruits each contains one smaller, rounder seed, each coffee fruit contains two seeds which are referred to as "beans" (despite the fact that they are not true beans).

In the United States, coffee is the second-most popular drink (water is first). Coffee drinkers drink an average of 3.1 9-ounce cups of coffee per day. Fifty-four percent of Americans at least 18 years old drink coffee every morning, and another 30% drink coffee occasionally. About 30% of people in the United States who drink coffee every day drink specialty drinks such as lattes and cappuccinos; only about 35% of coffee drinkers like their coffee black.

The average coffee-drinker spends $165 per year on coffee. Coffee drinkers who get their coffee from a coffeehouse stand in line there about 45 hours per year.

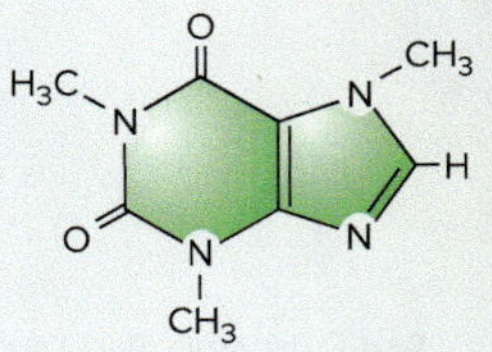

Caffeine

GFC Collection/Alamy Stock Photo

Figure 31.17 Coffee fruit (i.e., "beans") contain caffeine, which, in nature, is a natural pesticide that paralyzes and kills certain insects that attack coffee plants. In humans, caffeine is a stimulant that temporarily wards off drowsiness and restores alertness.

INQUIRY-BASED LEARNING

Grocery store botany: What plant organs are sold at your local grocery store?

Observation: The produce sections of grocery stores and markets contain a vast array of plant organs, each having a characteristic shape, taste, and texture. These plants have been the subject of intensive artificial selection over many generations, and today we enjoy the many varied products of that selection.

Question: What plant organ (e.g., modified root, stem, leaf, flower) are you purchasing when you buy items for a salad? Assume that your salad includes beets, radishes, sweet potatoes, celery, carrots, broccoli, asparagus, eggplant, lettuce, and onion.

a. Establish a working lab group and obtain Inquiry-Based Learning Worksheet 31 from your instructor.
b. Discuss with your group well-defined questions relevant to the preceding observation and question. Choose and record your group's best question for investigation.
c. Translate your question into a testable hypothesis and record it.
d. Outline on Worksheet 31 your experimental design and supplies needed to test your hypothesis. Ask your instructor to review your proposed investigation.
e. Conduct your procedures, record your data, answer your question, and make relevant comments.
f. Discuss with your instructor any revisions to your questions, hypotheses, or procedures. Repeat your work as needed.

Peter Fakler/Alamy Stock Photo

(a) A drupe is a fleshy fruit with a pit containing a single seed produced from a simple ovary.

D. Hurst/Alamy Stock Photo

(b) A berry is a fleshy fruit having seeds and pulp produced from a compound ovary.

Danny Smythe/Shutterstock

(c) A legume is a dry dehiscent fruit produced from a simple ovary.

Robert Llewellyn/Corbis/Getty Images

(d) A samara is a dry indehiscent fruit produced from a simple ovary.

MarcoFood/Shutterstock

(e) An aggregate fruit contains many fleshy fruits produced from simple ovaries of the same flower.

Ingram Publishing/Alamy Stock Photo

(f) A multiple fruit contains many fused fruits produced from simple ovaries of individual flowers.

Figure 31.18 Diagrams of fruit and their originating flowers.

Questions for Further Study and Inquiry

1. What is meant by "double fertilization"?

2. What features of seeds and fruits have enabled angiosperms to become so widespread?

3. What are the similarities and differences between cones and flowers?

4. What is the difference between a fruit and a vegetable?

5. In what ways does human welfare depend on seed plants?

6. How can you distinguish a monocot from a eudicot? If a plant has six stamens and leaves with parallel veins, is the plant a monocot or a eudicot?

7. How are insects such as bees important for the reproduction of angiosperms?

8. What advantages do flowers give angiosperms over gymnosperms?

9. How is microsporogenesis different from megasporogenesis?

10. What are the major parts (and their functions) of your favorite flower?

WRITING TO LEARN BIOLOGY

What are the functions of a flower? Describe how a flower of your choice is adapted to each function.

Summarizing What You've Learned

After each of your studies of plants (Exercises 28–31), complete the relevant parts of this table to help you summarize what you've learned.

Exercise Number	Phylum	Common Name	Representatives	Dominant Generation	How Propagated
28	Hepatophyta				
28	Bryophyta				
28	Anthocerophyta				
29	Pterophyta				
29	Lycophyta				
30	Cycadophyta				
30	Ginkgophyta				
30	Coniferophyta				
30	Gnetophyta				
31	Anthophyta				

EXERCISE 32

Plant Anatomy

Vegetative Structure of Vascular Plants

Learning Objectives

By the end of this exercise you should be able to:

1. Describe the functions of roots, stems, and leaves.
2. Distinguish between primary and secondary growth.
3. Describe the functional significance of the internal and external structure of roots, stems, and leaves.
4. Explain what causes growth rings in wood.

Please visit connect.mheducation.com to review online resources tailored to this lab.

The structure of plants varies greatly among species; compare, for example, an oak tree with a cactus. However, these structural differences are typically quantitative rather than qualitative; that is, the differences among roots, stems, and leaves result not from tissues unique to one species or another, but rather from different arrangements and proportions of the same tissues. These differences among plants represent a variety of adaptations that result in the universal evolutionary outcomes of survival and reproduction. This exercise will concentrate on the structure of roots, stems, and leaves of flowering plants.

ROOTS

During seed germination, a **radicle** or young **primary root** emerges from the seed and grows down. The primary root soon produces numerous **secondary roots** and forms a root system that absorbs water and minerals, anchors the plant, and stores food. Root systems have different morphologies. For example, a **taproot system** has a large main root and smaller secondary roots branching from it (e.g., carrot). In a **fibrous root system,** the primary and secondary roots are similar in size (fig. 32.1). Examine the displays of taproot and fibrous root systems available in the lab.

Primary growth of roots and all primary tissues is formed by **apical meristems.** A meristem is a localized area of cellular division. Apical meristems occur at the tips of roots and stems. Primary growth (i.e., growth in length) produces herbaceous (nonwoody) tissue. **Secondary growth** refers to growth in girth resulting from nonapical meristems, some of which are discussed later in this exercise.

Question 1

a. How do taproot systems and fibrous root systems help plants survive and reproduce?

b. Would one type of root system provide more adaptive advantages in a particular environment such as a rain forest? A desert? Explain your answer.

The Root Apex

Examine the root tips of two-day-old seedlings of radish (*Raphanus*) and corn (*Zea*) with your dissecting microscope. Refer to figure 32.2 and identify the **root cap, root apical meristem, zone of elongation,** and **zone of maturation.**

The cone of loosely arranged cells at the root cap perceives gravity and protects the root apical meristem. The root cap protects the root by secreting mucilage and sloughing cells as the root grows through the soil (fig. 32.3). The root apical meristem is behind the root cap and produces all of the new cells for primary growth. These cells elongate in the zone of elongation, which is 1–4 mm behind the root tip. This elongation produces primary growth.

(a) Peter Anderson/Getty Images

(b) redmal/E+/Getty Images

Figure 32.1 Two common types of root systems of vascular plants. (*a*) The taproot system consists of a prominent, thickened taproot. (*b*) The fibrous root system consists of many similarly sized roots. Fibrous root systems form extensive networks in the soil and minimize soil erosion.

Question 2

a. In the following space, sketch the root tips that you examined. In which area of a root tip are cells largest? In which area are they smallest?

b. Aside from their size, do all cells in the root tip appear similar? Why is this significant?

Re-examine the two-day-old radish seedling with your dissecting microscope. Note the many **root hairs** in the zone of maturation. Root hairs are outgrowths of epidermal cells and are short-lived. Root hairs increase the surface area of the root.

Question 3

a. Why do you think root hairs occur only in the zone of maturation? Why aren't they in the zone of elongation?

b. What is the function of root hairs?

Primary Tissues of the Root

The root apical meristem produces cells that differentiate into primary tissues of the root. The outer layer of cells is the **epidermis.** Just inside the epidermis is the **cortex,** whose cells contain numerous **amyloplasts,** which are starch-containing plastids. The inner layer of the cortex is the **endodermis,** which affects water flow and regulates the flow of nutrients to the vascular tissue in the center of the root. Immediately inside the endodermis is the **pericycle,** which can become meristematic and produce **secondary roots** (figs. 32.4 and 32.5).

Procedure 32.1 **Examine primary tissues of a root**

1. Examine a prepared slide of a cross section of a buttercup (*Ranunculus,* a eudicot) root (figs. 32.4 and 32.5). Sketch what you see. Label and state the function of each tissue that is present. (Note: Your

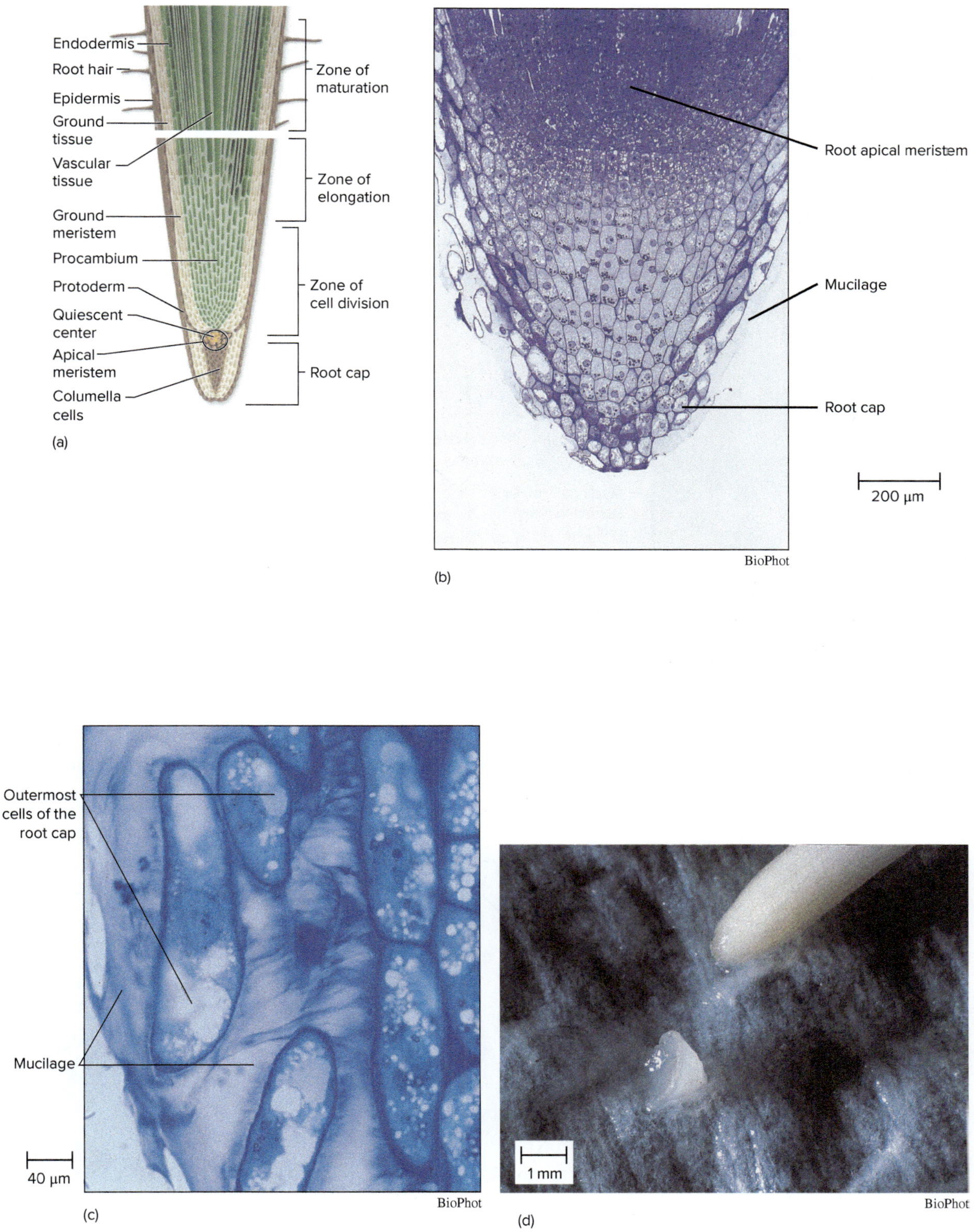

Figure 32.2 (*a*) Root tip showing root cap, root apical meristem, zones of structure and function. (*b*) The root apical meristem is covered by a thimble-shaped root cap that protects the meristem as the root grows through the soil (60×). (*c*) Mucilage produced by the root cap lubricates the root as it grows through the soil (see fig. 32.3) (420×). (*d*) In plants such as corn, the root cap can easily be removed from the rest of the root (6×).

instructor may have you examine a slide comparing cross sections of roots of monocots and eudicots.)

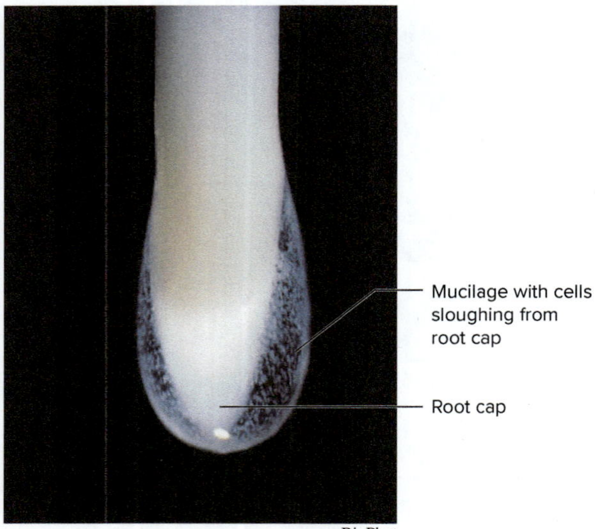

BioPhot

Figure 32.3 Tips of roots secrete large amounts of mucilage, a lubricant that helps the root force its way through the soil. The mucilage is secreted primarily by the root cap. Movement of the root through soil is also aided by the sloughing of root-cap cells. These sloughed cells are visible in the drop of mucilage on the tip of this root.

2. Examine a prepared slide labeled "lateral root origin." Locate the epidermis, cortex, pericycle, and newly formed secondary root. Sketch the lateral root and label its parts.

3. If time permits, also examine a prepared slide of a cross section of a corn (*Zea*, a monocot) root.

Question 4

a. Based on the presence of amyloplasts, what do you suppose is the primary function of the cortex?

b. Do secondary roots arise inside the primary root or on its surface?

c. How does the structure of a monocot root differ from that of a eudicot?

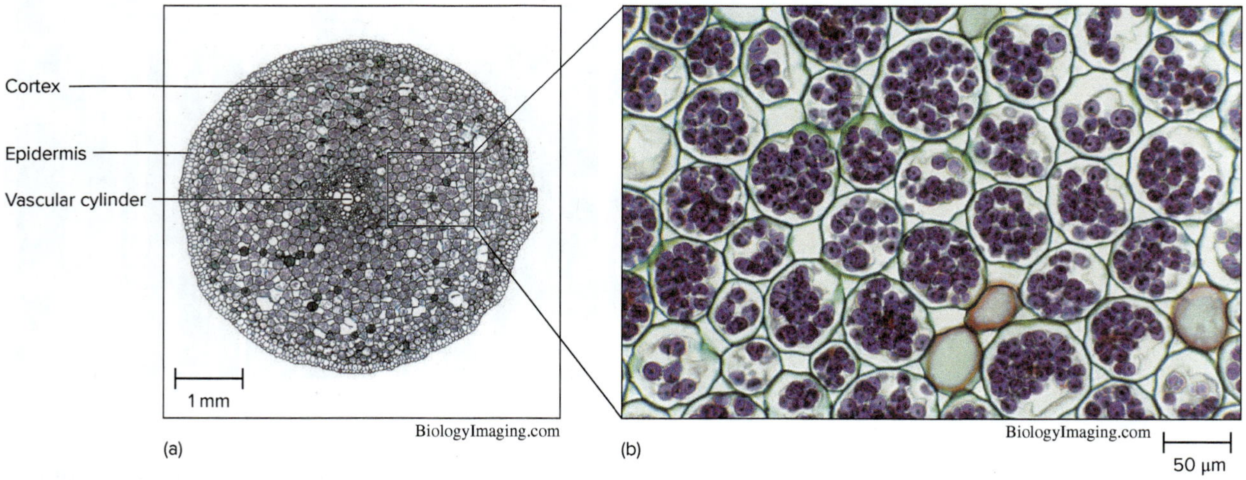

BiologyImaging.com BiologyImaging.com

Figure 32.4 Transverse sections of a root of a buttercup (*Ranunculus*), a eudicot. (*a*) Overall view of mature root. The vascular cylinder includes tissues specialized for long-distance transport of water and solutes, whereas the epidermis forms a protective outer layer of the root (16×). (*b*) Detail of cortex (250×). Each parenchyma cell in the cortex contains many amyloplasts, which store starch. In this image, the starch in the amyloplasts has been stained with iodine (Exercise 6).

In the center of the buttercup root is the **vascular** (fluid-conducting) **cylinder** composed of **xylem** and **phloem** (fig. 32.5). Xylem transports water and minerals; phloem transports water and most organic compounds in the plant, including carbohydrates. Water-conducting cells in the xylem of angiosperms are called **tracheids** and **vessel elements** and are dead and hollow at maturity. Tracheids are long, spindle-shaped cells with thin areas or apertures called **pits** where the cell walls of adjacent cells overlap (fig. 32.6). Water moves through these pits from one cell to the next. Vessels are stacks of cylindrical cells with thin or completely open end-walls. Water moves through vessel elements in straight, open tubes. These tubes are usually stained red or blue in slide preparations of buttercup roots.

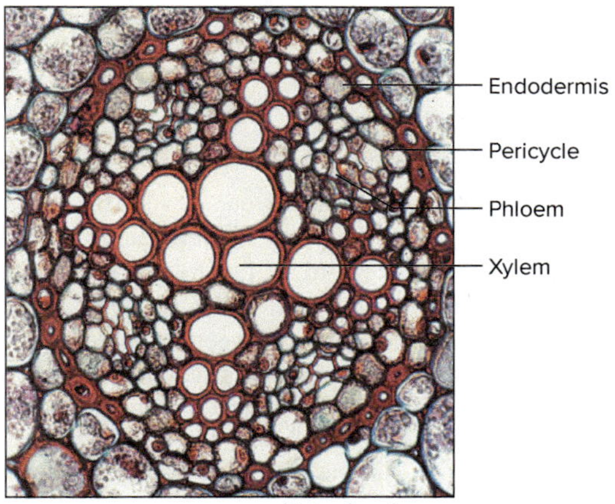

BiologyImaging.com

Figure 32.5 A cross section through the center of a root of a buttercup (*Ranunculus*), a eudicot (125×). The phloem and xylem are vascular tissues of the vascular cylinder shown in figure 32.4*a*.

Conducting cells in phloem are called **sieve cells** and **sieve tube members** and are alive at maturity. Phloem cells are small, thin-walled, and arranged in bundles that alternate with the poles of xylem (fig. 32.5). Sieve tube members are usually stained green.

Procedure 32.2 Examine carrot root

1. Prepare two thin cross sections of a carrot root.
2. Stain one slice with iodine (a stain for starch) and examine it with your microscope.
3. Stain the other section of carrot root with phloroglucinol. Phloroglucinol stains **lignin,** a molecule that strengthens xylary cell walls.

Phloroglucinol contains hydrochloric acid. Do not spill any on yourself or your belongings!

Question 5

a. Where is starch located in a carrot root?

b. What can you conclude from this observation?

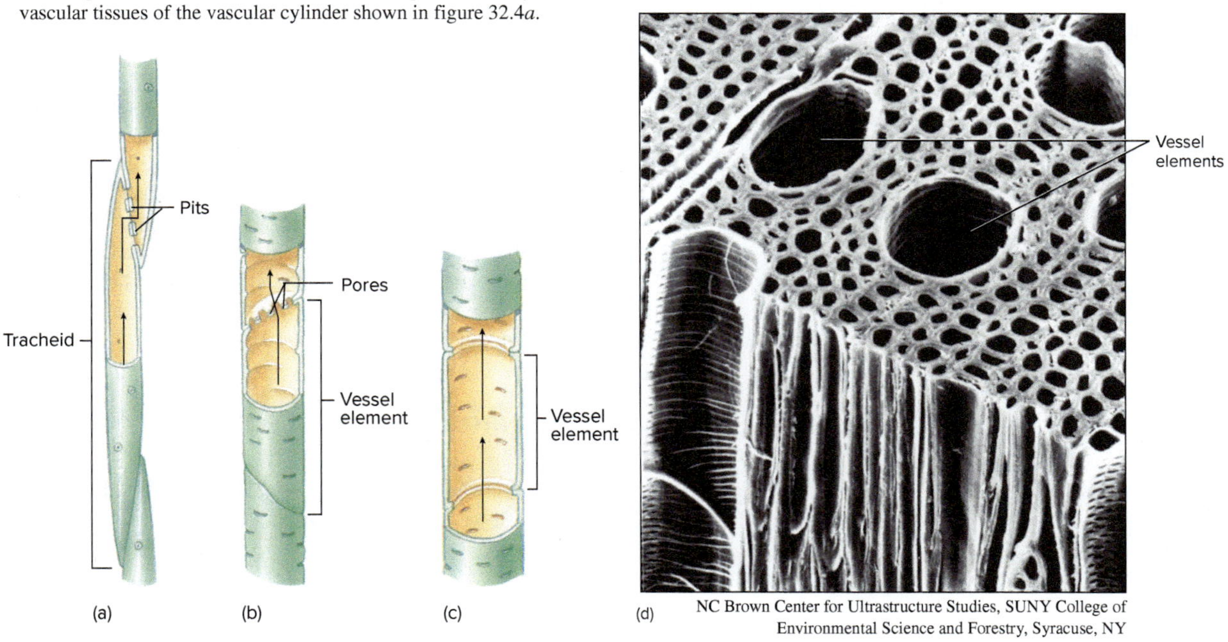

NC Brown Center for Ultrastructure Studies, SUNY College of Environmental Science and Forestry, Syracuse, NY

Figure 32.6 Comparison of vessel elements and tracheids. (*a*) In tracheids, water passes from cell to cell through pits, which are thin areas where cell walls overlap. (*b, c*) In vessel elements, water moves through pores, which may be simple or interrupted by bars. (*d*) The large, hollow tubes shown in this scanning electron micrograph of the wood of a red maple (*Acer rubrum,* a eudicot) are vessel elements (350×).

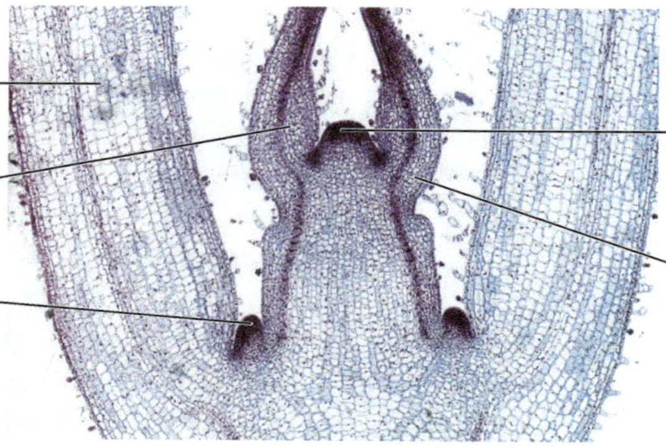

Figure 32.7 Shoot tip of *Coleus,* a eudicot. The apical meristem is the site of rapid cell division and primary growth. Young leaves are produced at the tip of the shoot. Shoot tips, unlike root tips, are not covered by a cap (20×).

STEMS

Stems are often conspicuous organs whose functions include support and the transport of water and solutes. Some stems (e.g., cacti) also photosynthesize and store food.

The Shoot Apex

Examine a living *Coleus* plant and note the arrangement of leaves on the stem. Keeping this observation in mind, examine a prepared slide of a longitudinal section of a shoot tip of *Coleus* (fig. 32.7). The dome-shaped **shoot apical meristem** is not covered by a cap as was the root. The shoot apical meristem produces young leaves (**leaf primordia**) that attach to the stem at a **node.** An **axillary bud** between the young leaf and the stem forms a branch or flower.

Question 6

a. How does the absence of a cap at a shoot apex differ from the apical meristem of a root?

b. Are all cells in the shoot apex the same size? Why is this significant?

External Features of a Mature Woody Stem

Examine the features of a dormant twig (fig. 32.8). A **terminal bud** containing the apical meristem is at the stem tip and is surrounded by **bud scales. Leaf scars** from shed leaves occur at regularly spaced **nodes** along the length of the stem. The portions of stem between the nodes are called **internodes. Vascular bundle scars** may be visible within the leaf scars. Axillary buds protrude from the stem just distal to each leaf scar.

Search for clusters of **bud scale scars.** The distance between clusters or from a cluster to the terminal bud indicates the length of yearly growth.

Primary Tissues of Stems

An **epidermis** covers the stem. The epidermis is coated with a waxy, waterproof substance called **cutin.** Below the epidermis is the **cortex,** which stores food. The **pith** in the center of the stem also stores food (fig. 32.9). In sunflower stems, the cortex is not uniform. Rather, the three to four cell-layers of the cortex just below the epidermis are smaller, rectangular cells with unevenly thickened cell walls. These are **collenchyma** cells; they support elongating regions of the plant (fig. 32.10). In stems, xylem and phloem are arranged in bundles (fig. 32.10).

Procedure 32.3 **Examine the primary tissues of stems**

1. Examine a prepared slide of a cross section of a sunflower (*Helianthus*) or alfalfa (*Medicago*) stem (fig. 32.9).

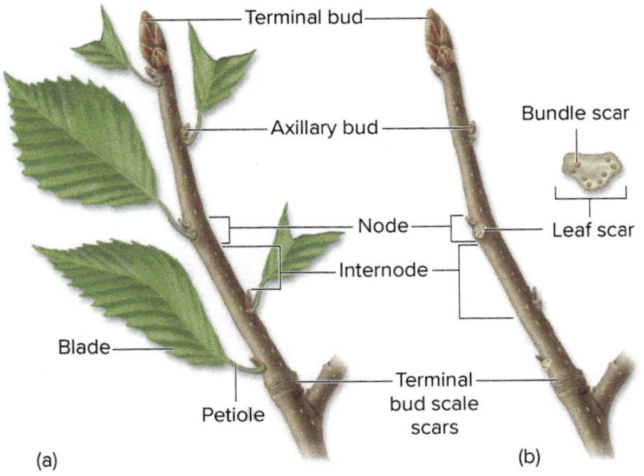

Figure 32.8 A woody twig. (*a*) In summer. (*b*) The twig in its dormant winter condition.

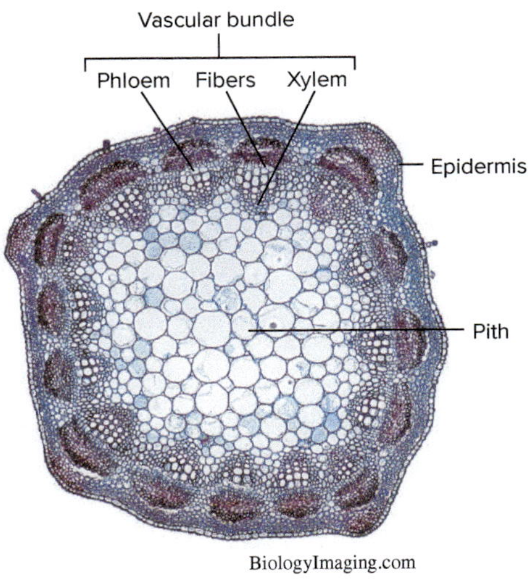

Figure 32.9 Cross-sectional view of a stem of alfalfa (*Medicago* sp.), a eudicot (10×). Note the ring of vascular bundles surrounding the pith. A high-magnification view of a vascular bundle from this stem is shown in figure 32.10.

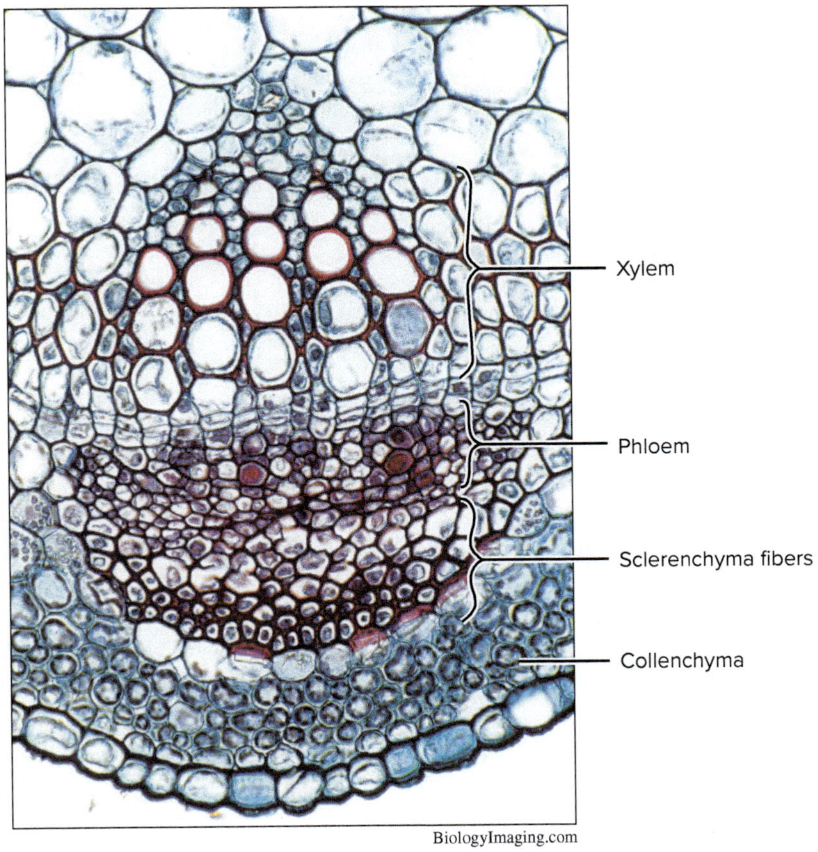

Figure 32.10 Cross section of a vascular bundle of a sunflower stem (485×). The sclerenchyma fibers support the stem and protect the vascular tissue. The walls of fibers are much thicker than those of adjacent cells.

2. On the slide you are examining, locate a vascular bundle. Sketch what you see.

Question 7

a. What is the significance of a coating of cutin on the epidermis?

b. How does the arrangement of xylem and phloem in stems differ from that in roots?

The darkly stained, thick-walled cells just outside the phloem in figure 32.10 are **sclerenchyma fibers,** which function in support. Sclerenchyma fibers, which can be more than a meter long in hemp and jute, are used to make linen, rope, and burlap.

The ring of vascular bundles in sunflower stems is typical of **eudicots,** which are flowering plants with two **cotyledons** (seed leaves; see Exercise 31). Examine a prepared slide of a cross section of a corn stem (fig. 32.11). Corn is a **monocot** (a flowering plant with only one cotyledon). In the following space, sketch a cross section of a sunflower stem and a corn stem. Note the distribution of the vascular bundles.

Question 8

How does the arrangement of vascular bundles differ in stems of monocots as compared to eudicots?

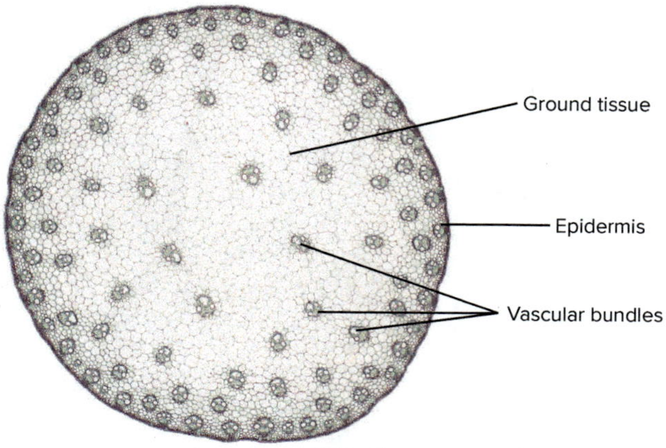

Ed Reschke

Figure 32.11 Cross section of a stem of corn (*Zea mays*), a monocot (5×). Unlike in eudicots such as alfalfa (fig. 32.9), bundles of vascular tissue in monocots occur throughout the ground tissue. The stem is surrounded by an epidermis.

Secondary Growth of Stems

Between the xylem and phloem and each vascular bundle in eudicot stems is a meristematic tissue called **vascular cambium.** The vascular cambium is a secondary meristem that produces secondary growth (i.e., growth in girth). The vascular cambium is cylindrical and produces secondary xylem to its inside and secondary phloem to its outside.

Question 9

How is secondary growth different from primary growth?

Procedure 32.4 **Examine secondary growth in woody stems**

1. Examine cross sections of 1-, 2-, and 3-year-old basswood (*Tilia*) stems. The vascular cambium is a narrow band of cells between the xylem and phloem (fig. 32.12).

2. Compare the structures of the three stems. The secondary xylem of older stems consists of concentric annual rings made of alternating layers of large and small cells. The large cells are formed in the spring, and the smaller cells are produced in summer.

3. In the following space draw cross sections of 1-, 2-, and 3-year-old stems.

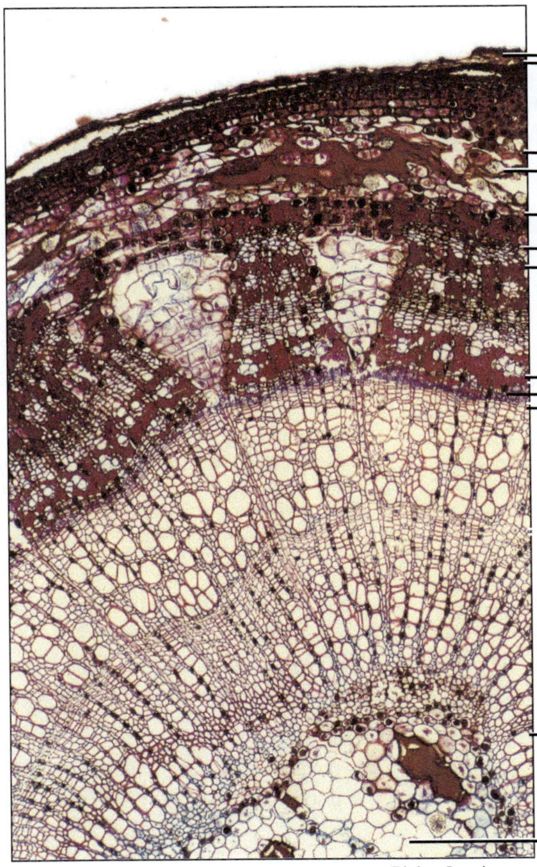

Figure 32.12 A cross section of a portion of a young linden (*Tilia*) stem showing secondary growth (30×). The vascular cambium produces secondary xylem (wood) to the inside and secondary phloem to the outside. Note the annual rings in the secondary xylem. A close-up of a growth ring from pine is shown in figure 32.13.

Question 10

a. How do you account for this seasonal production of different-sized cells?

b. What is the common name for secondary xylem?

c. What is "grain" in wood?

d. Aside from conducting water and minerals, what is another important function of secondary xylem?

4. Examine a prepared slide of a cross section of secondary xylem of pine (*Pinus*), a gymnosperm (fig. 32.13). In cone-bearing plants such as pine, the conducting cells of xylem are all tracheids. The absence of vessel elements gives wood of these plants a relatively uniform appearance.

5. Now examine a prepared slide of a cross section of secondary xylem of an angiosperm (fig. 32.14). Sketch what you see.

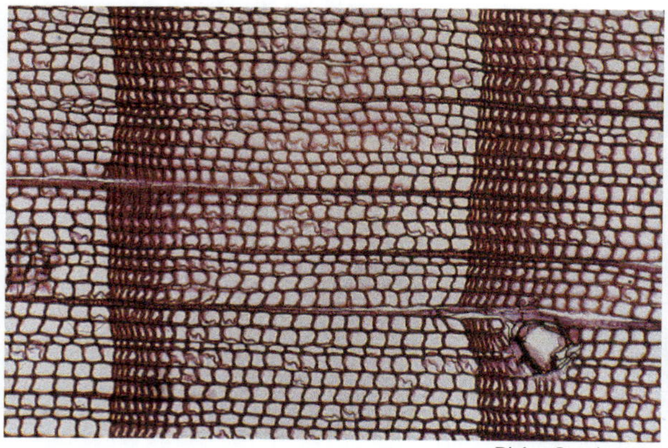

BiologyImaging.com

Figure 32.13 The wood of gymnosperms (such as this pine) consists almost exclusively of tracheids. These water-conducting cells are relatively small and help to support the plant. Although water moves slower through tracheids than through vessel elements, tracheids are less likely to be disabled by air bubbles that form in response to freezing and wind-induced bending of branches. The larger cells on the right side of this photo form during the wet days of spring and are called spring wood; the smaller cells at the left form during the drier days of summer and are called summer wood. The change in density between spring and summer wood produces a growth ring, which appears as "grain" in wood (280×).

Question 11

a. What are the large cells in the wood?

b. What is their function?

c. Which type of wood do you think transports more water per unit area, a gymnosperm or an angiosperm? Why?

Bark

Bark includes all tissues outside of the vascular cambium, including the secondary phloem (fig. 32.12). When viewed in cross section, secondary phloem consists of pyramidal masses of thick- and thin-walled cells. The thin-walled cells are the conducting cells.

The increase in stem circumference resulting from activity of the vascular cambium eventually ruptures the epidermis. The ruptured epidermis is replaced by a tissue called the **periderm** that, like the epidermis, functions to minimize water loss. Periderm consists of cork cells produced by another secondary meristem called the **cork cambium.**

Locate the cork cambium in cross sections of 1-, 2-, and 3-year-old basswood stems. The cork cambium is a band of thin-walled cells located beneath the epidermis. The cork cambium produces cork cells to the outside and cork parenchyma to the inside. Cork cells stain red because of the presence of suberin, a water-impermeable lipid.

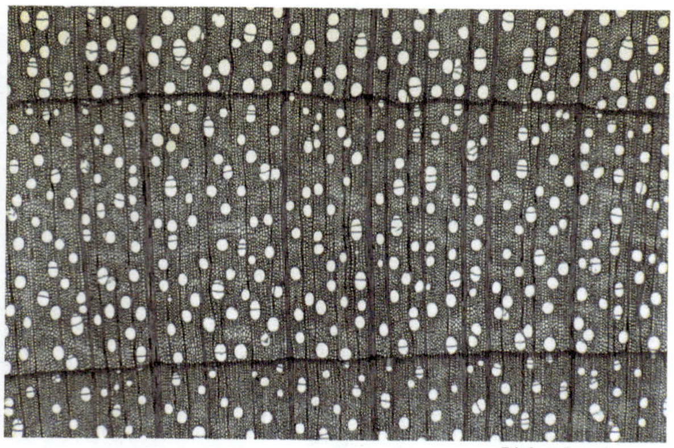

BiologyImaging.com

Figure 32.14 Unlike the xylem of gymnosperms, which contains only tracheids, the xylem of angiosperms (such as this oak) also contains vessel elements. These vessel elements are much wider than tracheids and appear in this micrograph as large circles. Vessel elements are an adaptation for increased rates of water flow in angiosperms (see fig. 32.6*c*) (40×).

Question 12

Is the amount of cork similar in 1-, 2-, and 3-year-old stems? If not, how does it differ? Why is this important?

As the stem diameter continues to increase, the original periderm ruptures and new periderms form in the underlying tissues. Tissues outside the new periderm die and form encrusting layers of bark.

Gas exchange through peridermal tissues occurs through structures called **lenticels** (fig. 32.15). Examine a prepared slide of a lenticel and locate lenticels on a mature woody stem.

Question 13

How does a lenticel differ from the remainder of the periderm?

LEAVES

With few exceptions, most photosynthesis occurs in leaves, although some may occur in green stems. Leaves typically consist of a **blade** and a **petiole.** The petiole attaches the leaf blade to the stem. **Simple leaves** have one blade connected to the petiole, whereas **compound leaves** have several

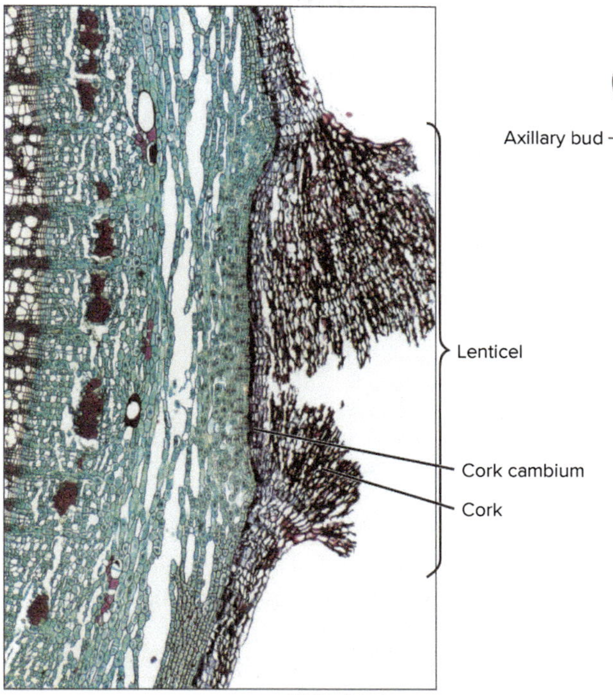

Figure 32.15 A lenticel in a cross section of part of a young stem of elderberry (*Sambucus*). Gas exchange across the cork layer of the stem occurs through lenticels (80×).

leaflets sharing one petiole (fig. 32.16). **Palmate** leaflets of a compound leaf arise from a central area, as your fingers arise from your palm. **Pinnate** leaflets arise in rows along a central midline. Examine the simple and compound leaves on display, and sketch a representative of each type of leaf.

Leaves are also classified according to their **venation** (i.e., arrangement of veins) (fig. 32.17). **Parallel veins** extend the entire length of the leaf with little or no cross-linking. **Pinnately veined** leaves have one major vein (i.e., a midrib) from which other veins branch. **Palmately veined** leaves have several veins each having branches. Veins of vascular tissue in leaves are continuous with vascular bundles in stems.

Examine the leaves on display in the lab and determine their venation. Sketch a few of these leaves to show their venation. List the names of common leaves on demonstration and indicate whether each is simple, pinnately compound, or palmately compound. Also indicate whether venation in each leaf is parallel, pinnate, or palmate.

Figure 32.16 Simple and compound leaves.

The arrangement of leaves on a stem is called **phyllotaxis** and characterizes individual plant species (fig. 32.18). **Opposite phyllotaxis** refers to two leaves per node located on opposite sides of the stem. **Alternate phyllotaxis** refers to one leaf per node, with leaves appearing first on one side of the stem and then on another. **Whorled phyllotaxis** refers to more than two leaves per node. Examine the plants on display in the lab to determine their phyllotaxis.

What Good Is Bark?

You've doubtless seen chips of bark used as garden mulch that is spread around plants. Because bark cells (secondary phloem and periderm) were once alive, they contain many nutrients that are released into the soil as the bark decays. However, sawdust (wood) has few nutrients in it and may result in nitrogen deficiency due to rapid decomposition. Consequently, sawdust is a poor mulch for plants.

Tree Girdling

You've probably heard of "tree girdling," which is the stripping of a ring of bark from a branch. (In the wild, porcupines also girdle trees.) Girdling removes secondary phloem from a tree, thereby leaving no pathway for photosynthate to move from the leaves to the roots. As a result, the shoot accumulates sugars and grows rapidly. Moreover, the roots cannot grow or provide water and nutrients to the shoot. The following spring it is impossible to send sugars from the roots to the girdled branch to renew growth. Thus, the branch or tree dies the year after it is girdled.

Figure 32.17 Palmate, pinnate, and parallel venation of leaves.

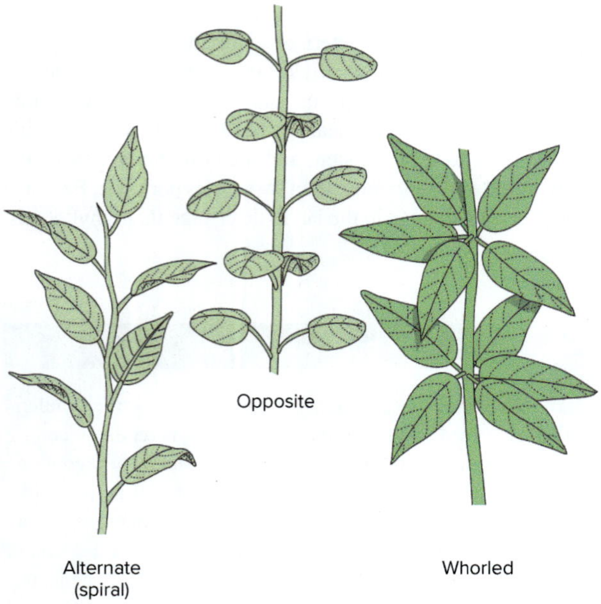

Figure 32.18 Patterns of leaf arrangement (phyllotaxis).

Internal Anatomy of a Leaf

Examine a cross section of a leaf of *Ligustrum* (privet) (figs. 32.19 and 13.9). The leaf is only 10–15 cells thick—pretty thin for a solar collector! The epidermis contains pores called **stomata,** each surrounded by two guard cells (you will study stomata again in the next exercise). Just below the upper epidermis are closely packed cells called **palisade mesophyll** cells; these cells contain about 50 chloroplasts per cell. Below the palisade layers are **spongy mesophyll** cells with numerous intercellular spaces. Examine and sketch a cross section of a corn leaf.

Question 14

a. What is the function of stomata?

b. Do epidermal cells of leaves have a cuticle? Why is this important?

c. What is the significance of chloroplasts being concentrated near the upper surface of the *Ligustrum* leaf?

d. What are the functions of air spaces near the lower surface of the leaf?

e. How is the internal anatomy of a corn leaf different from that of *Ligustrum*?

f. How is it similar?

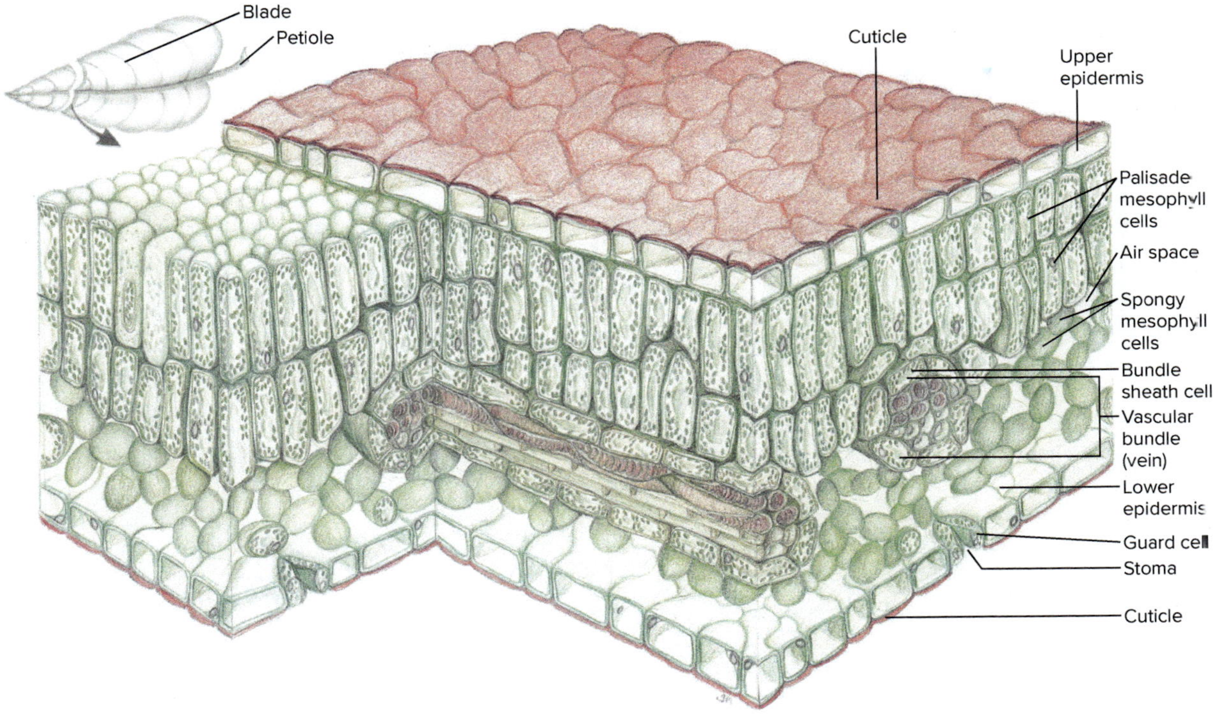

Figure 32.19 *Ligustrum* leaf, cross section. Most photosynthesis occurs in the densely packed palisade mesophyll cells, just beneath the upper epidermis of the leaf. Gas exchange occurs through stomata, which are usually most abundant on the lower side of the leaf. Water loss is minimized by the waxy cuticle that covers the leaf.

g. Based on the arrangement of vascular tissue, how could you distinguish the upper versus lower surfaces of a leaf?

h. Examine prepared slides of leaves of corn (*Zea mays,* a monocot) and pine (*Pinus,* a gymnosperm). What differences are there in the structures of these leaves? How do these structural differences correlate with functional differences?

INQUIRY-BASED LEARNING

How do plants sense and respond to light and gravity?

Observation: The ability of plant roots to grow downward and shoots to grow upward is adaptive because it increases the plants' chances of encountering water (by the roots) and light (by the shoot).

Question: When the roots of a corn (*Zea mays*) seedling grow downward, is this a negative response to light or a positive response to gravity?

a. Establish a working lab group and obtain Inquiry-Based Learning Worksheet 32 from your instructor.

b. Discuss with your group well-defined questions relevant to the preceding observation and question. Choose and record your group's best question for investigation.

c. Translate your question into a testable hypothesis and record it.

d. Outline on Worksheet 32 your experimental design and supplies needed to test your hypothesis. Ask your instructor to review your proposed investigation.

e. Conduct your procedures, record your data, answer your question, and make relevant comments.

f. Discuss with your instructor any revisions to your questions, hypotheses, or procedures. Repeat your work as needed.

Questions for Further Study and Inquiry

1. What is the function of xylem? Phloem? Vascular cambium? Epidermis?

2. How are the structure and function of stomata similar to lenticels? How are they different?

3. How is the internal anatomy of a stem different from that of a root?

4. How is primary growth different from secondary growth?

5. How is a leaf structurally adapted for its function?

6. Why does a stem typically contain more sclerenchyma and collenchyma than does a leaf?

7. An old friend tells you that 30 years ago she nailed a sign into a tree trunk at a height of 1 meter. She now says the sign is 25 meters up in the tree. Should you believe her? Why or why not?

8. Compare and contrast (*a*) monocot and dicot roots and (*b*) monocot and dicot stems.

9. What are three examples of plants growing on your campus that have (a) simple leaves and (b) compound leaves?

DOING BIOLOGY YOURSELF

Choose a couple of defined environments nearby, such as a vacant field or riverside. Survey the variety of leaf morphologies of the dominant plants. What can you conclude about the predominance of monocots versus eudicots in each environment?

WRITING TO LEARN BIOLOGY

Why is leaf abscission especially important for temperate plants?

WRITING TO LEARN BIOLOGY

Would you expect to find annual rings in wood of a tropical dicot tree? Why or why not?

EXERCISE 33

Plant Physiology

Transpiration

Learning Objectives

By the end of this exercise you should be able to:

1. Describe the structure and function of stomata.
2. Describe how environmental conditions such as wind and light influence stomatal opening.
3. Measure the transpiration rate of a plant.

Please visit connect.mheducation.com to review online resources tailored to this lab.

To survive and reproduce, land plants must cope with many environmental challenges. Among the biggest of these problems are obtaining water and avoiding desiccation (water loss). Solutions to these and other challenges require expenditure of energy and an efficient structure and physiology. Structural and functional solutions to problems such as maintaining water balance are called adaptations, and the study of these adaptations is among the most interesting aspects of biology (fig. 33.1).

In this exercise, you will study the "water physiology" of plants—how plants are adapted to control their water content and evaporative loss. Before starting today's exercise, review in your textbook the structure and function of xylem and stomata, and review how water moves through plants (fig. 33.2).

The movement of water through plants and evaporation from aerial plant parts is called **transpiration.** Most transpiration occurs through stomata of leaves. However, transpiration also occurs in flowers. Demonstrate this with the following procedure.

Magda Urbanova/Shutterstock

Figure 33.1 The droplets of water at the edges of these leaves of a strawberry plant are formed by guttation. Guttation results from root pressure (i.e., a positive pressure in the xylem) common in small plants growing in moist soil on cool, damp mornings. Although most water movement in plants results from evaporation of water from leaves, guttation may be an important way of moving water in short, herbaceous plants.

Procedure 33.1 **Visualize transpiration in flowers**

1. Obtain a small beaker containing a solution of food color from your lab instructor at the beginning of your laboratory period.
2. Working in groups of three or four, cut the stem of a white flower such as periwinkle (*Catharanthus*) or carnation (*Dianthus*) so that the cut is about 0.5 cm from the base of the flower. Do not cut all the way through the stem.
3. Remove the remainder of the stem about 2.0 cm below the partial cut.
4. Float the flower in the food color so that the partial cut is submerged.
5. Set the beaker containing the flower and dye in an illuminated area, and examine the flower after 2–3 h.

Question 1

a. In which part of the flower is the dye located?

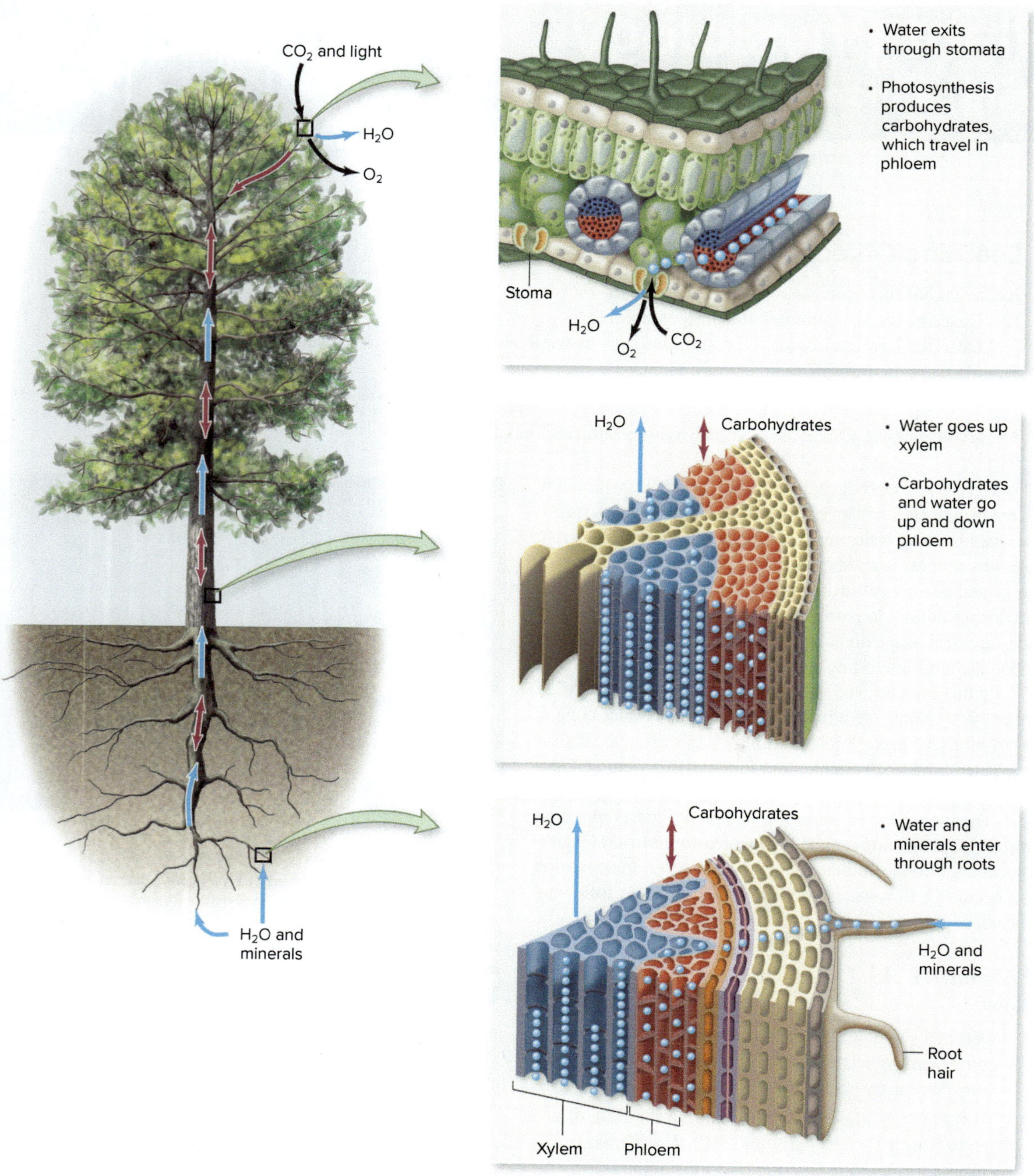

Figure 33.2 Flow of materials in a plant. Water with dissolved minerals enters the plant through root hairs and moves upward through the xylem of the vascular system. Most of this water is eventually lost through the leaves during transpiration. Dissolved minerals are used by plant cells for metabolic reactions. Carbohydrates made during photosynthesis move throughout the plant in the phloem. These carbohydrates are either stored or used directly as an energy source for metabolism and growth.

b. What do you conclude from this observation?

PLANT–WATER RELATIONS

Leaves of most plants have epidermal pores called **stomata** (sing., *stoma*), formed by the separation of a pair of guard cells (fig. 33.3). Together, the guard cells and the pore between them form a **stoma** (plural, *stomata*).

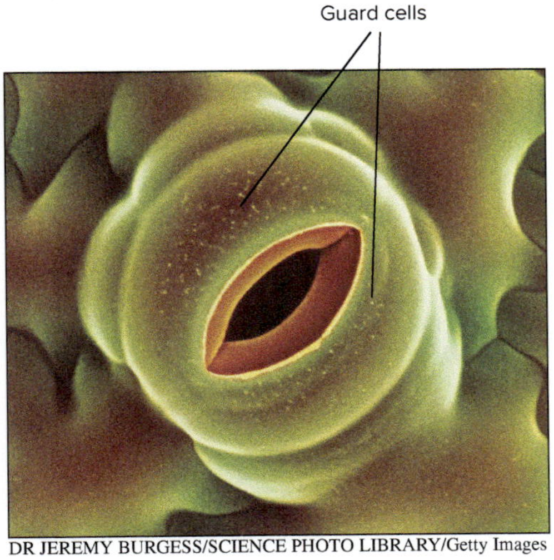

DR JEREMY BURGESS/SCIENCE PHOTO LIBRARY/Getty Images

Figure 33.3 A scanning electron micrograph of a stoma in a leaf of a tobacco plant. Guard cells change shape and, in doing so, open and close a pore. In dry conditions, stomata close and minimize water loss (100×). Water vapor passes out through the stomata, and carbon dioxide enters by the same portals. The mechanism of stomatal opening and closing is described in figure 33.4.

In Exercise 13, you learned that CO_2 is essential for photosynthesis. Carbon dioxide in the air reaches the photosynthetic cells of the leaf via stomata surrounded by guard cells If enough water is available, guard cells absorb water and become turgid. When guard cells are turgid, they expand and bend to form a stomatal pore through which CO_2 enters and water vapor exits a leaf. When water is scarce, guard cells lose their turgidity and shrink. As a result, guard cells touch each other and the stomatal pore closes (fig. 33.4). Thus, the water status of a plant determines whether stomata are open or closed and indirectly determines whether gases move in and out of the leaf.

STOMATAL STRUCTURE

Guard cells in dicots are bean-shaped and attached to each other at their ends. Microfibrils form transverse bands around the cell so that when water enters the cells they cannot expand around the middle. Instead, they lengthen and bow apart, thereby opening the stomatal pore. Use procedure 33.2 to examine living guard cells and how they respond to changing environmental conditions.

Procedure 33.2 **Stomatal structure**

1. Remove part of a leaf from a well-watered *Kalanchöe* or *Zebrina* plant that has been illuminated for 4–5 h. Bend the leaf until it snaps; then pull the surface portion of the leaf away from the rest of the leaf in a slow downward motion (your instructor will demonstrate this technique). A thin sheet of transparent epidermis will tear away from the leaf.
2. Quickly place the piece of epidermis in a few drops of distilled water on a microscope slide (do not let the tissue desiccate) and place a coverslip on the tissue. Be careful not to trap any air bubbles beneath the tissue or coverslip.

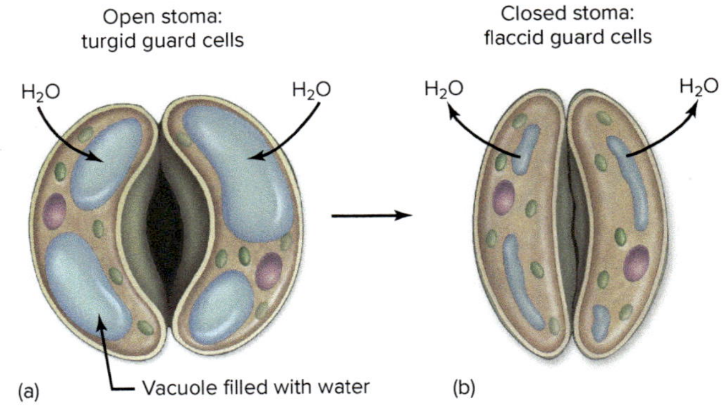

Figure 33.4 The opening and closing of stomata. (*a*) Two guard cells form each stoma on the surface of a leaf. Ions moving into guard cells increase the concentration of the solutes in the guard cells and cause water to diffuse into the cells by osmosis. This influx of water causes the cells to swell and bow apart, thereby forming a pore and opening the stoma. Water evaporates through these stomatal pores during transpiration. (*b*) When the solute pressure is low in the guard cells, water diffuses out, the guard cells become flaccid, and the stomatal pore closes. A variety of environmental factors cause stomata to open and close.

3. Examine the tissue with your microscope. Diagram the epidermis and its stomata.

4. Place a few drops of 10% NaCl at one edge of the coverslip covering the tissue.
5. Wick this solution under the coverslip by touching the solution at the opposite edge of the coverslip with a piece of paper towel.
6. Reexamine the stomata. Sketch what you see.

7. Observe the demonstration of an epidermal peel of *Zea mays* (corn), a monocot. Diagram cells of the epidermis of corn.

Question 2

a. Approximately how many stomata are in a mm^2 of the lower surface area of *Kalanchöe* and *Zebrina?*

b. Are the densities of stomata similar on the upper and lower surfaces of *Kalanchöe* and *Zebrina?*

c. When the guard cells of *Kalanchöe* and *Zebrina* are turgid, they form a stomatal pore. How?

d. Are stomatal densities similar in *Zea mays* and *Kalanchöe* and *Zebrina?*

e. Are stomata of the two plants arranged in a pattern, or do they occur randomly?

f. How is the structure of individual stomata different in *Zea mays* than in *Kalanchöe* and *Zebrina?*

Question 3

a. Did the NaCl solution cause stomata to close? Why or why not?

b. What is the advantage of closed stomata when water is in short supply?

c. What are the disadvantages?

Procedure 33.3 Stomatal density in other plants

Unlike *Kalanchöe* and *Zebrina,* not all plants have an epidermis that can be removed easily from their leaves. We can examine the stomata of these plants by creating a mold of the epidermis made with nail polish.

Before starting this procedure, examine leaves on the plants available in lab. For each plant, which surface of the leaf—that is, the upper or lower surface—do you think will have the most stomata? Which do you think will have the fewest? Why? Write your hypotheses here:

Test your hypotheses by making molds of the surfaces of the leaves, and then calculating the stomatal density of each surface. Here's how:

1. Obtain two leaves from a plant available in lab.

2. Paint a coating of clear nail polish onto the upper and lower sides of the leaves. The polish should cover 1–2 cm^2.

Handle the nail polish carefully; it can be toxic if ingested or inhaled.

3. Wait for the polish to dry completely. This will take 15–20 min. The dried polish is now a mold of the leaf surface. Commercially available "liquid bandage" also works for this procedure.

4. When the polish has completely dried (i.e., it is no longer sticky), apply a piece of clear cellophane tape to the patch of nail polish. (Do not use opaque tapes such as Scotch matte-finish "Magic" tape; instead, use clear carton-sealing tape.)

5. Lay the leaves flat on the table and use your thumb to gently press the hardened polish onto the leaf.

6. Peel the nail polish from the leaf by gently pulling on a corner of the tape. The patch of nail polish is the impression of the leaf surface that you will examine.

7. Tape your mold of the leaf surface to a clean microscope slide (with the sticky side of the tape down). If the mold doesn't come off the leaf cleanly, use forceps to remove the polish.

8. Examine the nail-polish mold with the high-power objective of your light microscope. Search for areas that are clean and intact and in which you can see stomata. Write the name of the plant and the leaf surface (i.e., upper or lower) you are examining on the slide.

9. Count the number of stomata in at least three different fields of view; then convert those numbers into stomatal densities (i.e., stomata mm^{-2}; see Exercise 3). Record your results here:

Name of plant ____________________________

Upper surface

Number of stomata per field of view ______ ______ ______

Density (stomata mm^{-2}) ______ ______ ______

Lower surface

Number of stomata per field of view ______ ______ ______

Density (stomata mm^{-2}) ______ ______ ______

Question 4

What is the average density of stomata on the upper versus lower surface of the leaf? Why is this significant?

10. If other plants are available in lab (e.g., monocots versus eudicots, aquatic plants versus desert plants), formulate hypotheses about the abundance of stomata on the upper and lower surfaces of their leaves. Write your hypotheses here:

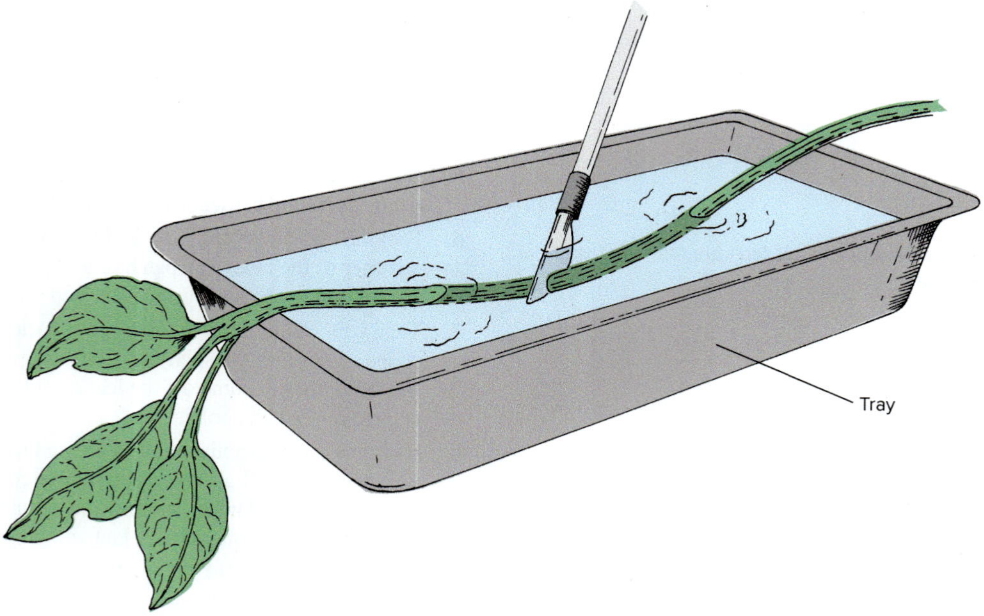

Figure 33.5 Method of cutting submerged stems. Cutting the stem underwater avoids introducing air into the vascular tissue.

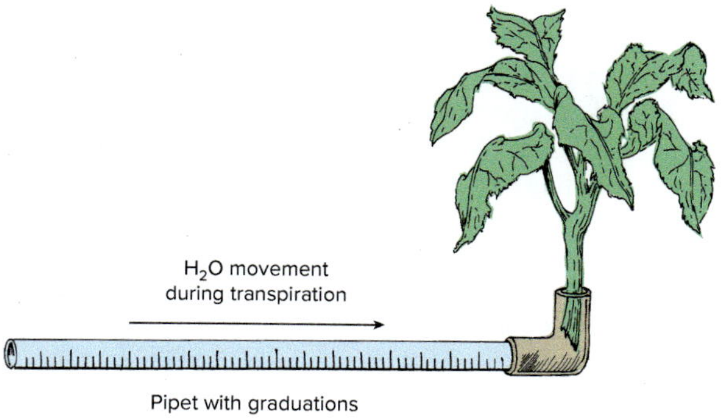

Figure 33.6 Experimental setup for measuring transpiration. As water is transpired from the leaves, the water moves through the pipet.

Table 33.1

Experimental Conditions for Measuring Transpiration

Condition	Transpiration Rate (mL H_2O h^{-1})
Control (standard illumination)	________
Darkness or reduced light	________
Light; mild breeze (created by a small fan)	________
Dark; mild breeze	________
Light; leaves coated with petroleum jelly	________
Light; all leaves removed	________

11. Test your hypotheses by making molds of the leaves' surfaces and calculating the density of stomata. Record your results here:

Question 5

What do you conclude about the density of stomata in the different types of plants you examined? For example, how did the density of stomata vary on the upper versus lower epidermis of the plants you examined?

TRANSPIRATION

Transpiration is the loss of water from plants; it is strongly influenced by environmental conditions such as the levels of CO_2, light, wind, and humidity. Use procedure 33.4 to quantify the effects of light and wind on transpiration.

Procedure 33.4 Quantify transpiration

1. Obtain a sunflower (*Helianthus*) plant from your lab instructor. If sunflower plants are not available, use the other stems available in the lab.
2. Submerge the stem in water and obliquely cut the stem with a sharp scalpel (fig. 33.5).
3. While it is still submerged in water, attach the severed stem to a water-filled tube connected to a 1.0-mL pipet as shown in figure 33.6. Keep the pipet in a horizontal position. Be sure that no air bubbles are in the system and that no water is on the leaves.
4. Place the stem in an illuminated area. Transpiration will move water through the pipet regardless of whether the stem is vertical or horizontal.
5. Allow about 10 min for the rate of water movement in the pipet to stabilize. Note the position of the meniscus in the pipet.
6. Allow transpiration to occur for 15 min.
7. Determine and record the volume of water (mL) transpired by observing the distance that the meniscus moved.
8. Multiply this value by 4 to convert to mL per hour. Record this rate of transpiration in table 33.1 as the control value.
9. Continue to determine the transpiration rate under the conditions in the order listed in table 33.1.

Question 6

State your conclusion about the influence of light, dark, breeze, clogged stomata, and removed leaves on water movement in plants.

Light:

Dark:

Breeze:

Clogged stomata:

Removed leaves:

INQUIRY-BASED LEARNING

How do transpiration rates vary in different species of plants?

Observation: In this lab you measured the rates of transpiration through individual parts of plants. All plants transpire, and their rates of transpiration are influenced by the environments in which they grow.

Question: Do transpiration rates vary among different species of plants?

a. Establish a working lab group and obtain Inquiry-Based Learning Worksheet 33 from your instructor.

b. Discuss with your group well-defined questions relevant to the preceding observation and question. Choose and record your group's best question for investigation.

c. Translate your question into a testable hypothesis and record it.

d. Outline on Worksheet 33 your experimental design and supplies needed to test your hypothesis. Ask your instructor to review your proposed investigation.

e. Conduct your procedures, record your data, answer your question, and make relevant comments.

f. Discuss with your instructor any revisions to your questions, hypotheses, or procedures. Repeat your work as needed.

Questions for Further Study and Inquiry

1. What structural features of plants minimize water loss?

2. Why was it important in today's exercise to sever the stems while they were submerged in water?

3. What adaptations help plants conserve water?

DOING BIOLOGY YOURSELF

Assume that a maple tree in your front yard has 100,000 leaves, each having an average area of 35 cm^2. If the transpiration rate of each leaf is 0.05 mL H_2O h^{-1} cm^{-2} leaf area, how many liters of water move through the plant in a day?

WRITING TO LEARN BIOLOGY

Describe any adaptations that would improve water economy in plants. In which environments might you expect to find the adaptations you described?

EXERCISE 34

Plant Physiology

Tropisms, Nutrition, and Growth Regulators

Learning Objectives

By the end of this exercise you should be able to:

1. Define the terms phototropism and gravitropism.
2. Describe symptoms of mineral deficiency in plants.
3. Explain how the quality and quantity of light affect seed germination.
4. Explain the modes of action of auxin and gibberellic acid.

Please visit connect.mheducation.com to review online resources tailored to this lab.

Plants respond to a variety of environmental stimuli such as light and gravity, and require several nutrients such as calcium and potassium. Plant growth and development are controlled by internal chemical signals called **growth regulators.** Plants have a small number of growth regulators that influence many processes and traits that we see. This is in contrast to animals, which have many more hormones and chemical regulators. Auxin and gibberellic acid are examples of growth regulators in plants.

In this exercise, you will study a variety of common physiological responses of plants. Throughout this exercise, remember that each of these responses is an adaptation for the survival and reproduction of plants in varied environments.

PLANT TROPISMS

A **tropism** is a movement in response to an external stimulus. The direction of movement is determined by the direction of the most intense stimulus. Two tropisms you will study in today's exercise are **phototropism,** which is directed growth in response to light (fig. 34.1), and **gravitropism,** which is directed growth in response to gravity (fig. 34.2).

Phototropism

Procedure 34.1 **Observe phototropism**

1. Obtain from your instructor six 10- to 14-day-old radish (*Raphanus*) seedlings. These seeds have been grown in diffuse, overhead light.
2. At the beginning of the lab period, place your seedlings approximately 25 cm from a 100-Watt light so that light strikes the shoots at a right angle.
3. Use a protractor to measure curvature of the seedlings every 30 min for 2 h.
4. Record your results in table 34.1.

Question 1

a. In which direction did the seedlings curve?

b. Is this curvature positive or negative phototropism?

c. What is the adaptive significance of phototropism?

d. Would you expect roots to react similarly to light? Why or why not?

Gaertner/Alamy Stock Photo

Figure 34.1 Sunflowers (*Helianthus annus*) get their common name because the flowers of some varieties track the sun across the sky, much like radiotelescopes track satellites. This "solar tracking" by sunflowers and other plants (e.g., cotton, alfalfa, beans) helps the plants regulate the amount of light that they absorb. Desert plants such as the "compass plant" use phototropism to orient their leaves parallel to the sun's rays and minimize the amount of light that they absorb; this prevents overheating and desiccation. Other plants orient their leaves perpendicular to the sun's rays to maximize the amount of light absorbed for photosynthesis.

Martin Shields/Alamy Stock Photo

Figure 34.2 Negative gravitropism by a stem. This plant was placed on its side in the dark 24 h before this picture was taken. The stems have curved away from the pull of gravity.

Table 34.1

Observations of Phototropism

Time (h)	Mean Curvature (degrees)
0.5	________
1.0	________
1.5	________
2.0	________

Gravitropism

Plants may perceive gravity by the movement of starch-laden amyloplasts within cells (fig. 34.3). Growing corn roots show their response to gravity by curving down (fig. 34.4).

Yesterday, your instructor placed *Zea mays* (corn) seedlings having roots approximately 1 cm long in a glass beaker. Seedlings labeled "H" had their root oriented

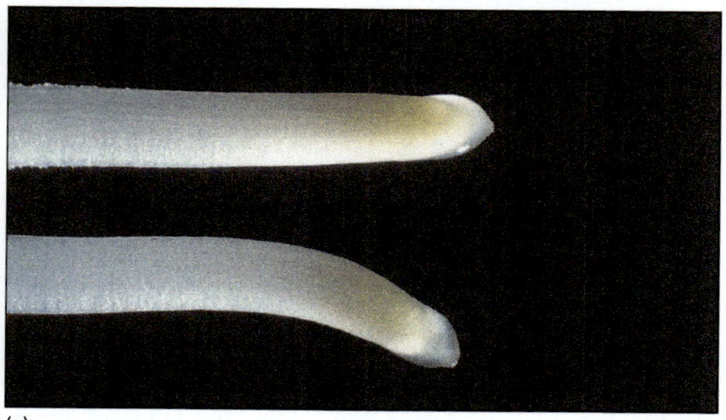

(a) BioPhot

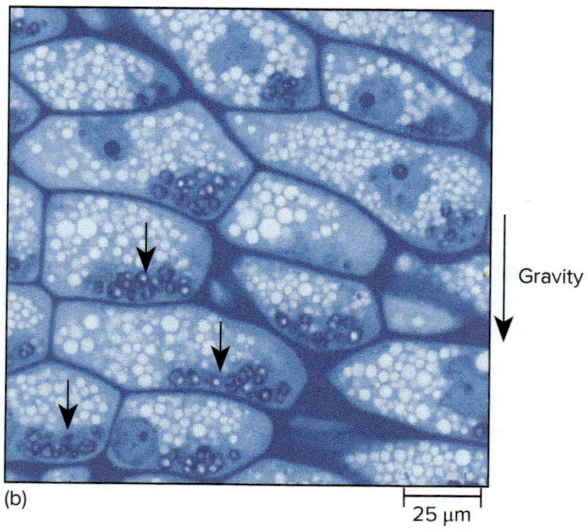

(b) BioPhot

Figure 34.3 Perception of gravity in root caps. (*a*) Cells in the center of a root cap contain numerous starch-laden amyloplasts located in the lower part of the cells (5×). (*b*) Amyloplasts (arrows) sediment to the bottom of the cell when roots are oriented horizontally (280×). This sedimentation of amyloplasts has long been thought to be the basis for how roots perceive gravity (upper photo) (see fig. 34.4). To see a higher-magnification view of the root tip and root cap, see figures 32.2*b* and 32.3.

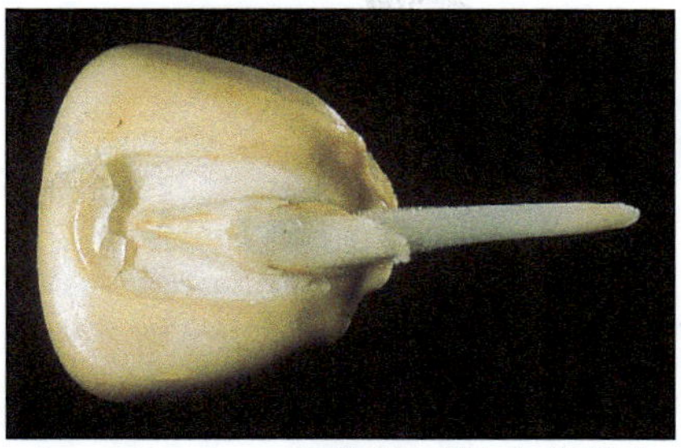

(a) Professor Malcolm B. Wilkins, Botany Dept., Glasgow University

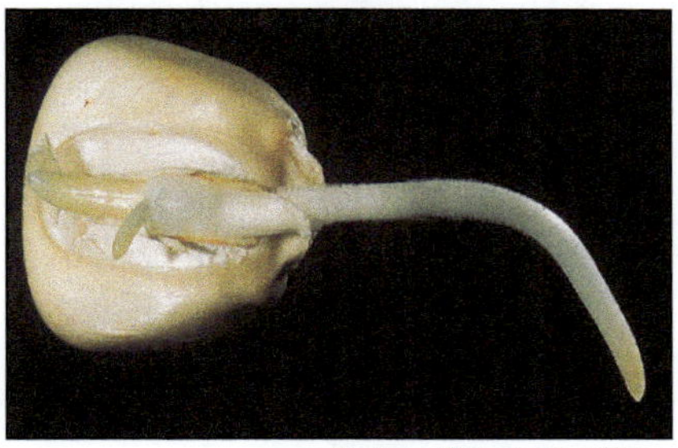

(b) Professor Malcolm B. Wilkins, Botany Dept., Glasgow University

Figure 34.4 Gravitropism by horizontally oriented roots of corn. Downward curvature (*a*) begins within 30 min and (*b*) is completed within a few hours. Curvature results from faster elongation of the upper side of the root than of the lower side.

horizontally, whereas those labeled "D" had their root pointing down. The terminal 3–4 mm was removed from the roots of seedlings labeled with an asterisk (*), whereas the other seedlings were not cut.

Procedure 34.2 Observe root gravitropism

1. Obtain the containers of corn seedlings oriented horizontally and vertically.
2. Examine the experimental setup and the direction of root growth.
3. Record your observations in table 34.2.

Question 2

a. Which roots grew down?

b. Which one(s) didn't?

c. What do you conclude from these observations?

Table 34.2

Observations of Gravitropism in Roots

Treatment or Orientation	Direction of Growth
Intact roots oriented vertically (V)	________
Intact roots oriented horizontally (H)	________
Detipped roots oriented vertically (V*)	________
Detipped roots oriented horizontally (H*)	________

d. Where in roots does the differential growth occur that produces gravitropism?

e. Are roots positively or negatively gravitropic?

f. What is the adaptive significance of gravitropism?

Procedure 34.3 Examine stem gravitropism

1. Obtain three containers of tomato plants (*Lycopersicon*) or sunflower (*Helianthus*) plants.
2. Turn the potted plants horizontally.
3. Measure the distance from the stem tip to the table's surface. Record your measurements in table 34.3.
4. Every 30 min remeasure the distance between the stem tips and table, and record your results.
5. Some of the plants available in the lab were turned horizontally and inverted yesterday; others were left upright.
6. Examine stem curvature after 24 h and record your observation in table 34.3.
7. Make a simple sketch of the seedling stem curvatures after 24 h.

Question 3

a. How are the stems oriented after 24 h?

b. What is the adaptive advantage of a gravitropic response?

Table 34.3

Observations of Gravitropism in Stems

Treatment–Time	Distance of Stem Tip from Table Surface
Horizontal—Time 0 min	________
Horizontal—Time 30 min	________
Horizontal—Time 60 min	________
Horizontal—Time 90 min	________
Horizontal—Time 120 min	________
Inverted—Time 24 h	Observation:

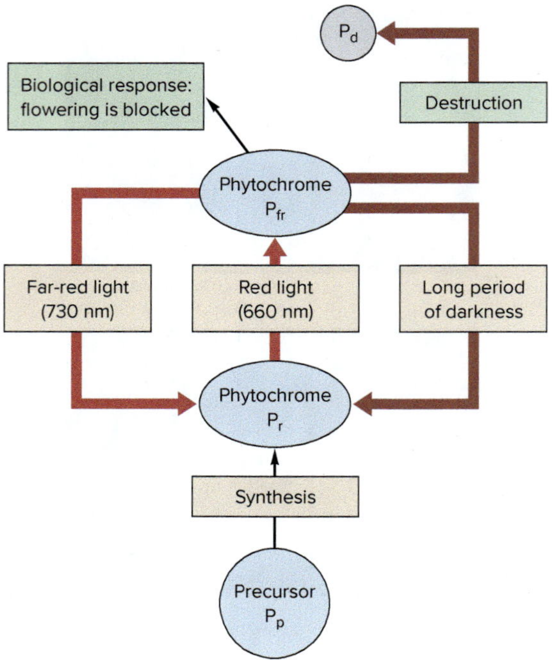

Figure 34.5 How phytochrome works. Phytochrome is synthesized in the P_r form from amino acids, designated P_p for phytochrome precursor. When exposed to red light, P_r changes to P_{fr}, which is the active form that elicits a response in plants. P_{fr} is converted to P_r when exposed to far-red light, and it also converts to P_r or is destroyed in darkness. The destruction product is designated P_d.

SEED GERMINATION

Several environmental factors affect seed germination. For example, seeds of many plants germinate only in response to certain types of light. The pigment that absorbs light affecting seed germination (and several other developmental responses) is **phytochrome.** Phytochrome alternates between two forms, depending on the light it has absorbed. Phytochrome is activated by red light (660 nm) and inactivated by far-red light (730 nm) and/or darkness (fig. 34.5). The activation and deactivation reactions are reversible, and the ultimate physiological effect induced by the phytochrome depends on which wavelengths were absorbed last. More than 50 different developmental processes are affected by phytochrome absorption of red and far-red light.

Grand Rapids lettuce seeds (*Lactuca sativa*) are excellent models for examining the effects of light on germination because they are sensitive to light and germinate quickly. The seeds are dormant when they are first shed, and germination is poor even with adequate oxygen and heat. Dormancy is broken by light absorption by phytochrome. The phytochrome usually exists in the red-absorbing form. Activation (absorption of red light) causes increased water absorption by the radicle cells, and the radicle grows and elongates. Germination begins.

Procedure 34.4 **Observe germination of lettuce seeds**

1. Obtain about 300 Grand Rapids lettuce seeds from your laboratory instructor.
2. Working in small groups, place 50 seeds in each of six petri dishes containing water-soaked filter paper.
3. Cover the petri dishes with lids and expose the seeds to the following treatments:

   Treatment 1: Continuous darkness (wrap the dish in metal foil)

   Treatment 2: Continuous light

   Treatment 3: Red light for 10 min, then darkness

   Treatment 4: Far-red light for 10 min, then darkness

   Treatment 5: Red light for 10 min, followed by far-red light for 10 min, then darkness

   Treatment 6: Red light for 10 min, far-red light for 10 min, red light for 10 min, then darkness
4. During your next laboratory period determine the percentage of germination resulting from each treatment.
5. Record your results in table 34.4.

Question 4

What do you conclude about the influence of light on germination of lettuce seeds?

Table 34.4

Germination of Lettuce Seeds under Six Different Light Treatments

Treatment	Percentage Germination
1. Dark	________
2. Constant room light	________
3. Red light, then dark	________
4. Far-red light, then dark	________
5. Red light, followed by far-red light, then dark	________
6. Red light, far-red light, red light, then dark	________

Table 34.5

Germination of Onion Seeds in Light and Dark

Treatment	Percentage Germination
Room light	________
Dark	________

Germinating seeds of different plants often respond differently to light. Germination of onion seeds, for example, is affected by light versus dark. Determine their sensitivity with procedure 34.5.

Procedure 34.5 **Observe germination of onion seeds**

1. Divide 50 onion seeds by placing 25 in each of two petri dishes containing water-soaked filter paper.
2. Put lids on the petri dishes and label the lids "light" and "dark." Place one petri dish in room light and the other dish in darkness.
3. Examine the seeds during your next lab period and record the percentage germination in table 34.5.

Question 5

a. Does light promote or inhibit germination of onion seeds?

b. Is this response different from that of lettuce?

c. How could seed germination be influenced by depth of planting? How might phytochrome be involved in such a response?

PLANT GROWTH IS AFFECTED BY LIGHT

Many seeds germinate in the dark and push their stems above the soil to receive light. If the seedlings do not receive light, they grow abnormally; the leaves grow longer and the plant appears pale and spindly. Such a plant is **etiolated.** Chlorophyll does not develop until the plant is exposed to light. Activation of phytochrome helps promote normal growth. Use procedure 34.6 to observe etiolation.

Procedure 34.6 **Examine etiolation**

1. Obtain two groups of bean (*Phaseolus*) seedlings. They were planted 10–14 days ago, with one group being grown in the dark and one group in light.
2. Complete the first two lines of table 34.6 with your observations.
3. After completing and recording your observations, reverse the two treatments and examine the plants during your next lab period.
4. Complete the last two lines of table 34.6 with your final observations.

Question 6

a. How did the seedlings manage to grow at all while in the dark?

b. Did etiolation include stem elongation?

c. What is the advantage of rapid stem elongation in a seed germinating in the dark?

PLANT NUTRITION

Growth of green plants requires suitable temperature and adequate amounts of carbon dioxide, oxygen, water, light, and a group of essential elements. Carbon, hydrogen, and

Table 34.6

Plant Growth under Different Light Treatments

Treatment	Leaf Size	Stem Diameter	Height	Color of Shoots	Stem Strength
Light					
Dark					
Light then dark					
Dark then light					

Table 34.7

Hoagland's Solution, a Common Nutrient Medium for Healthy Plant Growth

Macronutrients	Grams/Liter	Micronutrients	Grams/Liter
$Ca(NO_3)_2 \cdot 4H_2O$	1.18	H_3BO_3	0.60
KNO_3	0.51	$MnCl_2 \cdot 4H_2O$	0.40
$MgSO_4 \cdot 7H_2O$	0.49	$ZnSO_4$	0.05
KH_2PO_4	0.14	$CuSO_4 \cdot 5H_2O$	0.05
		$H_2Mo_4 \cdot 4H_2O$	0.02
		Ferric tartrate	0.50

oxygen compose 98% of the fresh weight of plants. The remaining 2% is composed of 13 other elements classified either as macronutrients or micronutrients:

- **Macronutrients** are nutrients needed in relatively large amounts (100–2000 ppm).[1] Nitrogen (N), phosphorus (P), calcium (Ca), potassium (K), magnesium (Mg), and sulfur (S) are macronutrients.
- **Micronutrients** are nutrients needed in relatively small amounts (<100 ppm). Iron (Fe), chlorine (Cl), copper (Cu), manganese (Mn), zinc (Zn), molybdenum (Mo), and boron (B) are examples of micronutrients.

Although plants need more macronutrients than micronutrients, all are essential for plant growth. The absence of any essential micronutrient or macronutrient will ultimately kill a plant. Before your lab period, seeds were germinated in distilled water and washed sand. Then the young seedlings were transferred to a variety of growth media. Control plants were watered with a solution containing all essential nutrients. Hoagland's Solution is a common mixture of nutrients used to provide for healthy growth (table 34.7). Your control plants were watered with a solution similar to Hoagland's. Other groups of seedlings were watered with solutions deficient in an essential element. Follow procedure 34.7 to observe the effects of nutrient deficiency.

Procedure 34.7 **Examine plants with symptoms of nutrient deficiency**

1. Examine the plants on demonstration that have been grown for their first month in nutrient solutions lacking Ca, N, P, Mg, K, S, and Fe.
2. Record the symptoms for deficiency of each nutrient in table 34.8. Do not refer to table 34.9 until you have completed your observations.
3. After examining all the seedlings, refer to table 34.9 and determine which deficiencies are most obvious and easily diagnosed.

1 One part per million (ppm) equals one second in 277 h (11.5 days) and one drop in 50 L (13.2 gallons).

Table 34.8

Symptoms of Nutrient Deficiency

Nutrient Solution	Symptom(s)
Complete (containing all required minerals)	
Complete solution minus Ca	
Complete solution minus N	
Complete solution minus P	
Complete solution minus Mg	
Complete solution minus K	
Complete solution minus S	
Complete solution minus Fe	

Table 34.9

Functions and Deficiency Symptoms of Some Major Elements

Element	Major Functions	Deficiency Symptoms
Calcium (Ca)	Component in pectin compounds of middle lamella. Present in organic acids bound to proteins. Plays a role in nitrogen metabolism and membrane integrity.	Deficiency may cause ion uptake imbalance, particularly with magnesium. Young leaves are affected first. Tips and margins of leaves become light green and later necrotic. Tips of leaves become limp. Terminal bud often dies.
Iron (Fe)	Electrons transported in cytochromes.	Effects localized on new leaves. Leaves chlorotic. Veins remain green.
Magnesium (Mg)	Constituent of chlorophyll. Important cofactor for enzymes in respiration and in phosphate metabolism.	Older leaves become chlorotic between the veins at the tips and margins. Usually not characterized by necrotic spots. Root system frequently overdeveloped. Leaf margins may cup upward.
Nitrogen (N)	Major component of amides, amino acids, and proteins. Present in membranes, organelles, and the cell wall. Balance of carbohydrates and nitrogenous substances necessary.	Leaves often more erect. Stem and leaves stunted with excess root development. Foliage, especially older leaves, chlorotic. Unable to flower.
Phosphorus (P)	Constituent of phospholipids, nucleic acids, and nucleoproteins. Important in respiration and energy transfer.	Small plants, narrow leaves, root system larger but fewer laterals. Accumulation of sugars in older leaves promotes the synthesis of purple anthocyanin pigments. Stiff but weak stems and leaves. Older leaves yellowed. Other leaves dark green.
Potassium (K)	Not known to be structurally part of organic compounds. Role is likely catalytic and regulatory. Needed to activate several enzyme systems.	Internodes short; stems weak. Localized chlorotic or molting of older leaves, particularly at the tips and margins. Later stages may have necrotic mottling. Leaf margins frequently curled under.
Sulfur (S)	Component of proteins. Component of iron-sulfur proteins of the electron transport system.	Younger leaves light green. Veins lighter than intervein area.

Question 7
How would you "cure" a plant suffering from these deficiency symptoms?

Commercial Fertilizers

Commercial fertilizer contains three primary nutrients: nitrogen, phosphorus, and potassium. It may also contain other macronutrients and/or micronutrients. Examine the bag of fertilizer in the lab—its contents are described by three numbers on the bag's label. For example, a bag labeled "10–12–8" is 10% nitrogen (as ammonium or nitrate), 12% phosphorus (as phosphoric acid), and 8% potassium (as mineral potash).

PLANT GROWTH REGULATORS

Plant growth and development are controlled by internal chemical signals called **growth regulators.** In this exercise you will study some common physiological responses of plants to growth regulators such as auxin and gibberellic acid (fig. 34.6). Remember that each of the plant responses is part of an overall design for survival and reproduction in varied environments.

Indoleacetic Acid (IAA)

Indoleacetic acid (IAA), also known as **auxin,** is the best-known growth regulator in plants. It is produced by shoot tips and dramatically affects cell growth (fig. 34.7). Auxin also inhibits development of axillary buds. That is, IAA promotes apical dominance.

Functions of Some of the Major Plant Hormones

Hormone		Major Functions	Where Produced or Found in Plant
Auxin (IAA)	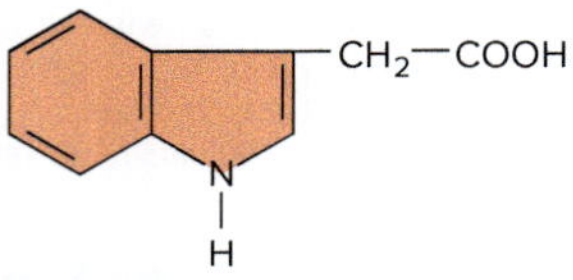	Promotion of stem elongation and growth; formation of adventitious roots; inhibition of leaf abscission; promotion of cell division (with cytokinins); inducement of ethylene production; promotion of lateral bud dormancy	Apical meristems; other immature parts of plants
Cytokinins	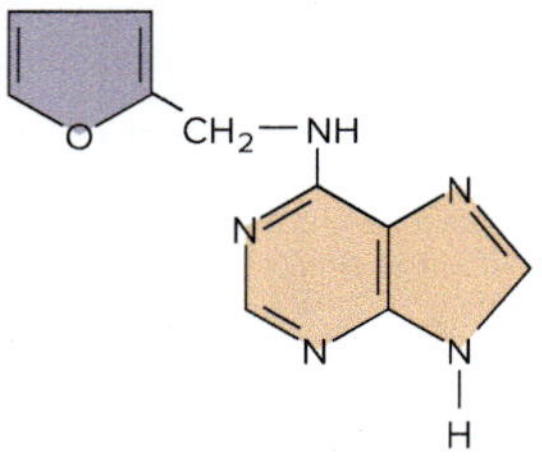	Stimulation of cell division, but only in the presence of auxin; promotion of chloroplast development; delay of leaf aging; promotion of bud formation	Root apical meristems; immature fruits
Gibberellins	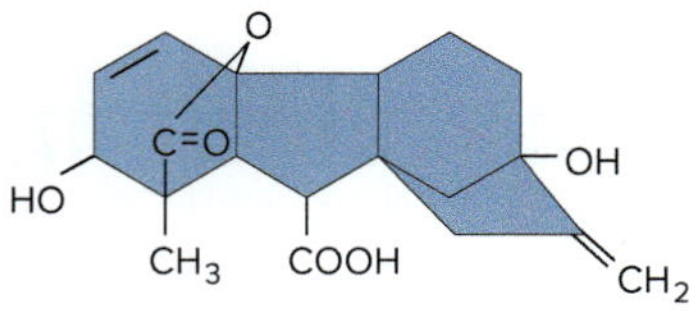	Promotion of stem elongation; stimulation of enzyme production in germinating seeds	Roots and shoot tips; young leaves; seeds
Ethylene	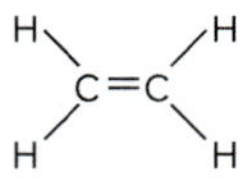	Control of leaf, flower, and fruit abscission; promotion of fruit ripening	Roots, shoot apical meristems; leaf nodes; aging flowers; ripening fruits
Abscisic acid	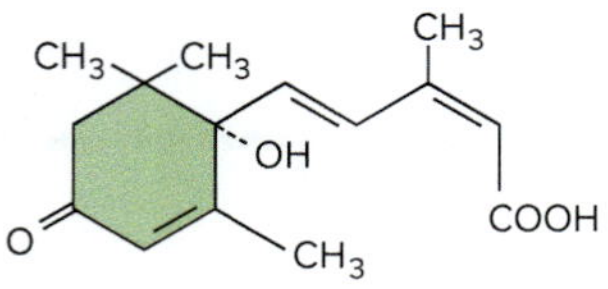	Inhibition of bud growth; control of stomatal closure; some control of seed dormancy; inhibition of effects of other hormones	Leaves, fruits, root caps, seeds

Figure 34.6 Functions of some of the major plant hormones.

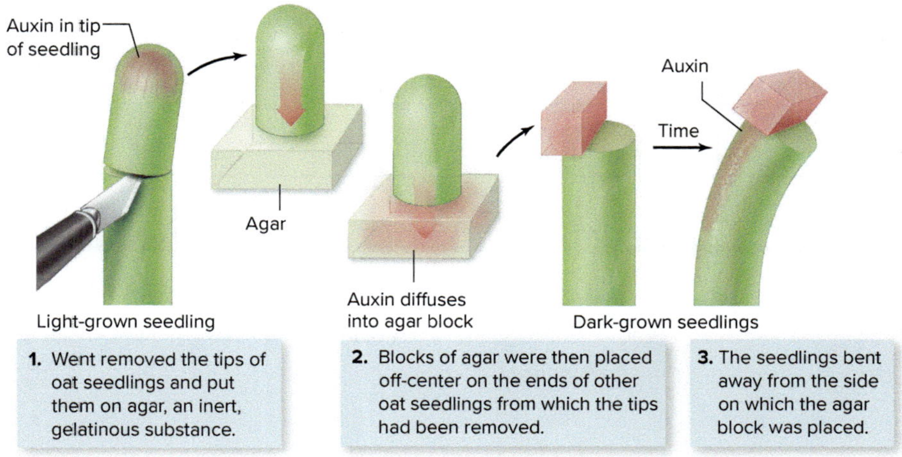

Figure 34.7 Went's demonstration of how auxin affects plant growth. Frits Went, a Dutch plant physiologist, discovered how auxin controls plant growth. Went concluded that auxin promoted cell elongation and that it accumulated on the side of a grass seedling bent away from the light.

Table 34.10

Effects of Auxin on *Coleus* Plants

Treatment	Observations
Control (no treatment)	
Shoot tip removed	
Shoot tip removed and the cut surface coated with 1% IAA in lanolin	
Shoot tip removed and the cut surface coated with lanolin alone	

Procedure 34.8 Examine the effects of auxin

1. Examine the *Zea* or *Coleus* plants subjected to the treatments described in table 34.10.
2. Record your observations in table 34.10.

Question 8

a. What is the effect of replacing the shoot tip with IAA?

b. Why was it necessary to apply only lanolin to one plant?

2,4-Dichlorophenoxyacetic Acid

2,4-dichlorophenoxyacetic acid (2,4-D) is a synthetic auxin used frequently in agriculture as a herbicide. 2,4-D is one of the most widely used herbicides in the world; it kills weeds in grass lawns by selectively eliminating broad-leaved dicots. 2,4-D is useful because it is synthetic and there are no enzymes in plants to detoxify it or break it down.

Table 34.11

Effects of 2,4-D on Monocots and Dicots

Type of Plant Treated with 2,4-D	Observations
Monocots	
Dicots	

Procedure 34.9 **Examine the effects of 2,4-D**

Be careful when handling 2,4-D; it can be especially irritating to your eyes.

1. Examine plants treated 2 weeks ago with 2,4-D. The pots originally contained a mixture of monocots (e.g., corn, oats) and dicots (e.g., peas, beans).
2. Record your observations in table 34.11.

Question 9
Which plants were affected most by the herbicide?

Gibberellic Acid (GA)

Dwarfism often results from a plant's inability to synthesize active forms of **gibberellic acid (GA),** another plant growth regulator. Gibberellic acid mobilizes enzymes during seed germination, promotes stem elongation, and is used commercially to increase the production of grapes (fig. 34.8).

(a) Omikron/Science Source

(b) Amnon Lichter, The Volcani Center

Figure 34.8 Gibberellins affect several aspects of plant growth and development. (*a*) Gibberellic acid stimulates cell elongation in cabbage. Cabbage (*Brassica oleracea*), a biennial native to the seacoast of Europe, will "bolt" when the heads are treated with gibberellin. (*b*) Applying gibberellic acid increases the space between grapes. As a result, larger grapes (right) develop because more space between grapes allows them to grow larger. This is important because it increases yields (and income) for farmers.

Table 34.12

Effects of Gibberellic Acid on Corn Plants

Plant and Treatment	Observations
Normal, untreated	
Dwarf, untreated	
Normal, treated with GA	
Dwarf, treated with GA	

Procedure 34.10 **Examine effects of gibberellic acid**

1. Examine four trays of bean or corn plants treated in the following ways:

   Tray 1: Normal plants—untreated

   Tray 2: Dwarf plants—untreated

   Tray 3: Normal plants—treated with two or three drops of GA on alternate days for two weeks

   Tray 4: Dwarf plants—treated with two or three drops of GA on alternate days for two weeks

2. Record your observations in table 34.12.

Question 10

a. What is the effect of GA on dwarfism?

b. Did applying GA to normal plants have any effect? If so, what?

INQUIRY-BASED LEARNING

How does light influence seed germination?

Observation: Seed germination is cued to a variety of environmental stimuli. Some of these stimuli are seasonal.

Question: Is seed germination of tropical species of plants more sensitive to light than seed germination of temperate species?

a. Establish a working lab group and obtain Inquiry-Based Learning Worksheet 34 from your instructor.

b. Discuss with your group well-defined questions relevant to the preceding observation and question. Choose and record your group's best question for investigation.

c. Translate your question into a testable hypothesis and record it.

d. Outline on Worksheet 34 your experimental design and supplies needed to test your hypothesis. Ask your instructor to review your proposed investigation.

e. Conduct your procedures, record your data, answer your question, and make relevant comments.

f. Discuss with your instructor any revisions to your questions, hypotheses, or procedures. Repeat your work as needed.

Questions for Further Study and Inquiry

1. How are phototropism and gravitropism similar? How are they different?

2. What is the adaptive significance of seed germination being influenced by environmental conditions such as light and water?

3. What comprises commercial "plant food"?

4. What are some commercial uses for plant growth-regulating substances?

5. Some seeds must pass through the digestive tract of an animal before they can germinate. What is the adaptive significance of this feature?

6. How are growth regulators in plants similar to hormones in animals? How are they different?

DOING BIOLOGY YOURSELF

Design an experiment to determine whether seed germination by tropical species of plants is more sensitive to light than that by temperate species.

WRITING TO LEARN BIOLOGY

Describe the characteristics of plants you would choose to culture were you to travel in space. Base your choices on nutrient requirements, phototropic sensitivity, and gravitropic sensitivity.

Survey of the Animal Kingdom

Phyla Porifera and Cnidaria

EXERCISE 36

Learning Objectives

By the end of this exercise you should be able to:

1. Describe how structures specific to poriferans and cnidarians help them survive and reproduce in their environment.
2. List the fundamental characteristics of members of phylum Porifera and phylum Cnidaria.
3. Recognize members of the three major classes of cnidarians.
4. Describe the body forms of cnidarians and describe reproduction of those species alternating between polyps and medusae.
5. Compare the feeding methods of sponges and jellyfish.
6. Examine the morphology of representative sponges and the major classes of cnidarians.
7. Discuss characteristics likely to promote long-term evolutionary success of sponges and cnidarians and relate the discussion to Learning Objective 1.

Please visit connect.mheducation.com to review online resources tailored to this lab.

Animals are uniformly characterized as **eukaryotic, multicellular,** and **ingestive-feeding heterotrophs.** Heterotrophs derive their energy from organic molecules made by other organisms. To many people, the most ancient animals, such as **phylum Porifera** (sponges) and **phylum Cnidaria** (sea jellies and corals), may appear primitive and unsophisticated. However, as we discuss in this and upcoming exercises the remarkable diversity of animals, remember that "primitive" refers to characteristics derived early in the phylogenetic history of an organism's evolution. Primitive does not imply simple, less important, or poorly functioning charactersitics. Instead, it refers to characteristics that have stood the test of time. The bodies and body cavities of sponges and jellyfish are structurally uncomplicated and lack the complex behavior and sensory capabilities of most higher animals. But don't let that simplicity fool you. What sponges and sea jellies (sometimes called jellyfish) lack in complexity is more than compensated for by their extraordinarily elegant design. After all, their overall body-plans (table 36.1) have persisted in changing environments for many millions of years; their "simple" morphology accomplishes the primary functions of food getting, reproduction, and adaptive responses to their environment as does the more-complex morphology of other more familiar animals.

PHYLUM PORIFERA

Sponges are the simplest of the major animal phyla and comprise 5150 species (fig. 36.1). Most sponges live in the ocean, but a few encrust rocks and wood in freshwater. Sponges lack tissues and organs and are typically **asymmetrical** assemblages of cells. Bodies of asymmetrical organisms have no symmetry or pattern such as left and right halves or anterior and posterior regions. Sponge cells are so loosely assembled and self-sufficient that if a sponge is forced through a mesh, the disassociated cells will survive. Even more remarkably, the disassociated cells of some species can reassemble as a functioning organism.

Grantia

Examine a preserved specimen of *Grantia*, one of the simplest sponges (fig. 36.2). At first glance, sponges such as *Grantia* appear plantlike because they are **sessile** (i.e., attached to a substrate). Some sponges even appear green because symbiotic algae live in their bodies. However, sponges are **filter-feeding** heterotrophs and have no photosynthetic pigments. Notice that *Grantia* is a tubular, open-ended chamber surrounded by a thin, porous, folded wall of cells. Phylum Porifera gets its name from the many pores in the chamber walls.

Table 36.1

Phyla Porifera and Cnidaria

Phylum	Typical Examples		Key Characteristics	Approximate Number of Named Species
Porifera (sponges)	Barrel sponges, boring sponges, basket sponges, vase sponges	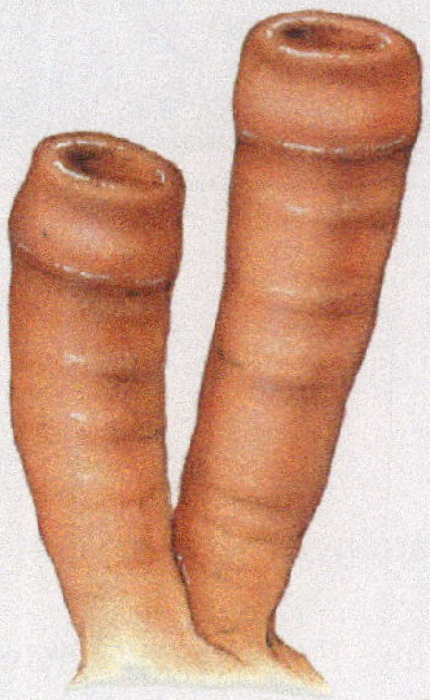	Asymmetrical bodies without distinct tissues or organs; saclike body consists of two layers breached by many pores; internal cavity lines with food-filtering cells called choanocytes; most marine (150 species live in freshwater)	5,150
Cnidaria (cnidarians)	Jellyfish, sea jellies, hydra, corals, sea anemones		Soft, gelatinous, radially symmetrical bodies whose digestive cavity has a single opening; possess tentacles armed with stinging cells called cnidocytes that shoot sharp harpoons called nematocysts; almost entirely marine	10,000

(a) Ocean Image Photography/Shutterstock (b) ArteSub/Alamy Stock Photo

Figure 36.1 Sponges (phylum Porifera) such as (*a*) the yellow tube sponge, *Verongia,* and (*b*) the red encrusting sponge, *Ophlitaspongia,* have a variety of colors. A seemingly inactive sponge may filter 1000 times its own volume of water per day through walls filled with small canals and chambers. There are up to 18,000 chambers per square millimeter of sponge. Flagellated cells line the chambers to circulate water and filter extremely small particles of food. Eighty percent of the organic matter captured by sponges is too small to be seen with a microscope. Use of this small-size fraction of dissolved matter has made sponges successful for hundreds of millions of years.

Carolina Biological Supply Company/Phototake

Figure 36.2 *Grantia* is a common tubular sponge with a folded body wall filled with pores for filtering water.

Question 1

Do any features of *Grantia* clearly distinguish this organism as an animal? If so, which ones?

Structure of Sponges

Sponge walls filter seawater and remove food particles. In the simplest sponges this wall is lined on the outside by an **epithelial layer** of flat cells (fig. 36.3). Inside the sponge is the **spongocoel,** a central cavity lined by flagellated cells called **choanocytes** (sometimes called collar cells). Their moving flagella draw water through pores within **pore cells** into the spongocoel and across the collars of the choanocytes to trap food particles (fig. 36.3). The flagella of choanocytes produce a constant flow of water into a sponge. Eventually the filtered water exits through a large hole in the end of the sponge called the **osculum.**

Question 2

Consider Learning Objective 1 listed at the beginning of this exercise. Are choanocytes significant to a fundamental process for sponges? What is the process and how are choanocytes significant?

Grantia has a folded wall. Examine a prepared slide of a cross section through *Grantia* (fig. 36.4). These folds form **incurrent canals** opening to the outside and **flagellated canals** opening to the central spongocoel. The flagellated canals are lined with choanocytes that move their flagella to draw water into the incurrent canals. From there, the water moves though openings formed by pore cells, through the flagellated canals, and on to the **spongocoel,** finally exiting by way of the **osculum.** A specialized collar of microvilli surrounding the flagellum of a choanocyte traps food particles that are engulfed by the cell body. Digestion is **intracellular,** meaning that it occurs inside cells. Some sponges are more complicated than *Grantia* and have highly folded walls or a complicated series of small chambers lined with choanocytes. A spongocoel may be difficult to distinguish in these complicated sponges (see fig. 36.1*b*).

The wall of a sponge is reinforced by proteinaceous **spongin** fibers and crystalline supportive structures called **spicules** (fig. 36.5). Spicules are usually calcareous (containing calcium) or siliceous (containing silicon). The wall of a sponge also contains **amoebocytes**. As their name implies, amoebocytes are creeping, motile cells having many functions, including digestion and a rather amazing ability to differentiate into other cell types as needed.

SAFETY FIRST Before coming to lab, you were asked to read this exercise so you would know what to do and be aware of safety issues. In the space below, briefly list the safety issues associated with today's procedures. If you have questions about these issues, contact your laboratory assistant before starting work.

Procedure 36.1 Examine spicules

1. Examine a prepared slide of cleaned and isolated spicules of sponges.
2. Also locate spicules in a cross section of *Grantia.*
3. If enough preserved sponge is available, make a wet mount of spicules in a depression slide by gently crushing a small piece of sponge on a slide with a coverslip.
4. If bleach is available, add one or two drops to dissolve the protein and expose the spicules.
5. Examine the preparation under low, then high, magnification. Sketch what you see.

Question 3

How do spicules help sponges survive in their environment (see Learning Objective 1)?

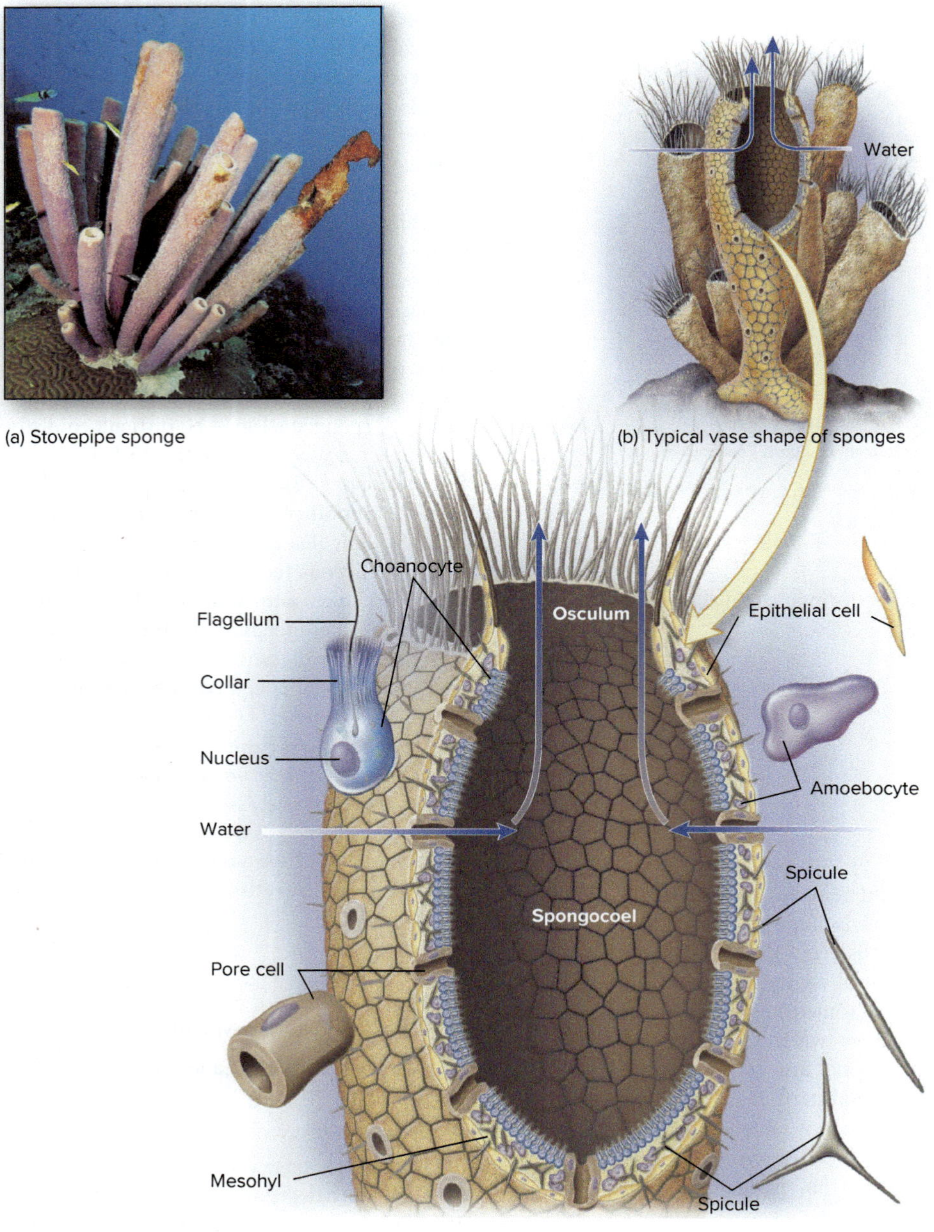

(a) James R. D. Scott/Moment/Getty Images

Figure 36.3 Sponge morphology. (*a*) The stovepipe sponge (*Aplysinaarcheri*) is a common sponge found on Caribbean reefs. (*b*) Many sponges have a vaselike shape. (*c*) A cross section reveals that sponges are multicellular animals, having various cell types but no distinct organs.

Examine the fused lattice of spicules of *Euplectella* (fig. 36.6). Spicules of different sponges have many shapes and may be fused in an ornate lattice. The beautiful lattice of the sponge *Euplectella* is unusual because it often houses several species of shrimp. Interestingly, when a male and female shrimp enter the spongocoel, they may grow too large to escape. A dried specimen of *Euplectella* containing remnants of its permanent residents was formerly used in Japan as a wedding present to symbolize the permanent bond of marriage.

Spongia

Examine some large, dried sponges on display. *Spongia* has a more complex arrangement of chambers than does *Grantia.*

Examine a prepared slide of spongin fibers. Fibers of spongin (made of protein) compose the skeleton of common bath sponges such as *Spongia.* Spongin fibers are flexible and are not like crystalline spicules. In the past, natural sponges were gathered, dried, and sold easily in the Caribbean and Mediterranean. Today, natural sponges are rare and have been replaced by synthetic products.

Question 4

a. Do sponges appear to have any organs or organ systems?

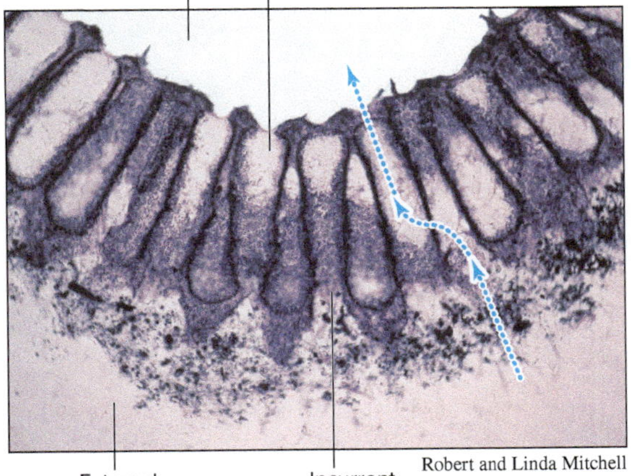

Robert and Linda Mitchell

Figure 36.4 Cross section through the folded body wall of *Grantia* (150×). Arrows show the path of water flow. The beating flagella of choanocytes draw water containing food particles from the external environment into an incurrent canal, through pore cells in the folded wall that is lined with choanocytes, into the flagellated canal, into the spongocoel, and out the osculum.

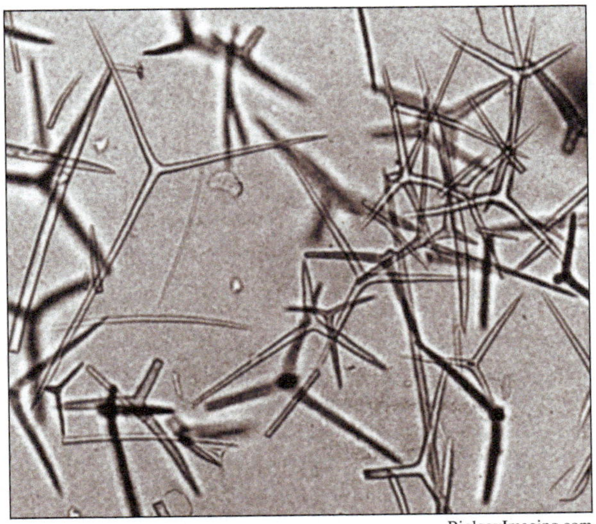

BiologyImaging.com

Figure 36.5 This micrograph shows a variety of sponge spicules (150×).

b. What is the advantage of a folded or convoluted wall in sponges?

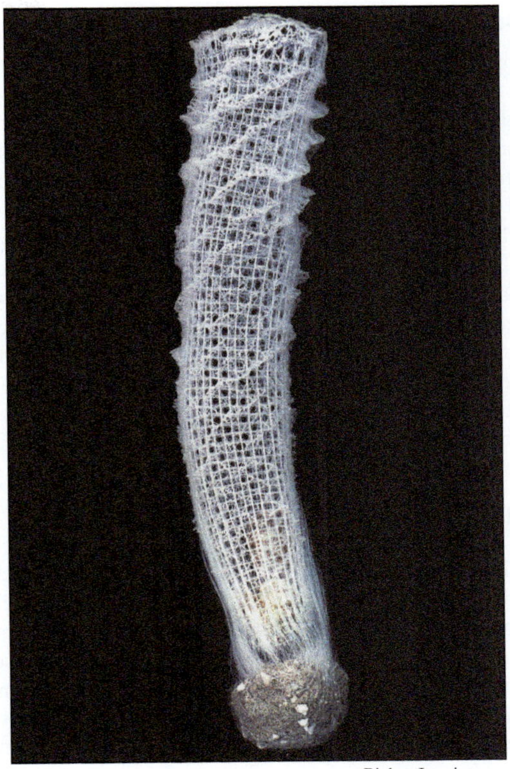

BiologyImaging.com

Figure 36.6 The fused glass (silicon) spicules of *Euplectella* form a beautiful, intricate lattice. The living tissue has been removed from this dried specimen.

c. What function other than support might spicules serve?

d. How many prongs do spicules of *Grantia* have?

e. What characteristics of *Spongia* make them useful as a household sponge?

f. Consider Learning Objective 1 listed at the beginning of this exercise. Are spicules significant to a fundamental process for sponges? In what way?

Sponge Reproduction

Sponges reproduce asexually and sexually. Asexual reproduction includes budding and the release of stress-resistant aggregates of amoebocytes called **gemmules.** In favorable conditions, amoebocytes in a gemmule can grow into a mature organism. During sexual reproduction, choanocytes and amoebocytes differentiate into gametes. Eggs remain in the gelatinous, interstitial (between cells) **mesenchyme,** but sperm are released into the water and are captured by choanocytes or amoebocytes of other sponges. The captured sperm are transported to eggs and fertilization occurs. After a brief development, the embryo is expelled from the sponge. Most species of sponges are hermaphroditic (i.e., produce both male and female gametes), but they usually produce eggs and sperm at different times.

PHYLUM CNIDARIA (COELENTERATA)

Phylum Cnidaria (also called coelenterates) includes **class Hydrozoa** (hydras), **class Scyphozoa** (sea jellies), **class Cubozoa,** and **class Anthozoa** (anemones and corals). Almost all cnidarians are marine carnivores. Their bodies are **radially symmetrical** and more complex than sponges (fig. 36.7). Radial symmetry describes a body plan with repetitive body areas arranged in a circle around a central point like the pieces of a pie. Sensory organs are exposed to the environment in all directions around the perimeter. This is evolutionarily adaptive for a slow or nonmotile organism.

The body wall of cnidarians has two cellular layers, an **ectoderm** on the outside and an **endoderm** (sometimes called the gastroderm) lining the interior gastrovascular cavity (fig. 36.8). A gelatinous **mesoglea** separates the two body layers. Cells of cnidarians are organized into true tissues (nervous, muscular, and reproductive), but organs are organized loosely, at best.

Cnidarians have two basic body-plans: **polyps** and **medusae** (fig. 36.8). Polyps are cylindrical animals with a mouth surrounded by tentacles atop the cylinder (i.e., the end facing away from the substrate). Polyps are usually attached to the substrate and may be solitary or colonial. In contrast to polyps, medusae are usually free-floating and umbrella-shaped. Their mouths point downward and are surrounded by hanging tentacles. The classes of cnidarians are distinguished primarily by the relative dominance of the polyp stage or the medusa stage in the life cycle. Some cnidarians occur only as polyps, others only as medusae, and still others alternate between these two forms. This alternation is a form of **polymorphism,** which means "many forms."

The life cycle of many cnidarians is characterized by alternation between polyp and medusa (i.e., polymorphism; fig. 36.8). During the life cycle, medusae produce and release eggs and sperm into water for fertilization, although some species retain their eggs. After fertilization the zygote develops into a swimming mass of ciliated cells called a **planula larva.** A planula eventually settles to the substrate and develops into a polyp. The polyp may reproduce asexually by budding other polyps or may continue the sexual cycle by budding immature medusae called **ephyrae.** An ephyra develops into a mature medusa.

All cnidarians are carnivores that capture their prey (small fishes and crustaceans) with tentacles that ring their mouth. Captured prey are pushed through the mouth into the **gastrovascular cavity** (GVC), where a combination of intracellular and **extracellular digestion** occurs followed by phagocytosis of small food particles and intracellular digestion (fig. 36.9). These tentacles are armed with stinging cells called **cnidocytes** containing small, barbed harpoonlike structures called **nematocysts** (fig. 36.10). Notice the transition from intracellular digestion within choanocytes of poriferans to a combination of intra- and extracellular processing in cnidarians. The phyla discussed in upcoming exercises rely almost exclusively on extracellular digestion.

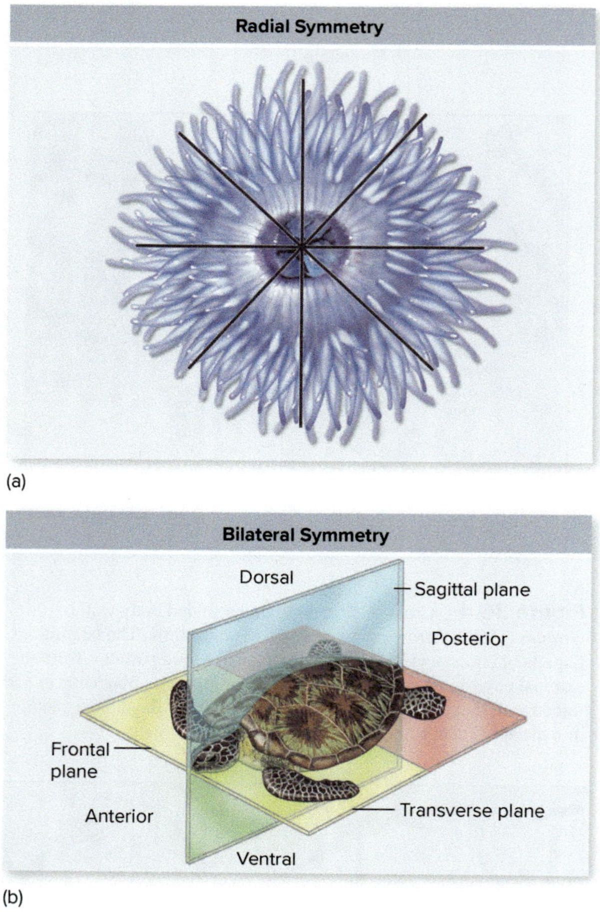

Figure 36.7 A comparison of radial and bilateral symmetry. (*a*) Radially symmetrical animals, such as this sea anemone, can be bisected into equal halves in any two-dimensional plane. (*b*) Bilaterally symmetrical animals, such as this turtle, can only be bisected into equal halves in one plane (the sagittal plane).

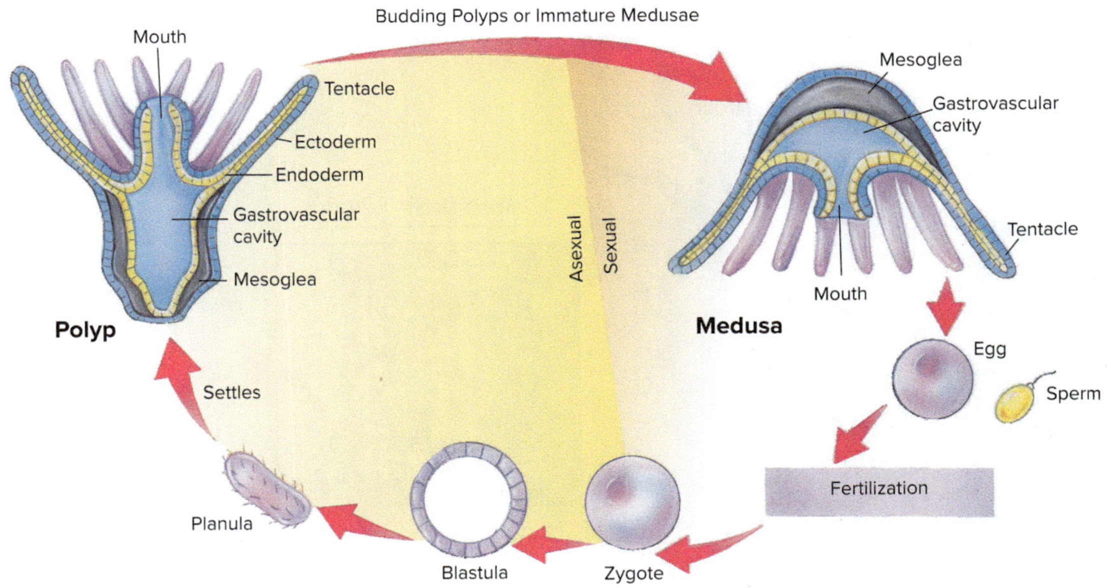

Figure 36.8 A generalized life cycle for Cnidaria. Cnidarians alternate between medusa and polyp forms. Male and female medusae use meiosis to produce gametes that, in water, are fertilized. These gametes are the only haploid (n) stage; all other stages are diploid ($2n$). After a short period of free swimming, the planula larva settles to the substrate and forms a polyp. When the polyp buds (an asexual process), additional polyps and medusa buds form. Medusae separate from the polyp and swim away. The polyp or medusa stage has been lost or reduced in many cnidarians, such as anemones and corals.

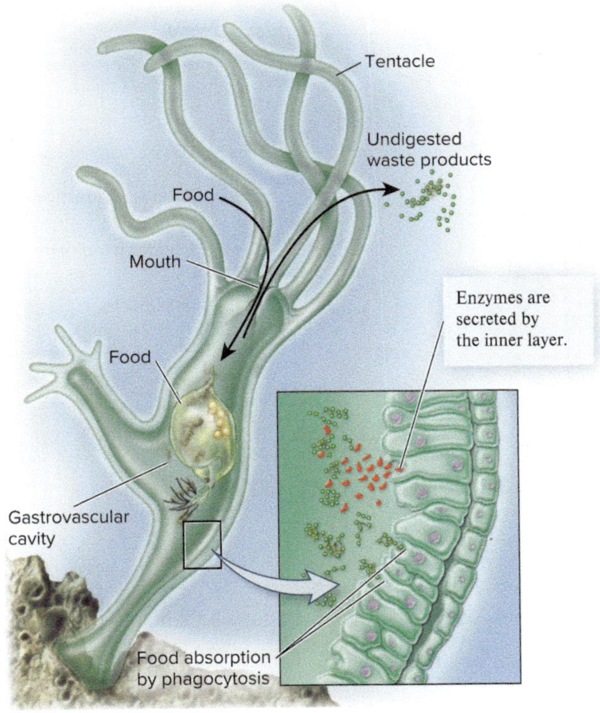

Figure 36.9 Extracellular digestion. In animals with gastrovascular cavities, such as the cnidarian *Hydra,* most digestion occurs extracellularly. Food sources are trapped by tentacles and delivered to the mouth. Digestion occurs within the gastrovascular cavity. Digested food products are phagocytosed directly into the cells that line the cavity, and wastes are excreted out the same opening in which food entered.

Question 5

a. Consider Learning Objective 1 listed at the beginning of this exercise. Are cnidocytes significant to fundamental processes for cnidarians? In what ways?

b. Consider Learning Objective 1 listed at the beginning of this exercise. How could polymorphism contribute to the evolutionary success of cnidarians in their environment?

Class Hydrozoa

The polyp stage dominates the hydrozoan life cycle, although both polyps and medusae occur in most species. The outer layer of cells, the ectoderm, and the inner layer, the endoderm, surround the gastrovascular cavity; these layers are separated by the gelatinous, acellular mesoglea. Amoeboid cells circulate in the mesoglea. Ectodermal cells include cnidocytes and muscular contractile cells. Endodermal and glandular cells secrete enzymes into the gastrovascular cavity for extracellular digestion. The gastrodermis lacks cnidocytes.

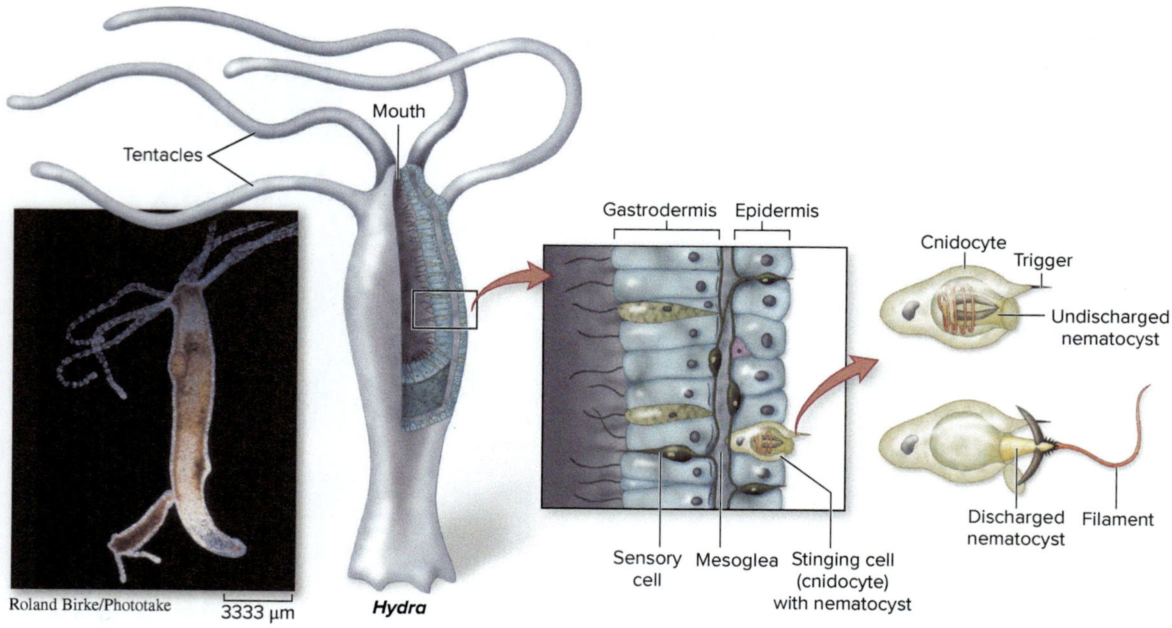

Figure 36.10 Phylum Cnidaria: cnidarians. The cells of a cnidarian such as this *Hydra* are organized into specialized tissues. The interior gut cavity is specialized for extracellular digestion—that is, digestion within a gut cavity rather than within individual cells. Cnidarians are radially symmetrical, with parts arranged around a central axis like the petals of a daisy. The epidermis includes stinging cells called cnidocytes, and each cnidocyte can discharge a harpoonlike nematocyst. Cnidocytes are scattered on the body wall and dense on the tentacles.

Hydra

Hydra are small, common hydrozoans that live in shallow, freshwater ponds (figs. 36.9 and 36.10). They are usually less than 1 cm tall and prey on smaller invertebrates among the filaments and leaves of freshwater algae and plants. *Hydra* have no medusa stage.

Procedure 36.2 Observe Hydra

1. Obtain a living *Hydra;* observe it in a small petri dish with a dissecting microscope. Polyps of *Hydra* are solitary and occasionally hang from the water's surface with their **basal disks** adhering to the surface of the water. More often, they attach to a hard substrate with their basal disk. However, *Hydra* can detach themselves and move, not by swimming but by somersaulting along their substrate.
2. Allow a few minutes for the animal to relax in the petri dish; then tap the edge and observe the animal's response.
3. If small, living crustaceans such as *Daphnia* or *Artemia* (brine shrimp) are available, place some of these organisms near the tentacles of a *Hydra.* When a prey item touches a tentacle, it sticks tightly. The nematocysts help entrap the *Daphnia.*
4. Observe the cellular structure of *Hydra* by studying a prepared slide of a cross section of the organism (figs. 36.10 and 36.11).

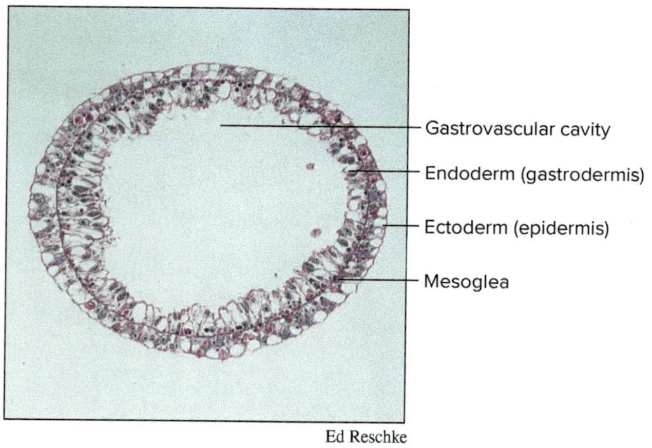

Figure 36.11 Micrograph of a cross section of a *Hydra,* a hydrozoan (100×).

Question 6

a. How do *Hydra* respond to a tap on their substrate?

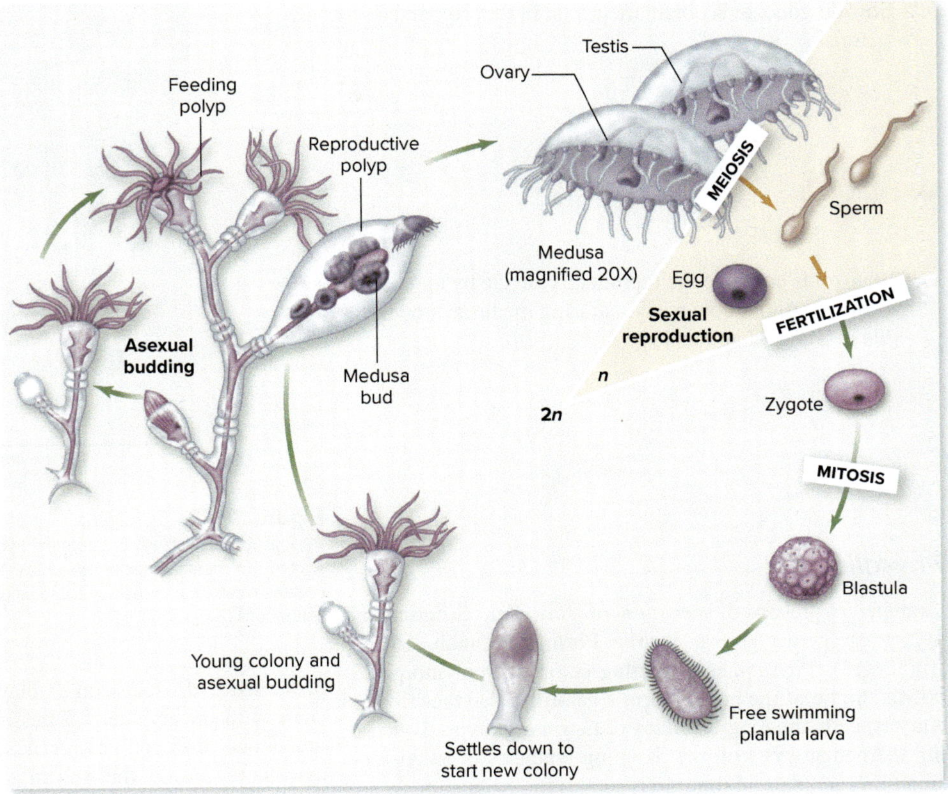

A medusa of *Obelia*, a marine hydrozoan.

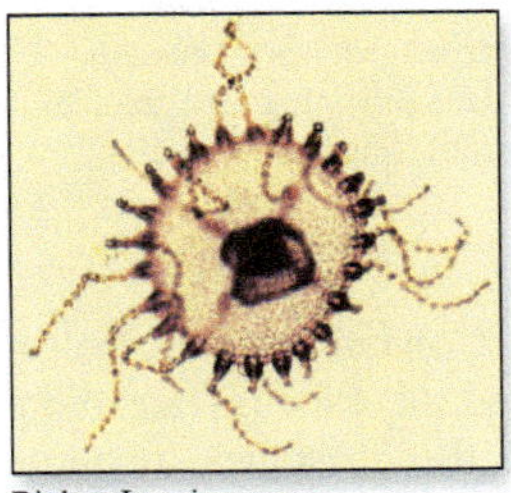

BiologyImaging.com

Figure 36.12 The life cycle of *Obelia,* a marine colonial hydroid. Polyps reproduce by asexual budding, forming colonies. Reproductive polyps may also give rise to medusae, which reproduce sexually via gametes. These gametes fuse, producing zygotes that develop into planulae, which in turn settle down to produce polyps. The inset photo shows a young recently budded medusa.

b. What tissues must exist for this response?

c. *Hydra* are predators. How actively do *Hydra* stalk their food?

d. What specialized cells of tentacles aid in capturing prey?

Obelia

Examine a prepared slide of another hydrozoan, *Obelia* (fig. 36.12). Examine a prepared slide of the small medusae of *Obelia* in addition to the polyp colony. *Obelia* typifies most hydrozoans because it has colonial polyps and free-swimming medusae. These colonial polyps appear plantlike and branch from a tube. Polyps of *Obelia* are polymorphic because some are specialized feeding polyps called **gastrozooids;** others are reproductive polyps called **gonozoids.**

Question 7

a. What structures determine whether a polyp of *Obelia* is a gastrozooid (feeding polyp) rather than a gonozoid?

b. How do gonozoids obtain their food in this colonial organism?

c. Gonozoids continue the reproductive cycle by budding medusae. About how many maturing medusae are visible in a typical gonozoid?

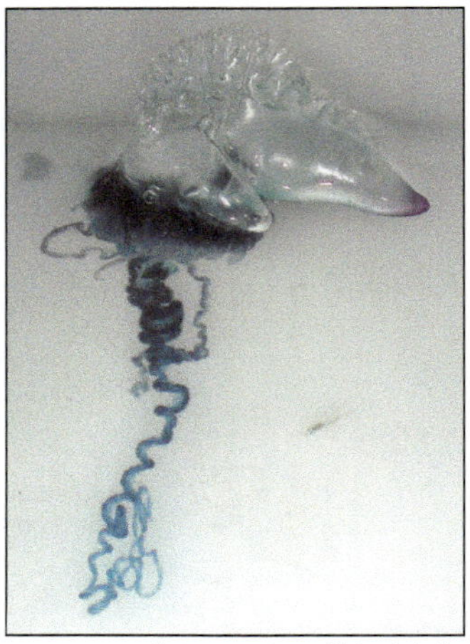

Image courtesy of Islands in the Sea 2002, NOAA/OER

Figure 36.13 Portuguese man-of-war, *Physalia utriculus.* The Portuguese man-of-war is a colonial hydrozoan that has adopted the way of life characteristic of sea jellies. This highly integrated colonial organism can ensnare fish by using its painful stings and tentacles, sometimes over 15 m long.

Physalia

Examine a preserved specimen of *Physalia*, a common hydrozoan better known as the Portuguese man-of-war (fig. 36.13). *Physalia* is a floating colony of polymorphic polyps. Some of the polyps form a gas-filled sac that floats and suspends the long tentacles of nutritive polyps. Touching the nematocysts on the dangling tentacles is lethal for small fish and painful (but rarely lethal) for swimmers.

Gonionemus

Examine a preserved specimen of *Gonionemus* with your dissecting microscope (fig. 36.14). Examine any other preserved hydrozoans on display. *Gonionemus* is a hydrozoan with an atypically large medusa. Medusa are more gelatinous than polyps because the mesoglea is more extensive. Locate the **velum** on the inner periphery of the medusae and the mouth at the end of the **manubrium.** The **gastrovascular cavity** radiates from the center as **ring canals** connected by a **circular canal** around the perimeter. The **gonads** (tissue that produces gametes) attach to the radial canals and appear similar in males and females. The **tentacles** have a rough surface.

INQUIRY-BASED LEARNING

How well do Hydra *detect nearby prey?*

Observations: *Hydra* are predators capable of relatively complex behavior. They readily detect vibration as well as changes in water chemistry produced by their potential prey.

Question: How do *Hydra* sense their prey?

a. Establish a working lab group and obtain Inquiry-Based Learning Worksheet 36 from your instructor.
b. Patiently observe *Hydra* movement and their responsiveness. Discuss with your group a well-defined question relevant to *Hydra* response to prey movement, to dissolved molecules from their prey, or both. Record the question on Worksheet 36.
c. Translate your question into a testable hypothesis and record it.
d. Devise a procedure to determine if *Hydra* are more responsive to prey movement, to dissolved molecules from their prey, or both.
e. Outline on Worksheet 36 your experimental design and supplies needed to test your hypothesis. Ask your instructor to review your proposed investigation.
f. Conduct your procedures.
g. Record your data, answer your question, and make relevant comments.
h. Discuss with your instructor any revisions to your questions, hypotheses, or procedures. Repeat your work as needed.

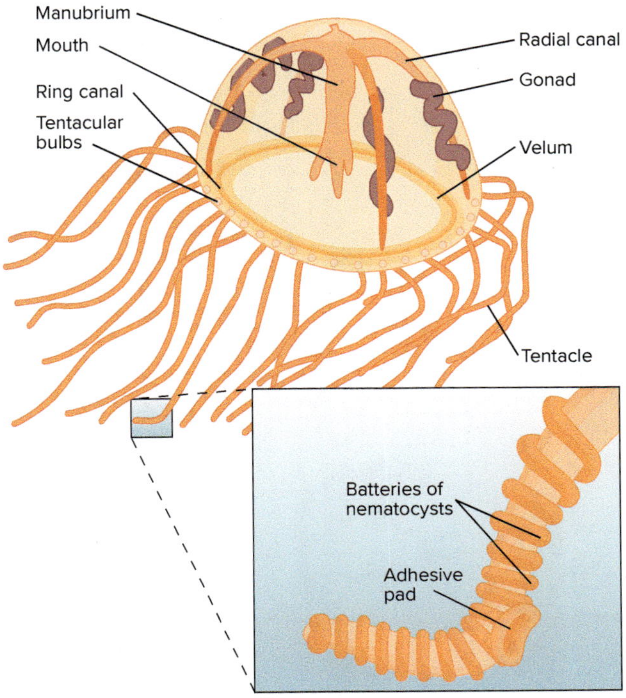

Figure 36.14 Diagram showing the structure of the medusa of *Gonionemus,* a hydrozoan.

Question 8

a. Which cells give tentacles of *Gonionemus* their rough surface?

b. What functions might radial and ring canals serve?

Class Scyphozoa

Observe preserved medusae of *Aurelia* and *Cassiopeia.* Scyphozoans are commonly called sea jellies (jellyfish) because the gelatinous medusa dominates their life cycle. The polyp is reduced to a small larval stage. The mesoglea has amoeboid cells. The GVC is divided into four radiating pouches, and the gastrodermis has cnidocytes. This group also includes one of the largest invertebrates in the world, *Cyanea capillota. Cyanea* lives in the North Sea and can exceed 2 m in diameter. *Aurelia* (2–6 cm) is a more typical sea jelly for you to examine (fig. 36.15).

BiologyImaging.com

Figure 36.15 *Aurelia,* the moon sea jelly (1×). The gelatinous medusa is the dominant body form in the life cycle of these cnidarians.

Examine prepared slides of (1) planula larvae produced by sexual reproduction of medusae, (2) the polyp stage called a **scyphistoma**, and (3) ephyra (immature medusae) budded from the polyp. Review the life cycle of *Aurelia* (fig. 36.16).

Question 9

How do medusae of *Aurelia* and *Gonionemus* differ in size, arrangement of tentacles, and shape of manubrium?

Class Anthozoa

Anthozoans (anemones and corals) form the largest class of cnidarians with more than 6000 species (fig. 36.17). Anthozoan polyps are solitary or colonial, and there is no medusa. The mouth leads to a tubular pharynx and to a gastrovascular cavity with septate compartments. Gonads are gastrodermal.

Metridium

Obtain a specimen of the common anemone (pronounced ah-NEH-moh-nee) *Metridium*, and find its mouth and tentacles. Make a cross section through the body of *Metridium* and expose the gastrovascular cavity (this dissection may be on demonstration). Locate the structures shown in figure 36.18. Anemones are sessile and attach themselves to a substrate with their flat and sticky basal disk. However, this attachment is not permanent, and anemones can slowly slide on a film of mucus. When pieces of the basal disk tear

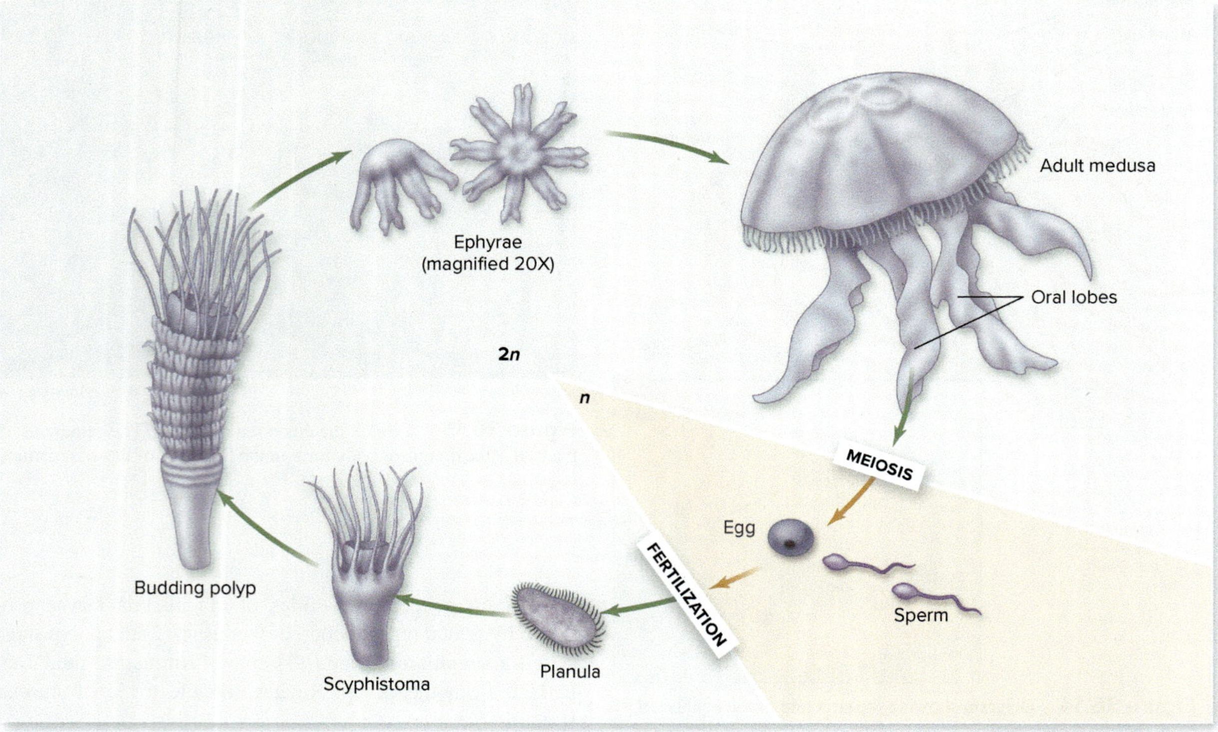

Figure 36.16 *Aurelia* life history. For the dioecious *Aurelia,* like all scyphozoans, the medusa stage dominates the life cycle. Male and female medusae produce gametes for fertilization in the water. These gametes are the only haploid (*n*) stage of the life cycle; all other stages are diploid (2*n*). The zygote develops into a planula larva that settles to the substrate and forms a scyphistoma (polyp) that produces ephyrae (immature medusae) by budding. Medusae separate from the polyp and swim away.

Figure 36.17 Sea anemone, a common anthozoan.

away from a moving anemone, the pieces form a new individual. This type of asexual reproduction is **fragmentation.**

The gastrovascular cavity of an anthozoan polyp is partitioned by thin septa. These septa distinguish anthozoan from scyphozoan and hydrozoan polyps. Along the edges of the septa are gonadal tissue and threads of tissue called

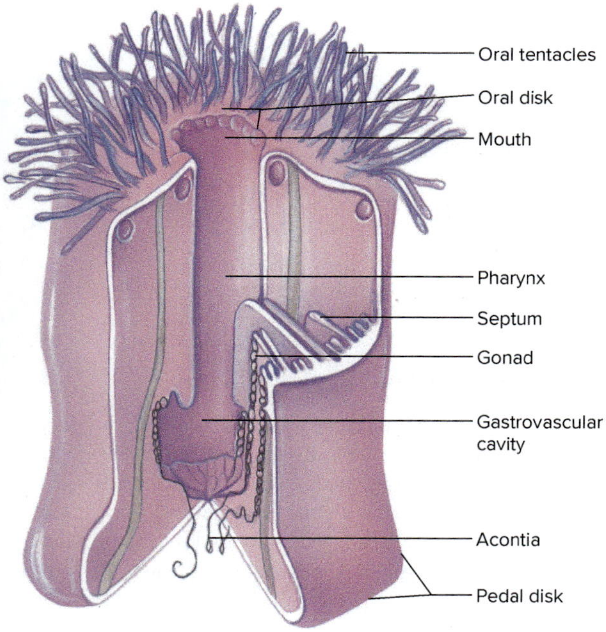

Figure 36.18 Class Anthozoa. The structure of the anemone, *Metridium.* Anthozoans have no medusa stage.

BiologyImaging.com

Figure 36.19 Calcium carbonate skeleton of a stony coral.

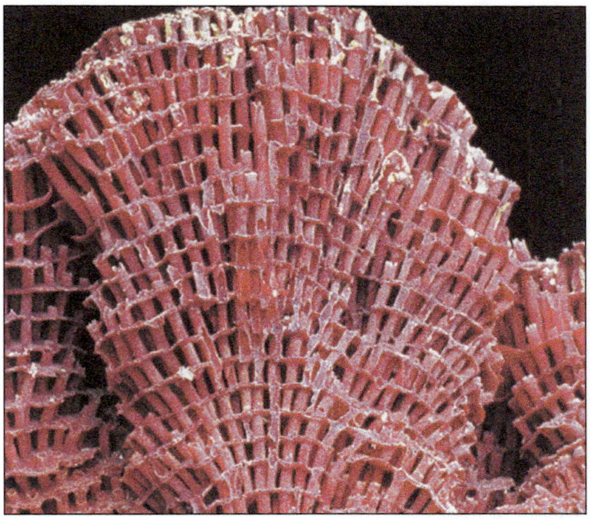

BiologyImaging.ccm

Figure 36.20 *Tubipora,* organ pipe coral.

acontia. Acontia threads bear dense batteries of cnidocytes that can protrude through the mouth or through pores in the body wall to overcome prey or provide defense.

Corals

Examine a piece of dry, calcareous coral and look for small depressions (fig. 36.19). Coral polyps are structurally similar to anemones, but corals are usually colonial and much smaller. Most corals secrete a hard skeleton of calcium carbonate with many small cups surrounding the polyps. The small, fragile polyps are probably absent from the specimen you are examining, but the depressions in which they lived are numerous.

Examine a piece of *Tubipora* (fig. 36.20) and any other anthozoans on display. The tropical organ pipe coral, *Tubipora,* is organized differently. Long parallel polyps are encased in calcareous tubes connected at intervals by transverse plates. The calcareous tubes are impregnated with iron salts that give the colony an attractive color.

Unfortunately, the calcium skeletons forming the bulk of coral reefs are highly vulnerable to increased levels of atmospheric CO_2 produced by our growing use of fossil fuels. Increasing amounts of CO_2 dissolve in ocean water and form carbonic acid, which acidifies the water and weakens the calcareous coral structure. In addition, rising temperatures associated with global warming causes coral polyps to expel their symbiotic algae and die. This is known as "coral bleaching" because it leaves expanses of white, lifeless coral skeletons behind.

Question 10

a. Locate the radial ridges within each depression on a piece of coral. What structures within the polyp did they support?

b. What is the advantage of a partitioned gastrovascular cavity?

c. Consider Learning Objective 1 listed at the beginning of this exercise. How does fragmentation contribute to the evolutionary success of anthozoans in their environment?

Question 11

a. In your notes draw and describe the life cycle of a cnidarian. What is the difference between a polymorphic life cycle and the typical life cycle of other animals?

b. How does the feeding method of a sponge compare with that of a coral?

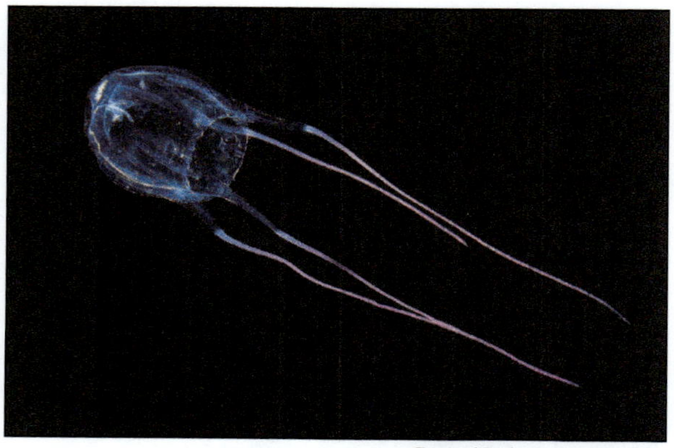

Bernard Radvaner/Getty Images

Figure 36.21 Class Cubozoa. *Chironex fleckeri,* a box jelly (1×).

Class Cubozoa

As their name implies, cnidarians of class Cubozoa are box-shaped medusae; the polyp stage is inconspicuous and in many cases not known. Most medusae are only a few centimeters in height, although some are 25 cm tall. A tentacle or group of tentacles is found at each corner of the box (fig. 36.21).

Box jellies are strong swimmers and voracious predators of fish in tropical and subtropical waters. The stings of some species can be fatal to humans. Examine a preserved specimen of a box jelly. How does its shape compare to that of scyphozoans?

Questions for Further Study and Inquiry

1. Which group within kingdom Protista probably gave rise to sponges? On what evidence do you base your answer?

2. Why are spicules used as a primary characteristic to classify sponges?

3. Sponges and cnidarians have no lungs or gills. How do they exchange gases with the environment? Are humans better off having lungs? If so, how?

4. Why are sponges considered to be an evolutionary dead end?

5. Explain how cnidocytes with their nematocysts function in food capture and defense.

6. Discuss how polymorphism in the cnidarians might have influenced adaptive radiation of the group.

7. How does digestion in cnidarians differ from digestion in sponges?

8. Prepare a simple table of all of the taxonomic groups, their common names, their distinguishing characteristics, and all representative genera covered in this exercise. Keep this table with your study notes.

WRITING TO LEARN BIOLOGY

What are the advantages and disadvantages to a solitary versus colonial existence?

Survey of the Animal Kingdom

Phyla Platyhelminthes and Mollusca

EXERCISE 37

Learning Objectives

By the end of this exercise you should be able to:

1. Describe how the prominent characteristics of platyhelminths and mollusks promote their survival and reproduction.
2. Describe the general morphology of flatworms in phylum Platyhelminthes and mollusks in phylum Mollusca.
3. List characteristics that phyla Platyhelminthes and Mollusca have in common with phyla Porifera and Cnidaria.
4. List characteristics of flatworms and mollusks more advanced than those of more primitive phyla.
5. List examples of each major class of flatworms and mollusks.
6. Understand the differences between acoelomates and coelomates and know which phyla are associated with each.
7. Discuss characteristics that have most likely contributed to flatworm and molluscan success over millions of years and relate the discussion to Learning Objective 1.

Please visit connect.mheducation.com to review online resources tailored to this lab.

Flatworms of **phylum Platyhelminthes** and **mollusks** of **phylum Mollusca** are remarkably successful if we measure evolutionary success by high diversity and persistence in the environment for 400 million years. Both phyla occur in marine, freshwater, terrestrial, and parasitic environments. Their morphology is considerably more complex than that of sponges and sea jellies (table 37.1). For example, flatworms and mollusks are **triploblastic** meaning that they have three embryological germ layers: **ectoderm**, **mesoderm**, and **endoderm**. They also have **organs** made of interdependent tissues and flatworms are the most ancient animals having **bilateral symmetry** with distinct anterior and posterior ends (see fig. 36.7).

Flatworms are **acoelomate**, meaning without a coelomic cavity. Their mesoderm is a solid mass of tissue with no internal cavity (figs. 37.1 and 37.2). In contrast, mollusks and the remaining phyla we will discuss are **coelomate**. They have an internal cavity lined with mesoderm to accommodate complex internal organs. This spacious internal cavity likely enhances the ability of internal organs to move, contract, and otherwise function independently. Note that the coelom of mollusks is greatly reduced in modern species.

Coelomates are further divided into **protostomes** and **deuterostomes**. Protostomes have well-developed nervous, circulatory, excretory, reproductive, and digestive systems. Protostomes include phyla Platyhelminthes, Mollusca, and Annelida, all grouped as lophotrochozoa because they commonly have a trochophore larval stage (fig. 37.3).

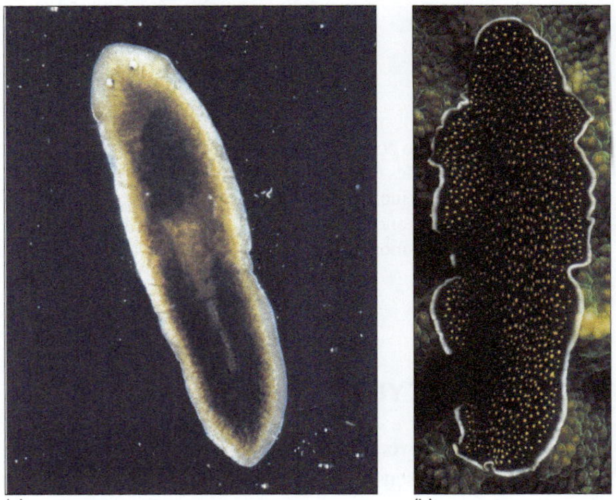
(a) (b)

(a) blickwinkel/Hecker/Alamy Stock Photo (b) R. Aaron Raymond/Radius Images/Getty Images

Figure 37.1 Flatworms (phylum Platyhelminthes). (*a*) A common freshwater flatworm (*Dugesia dorotocephala*) (20×). (*b*) A free-living marine flatworm (*Thysanozoon nigropapillosum*) (5×).

Protostomes also include phyla Nematoda and Arthropoda grouped as ecdysozoa because they commonly molt (shed) their outer cuticle or exoskeleton (see Exercises 38 and 39). Deuterostome phyla include Echinodermata and Chordata (see Exercise 40). A detailed comparison of protostomes and deuterostomes is presented in Exercise 40.

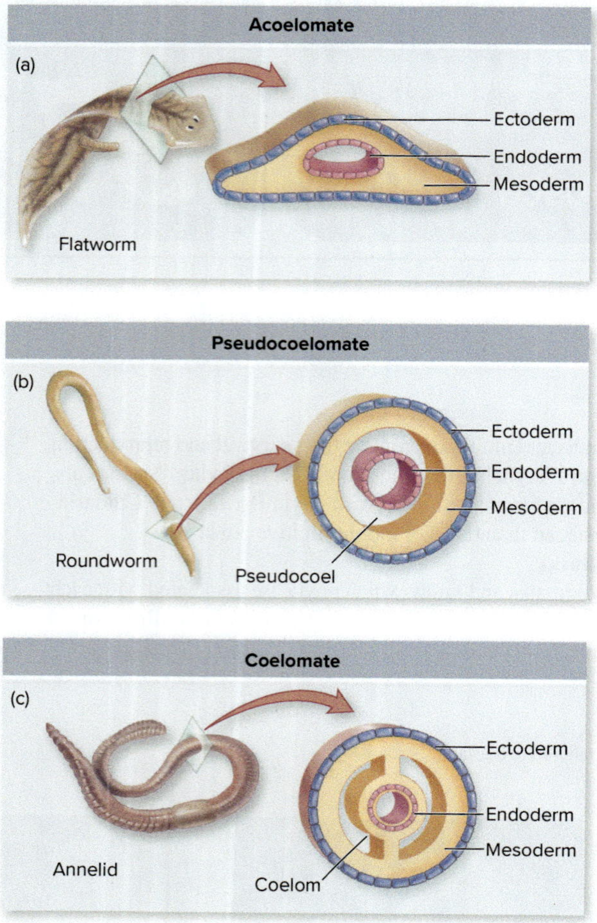

Figure 37.2 Three body plans for bilaterally symmetrical animals. (*a*) Acoelomates (including flatworms) have no body cavity. (*b*) Pseudocoelomates (including nematodes) develop a body cavity between the mesoderm and endoderm. (*c*) Coelomates (including mollusks, annelids and more advanced phyla) have a body cavity bounded by mesoderm.

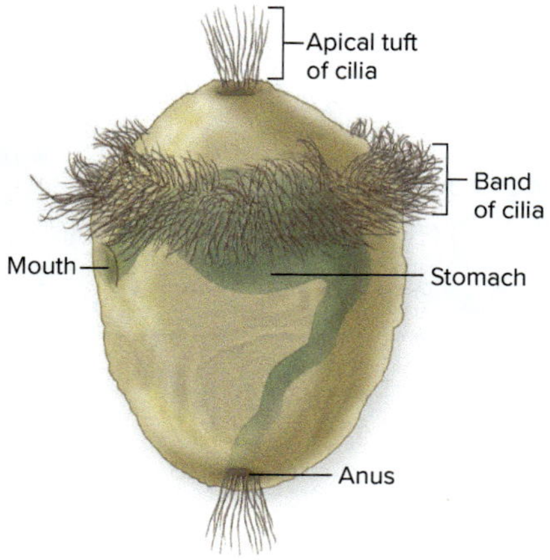

McGraw Hill

Figure 37.3 Trochophore larval form. Protostome phyla include Platyhelminthes, Mollusca, and Annelida. They share a close common ancestry as indicated by many of their species having a similar larval stage called a trochophore. These phyla are grouped as lophotrochozoans.

PHYLUM PLATYHELMINTHES

Flatworms are dorsoventrally compressed, and free-living species have primitive sense organs (fig. 37.1). They also have a gastrovascular cavity with one opening that is both mouth and anus. Their nervous system is more advanced than that of cnidarians (see Exercise 36) and consists of a ladderlike arrangement of nerve cords extending the length of the body.

Subphylum Turbellaria

Turbellarians (3000 species) are free-living flatworms inhabiting freshwater, saltwater, and moist terrestrial environments. Turbellarians scavenge and prey on small animals and are **hermaphroditic,** meaning that individuals have both male and female sex organs.

Dugesia

Dugesia, often called planaria, is a common freshwater turbellarian with typical characteristics of flatworms (fig. 37.4). The head has lateral lobes and sensory organs called **eyespots.** *Dugesia* feed by sucking food through its mouth and into a tubular **pharynx** leading to the gastrovascular cavity. The muscular pharynx is usually retracted in the body but can be everted through an opening in the midventral epidermis (middle of the lower surface). Most digestion in the gastrovascular cavity is extracellular, but phagocytic cells line the cavity and complete digestion of small particles intracellularly.

The mesodermal tissue of planaria is the loose mass of cells between the ectoderm on the surface and the endoderm lining the gastrovascular cavity and pharynx (fig. 37.4). This body plan allows for remarkable powers of regeneration. A chopped planarian can regenerate a complete new head or regrow the entire right side of its body if cut in half lengthwise. Wouldn't it be remarkable if humans could rekindle the ability to regenerate lost appendages, an ability lost in more advanced phlya?

SAFETY FIRST Before coming to lab, you were asked to read this exercise so you would know what to do and be aware of safety issues. In the space below, briefly list the safety issues associated with today's procedures. If you have questions about these issues, contact your laboratory assistant before starting work.

Procedure 37.1 **Observe planaria**

1. Obtain a living *Dugesia* and examine its morphology with a dissecting microscope.
2. Place the animal in the center of a petri dish and follow its movements for a few minutes.

Table 37.1

Phyla Platyhelminthes and Mollusca

Phylum	Typical Examples		Key Characteristics	Approximate Number of Named Species
Platyhelminthes (flatworms)	*Planaria*, tapeworms, liver flukes	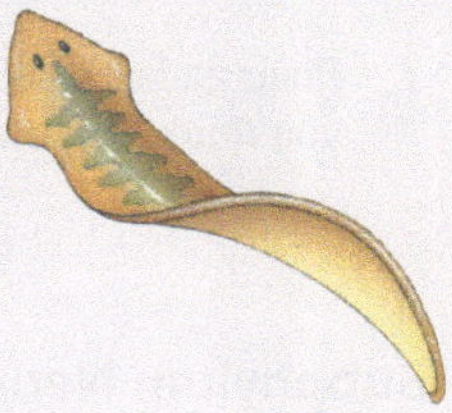	Solid, unsegmented, bilaterally symmetrical worms; no body cavity; digestive cavity, if present, has only one opening	20,000
Mollusca (mollusks)	Snails, oysters, octopuses, nudibranchs		Soft-bodied coelomates whose bodies are divided into three parts: head-foot, visceral mass, and mantle; many have shells; almost all possess a unique rasping tongue, called a radula; 35,000 species are terrestrial.	110,000

3. Gently touch the animal with a probe and watch how it responds.
4. If a spotlight is available, determine how *Dugesia* may respond to strong light.
5. Offer *Dugesia* a small piece of liver, boiled egg, or cat food; watch it eat.
6. After a few minutes, gently roll the animal from the food and find its protruded pharynx.
7. Examine a whole mount of a stained planaria and a prepared slide of a cross section through a planaria. Specimens vary. Be sure to examine more than one slide, and do not rely on photographs.
8. Locate the structures shown in figure 37.4.
9. Examine cross sections taken through the region of the pharynx and away from the pharynx. Draw and label these cross sections, and have your instructor check them for accuracy.

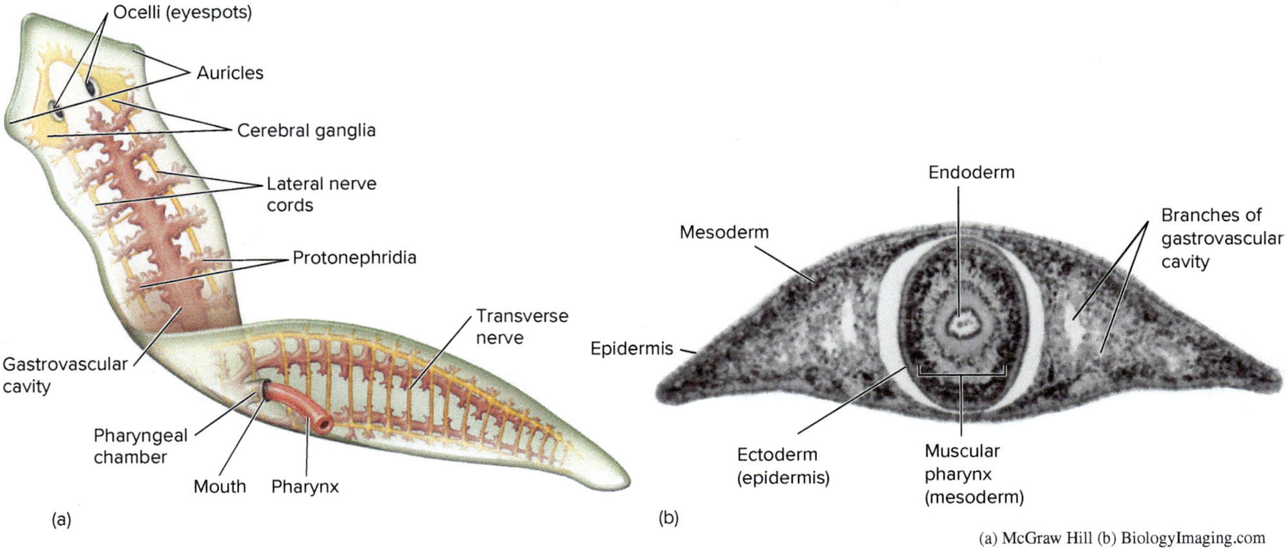

(a) McGraw Hill (b) BiologyImaging.com

Figure 37.4 Anatomy of a free-living flatworm, *Dugesia*. (*a*) External structure. (*b*) Cross section of a planarian taken through the pharynx region (40×).

Question 1

a. What features of *Dugesia* distinguish its head from its tail?

b. What is the difference between the eyes of most animals you are familiar with and the eyespots of *Dugesia?*

c. How does the head of *Dugesia* move differently from the tail?

d. Does *Dugesia* move randomly or in an apparent direction?

e. How is *Dugesia* adapted for directional movement?

f. How does a flatworm respond when touched with a probe?

g. Do the planaria move toward or away from light?

h. Where is the feeding tube located? Why is this unusual for bilaterally symmetrical organisms?

i. Is the gastrovascular cavity of *Dugesia* a simple sac? How is it divided and what advantage do these divisions offer?

j. Consider Learning Objective 1 listed at the beginning of this exercise. How could being monoecious contribute to evolutionary success of flatworms in their environment?

k. Planarians have a head. In biological terms, what constitutes a "head"? How does it relate to Learning Objectives 1, 4, and 7?

l. Planaria lack specialized gas-exchange organs. How do you think planaria accomplish this task?

Subphylum Neodermata

All parasitic flatworms are in subphylum Neodermata. This name refers to "new skin" and refers to the animal's outer surface, termed the neodermis. Neodermata contains two subgroups: Trematoda, flukes, and Cercomeromorpha, tapeworms and their relatives.

Subgroup Trematoda

Trematodes, commonly called flukes, are parasites, and their oval bodies are usually a few millimeters long. Flukes infect vertebrates and include both **endoparasites** (parasites inside their host) (fig. 37.5) and **ectoparasites** (parasites on the surface of their host). Trematodes lack an epidermis and are covered by an acellular but metabolically active **epicuticle.** This epicuticle is made of protein and lipids secreted by mesodermal cells and resists digestive enzymes. The epicuticle helps in respiration and absorbing nutrients. The ventral surface of a fluke usually has two adhesive organs (suckers). The oral sucker surrounds the mouth.

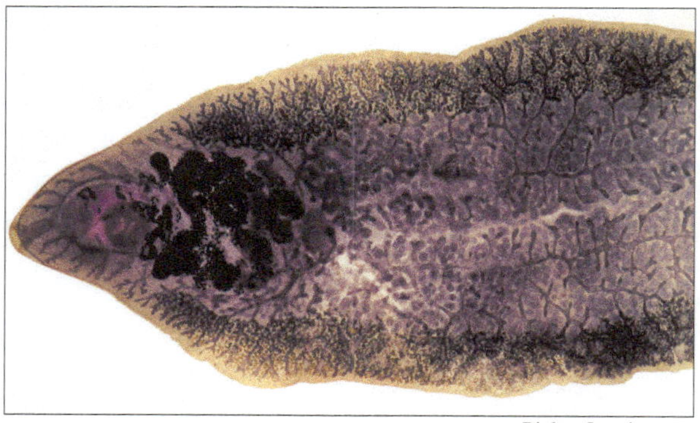

BiologyImaging.com

Figure 37.5 *Fasciola* sp. is a parasitic fluke that infects the small intestine of vertebrates (10×). Humans acquire this parasite by eating water chestnuts and other aquatic plants with larvae attached. The larvae develop in intermediate hosts such as snails and fish. A major adaptation of flukes and other parasites is the prolific production of eggs—as many as 3000 eggs per day for a lifetime of many years. The dark, coiled uterus shown here contains hundreds of eggs, but most of these eggs will never be engulfed by an intermediate host and hatch. To complete a complicated life cycle through multiple hosts, a fluke must produce immense numbers of eggs to ensure that a few are successful. To appreciate the structure of the entire organism, see figure 37.8.

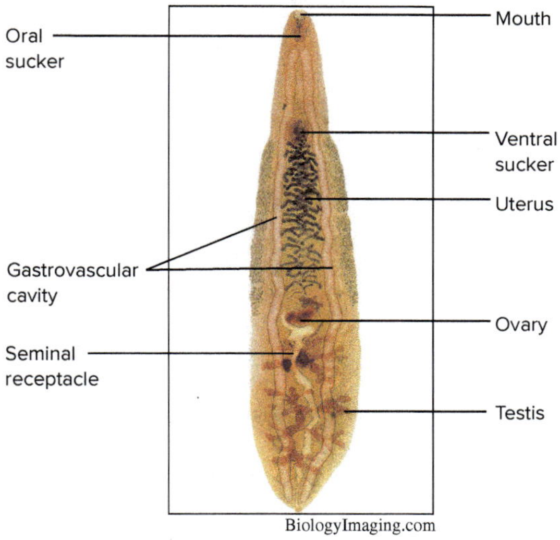

Figure 37.6 Internal structure of *Opisthorchis,* the oriental liver fluke (30×).

Opisthorchis

Use a dissecting microscope to examine a prepared slide of *Opisthorchis* and locate the structures shown in figure 37.6. *Opisthorchis* (*Clonorchis*) *sinensis,* the oriental liver fluke, often parasitizes humans in Japan and China. This hermaphroditic, adult fluke attaches to the bile duct and releases eggs that move through the digestive system of the host and exit with the feces. Larvae of flukes typically develop in snails and fish. Humans are infected when they eat raw or poorly cooked fish (fig. 37.7).

Question 2

a. How does the shape of the digestive sac of *Opisthorchis* compare with that of *Dugesia?*

b. How does the position of the mouth of *Dugesia* and flukes compare?

Fasciola

Examine figure 37.5 and read its caption carefully. Use a dissecting microscope to examine a prepared slide or plastic mount of *Fasciola* and locate the features shown in figure 37.8. *Fasciola hepatica,* the sheep liver fluke, infects sheep, other vertebrates, and (rarely) humans. It is much larger than *Opisthorchis* (fig. 37.6) but similar in structure. *Fasciola* sucks food (blood, mucus, and cells) through a muscular pharynx located at the mouth.

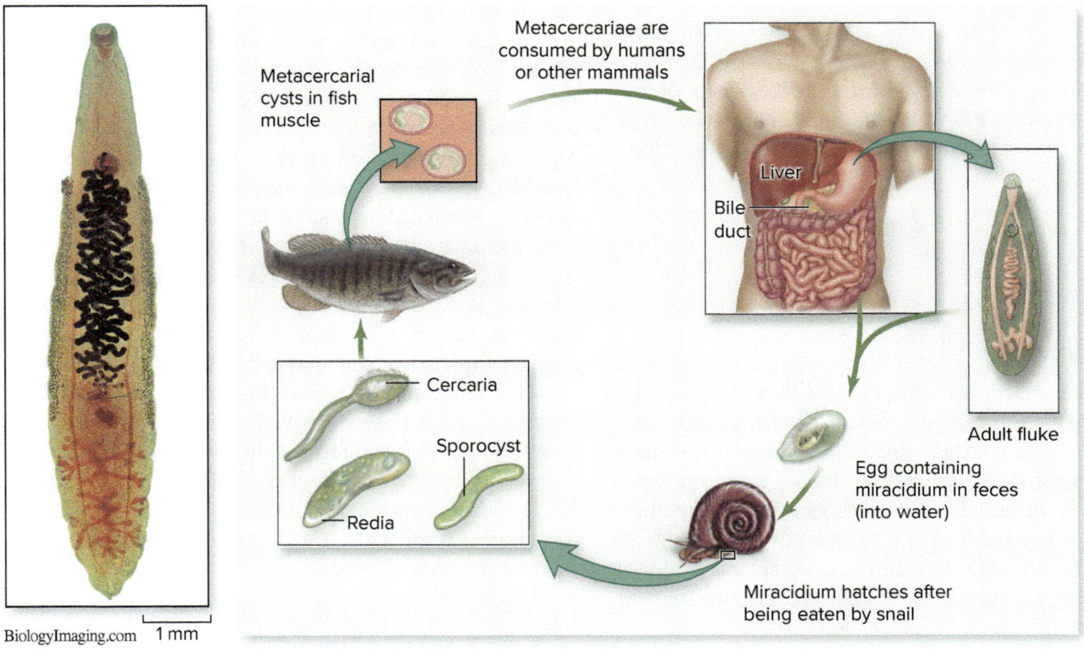

Figure 37.7 Life cycle of the oriental liver fluke, *Opisthorchis sinensis.*

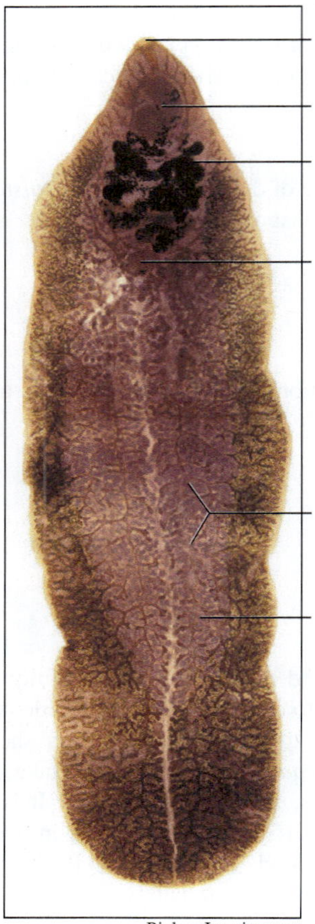

Figure 37.8 Internal structure of *Fasciola,* the sheep liver fluke (2×).

Question 3
Consider Learning Objective 1 listed at the beginning of this exercise. How could the production of large numbers of eggs contribute to the evolutionary success of flatworms in their environment?

Schistosoma

Examine a demonstration slide of *Schistosoma,* and draw its body shape. *Schistosoma,* a blood fluke, inhabits the intestinal veins and other organs of many vertebrates (including humans) and causes the disease schistosomiasis (fig. 37.9). *Schistosoma* is socioeconomically important because it infects more than 200 million people in countries having tropical and temperate climates. Symptoms of infection include an enlarged liver, spleen, and bladder, as well as nutritional deficiency. Some of these symptoms, such as bleeding ulcers in the intestine and other organs, are inflamed by large deposits of eggs (up to 3500 per day from one female).

Snails are **intermediate hosts** for *Schistosoma.* An intermediate host is an organism harboring immature stages of a parasite, whereas a **definitive host** contains sexually mature, egg-laying stages of the life cycle. Irrigation ditches with snail populations enhance the spread of schistosomiasis in underdeveloped agricultural areas. Immature larvae of blood flukes released from the snails burrow through skin and infect people wading in these ditches.

Unlike most trematodes, *Schistosoma* is dioecious (fig. 37.9*a*). The male has a ventral groove along the length of his body into which the slender female cradles for copulation. Often the female remains in this groove for the remainder of her lifetime and extends slightly to lay eggs. Your prepared slide may include schistosomes in this position for copulation.

Question 4
How does the shape of *Schistosoma* differ from that of other flukes you have studied?

Subgroup Cercomeromorpha

Tapeworms are the most specialized platyhelminths. They are endoparasites of the gut of vertebrates and are covered by a cuticle similar to that of trematodes. However, tapeworms lack a mouth or digestive tract and have a unique body plan (fig. 37.10). Their cuticle efficiently absorbs nutrients from their host. The anterior end, or **scolex,** adheres to the host's intestinal wall with hooks or suckers. Behind the scolex is the **neck** followed by a series of segments called **proglottids.** The scolex and neck are small, but the chain of proglottids may be 10–15 m long.

A tapeworm grows as the scolex and neck continually produce a chain of proglottids as self-contained packets of male and female reproductive organs (fig. 37.11). Self-fertilization occasionally occurs in a proglottid, but cross-fertilization by copulating proglottids of adjacent worms is more common. **Gravid** (egg-carrying) proglottids eventually break from the end of the worm and pass out of the host with its feces.

Examine a prepared slide or whole specimen of a pork tapeworm, *Taenia solium.* Then examine a prepared slide of a scolex. Draw the basic shape of the scolex and note any hooks or suckers. Also examine and sketch from prepared slides young, mature, and gravid proglottids.

An adult *Taenia solium* may be up to 10 m long. Humans infect themselves with *Taenia* by eating uncooked meat from pigs, often intermediate hosts (fig. 37.12).

Compare a mature proglottid with that shown in figure 37.11. Although each proglottid has a complete reproductive system, the excretory ducts and longitudinal nerves are continuous between proglottids.

Examine a demonstration specimen of *Dibothriocephalus latus* if it is available. *Dibothriocephalus latus* is the largest tapeworm to infect humans and can be up to 20 m long. Its intermediate hosts are small crustaceans and fish.

SPL/Science Source

Figure 37.9 Life cycle of a blood fluke, *Schistosoma.* (*a*) Micrograph of *Schistosoma* (30×). (*b*) *Schistosomiasis,* an infection of humans caused by the blood fluke *Schistosoma,* is an extremely prevalent disease in Egypt—especially since the building of the Aswan High Dam. Standing water in irrigation ditches, combined with unsanitary practices, has created the conditions for widespread infection.

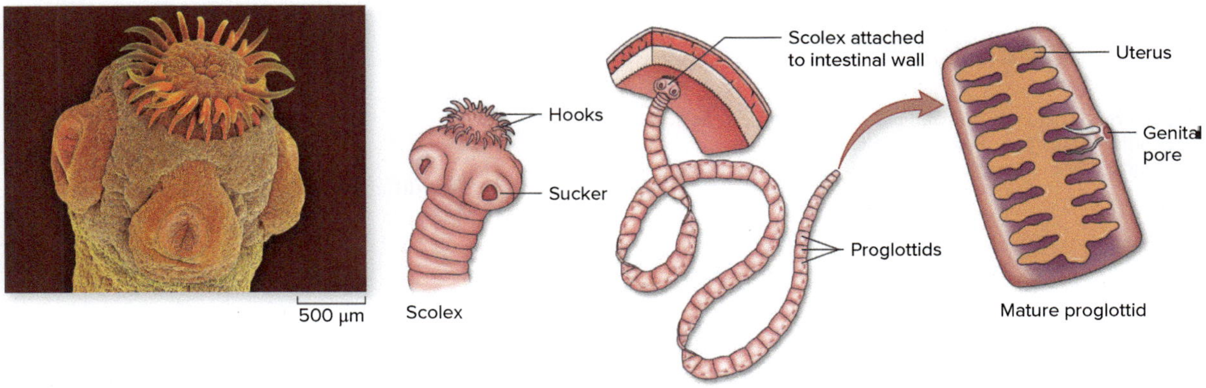

Dennis Kunkel Microscopy, Inc./Phototake

Figure 37.10 Body plan of a tapeworm (subgroup Cercomeromorpha). A mature proglottid has mature reproductive organs; a gravid proglottid contains many eggs.

INQUIRY-BASED LEARNING

Can the primitive nervous system of planarians detect the chemistry of the water surrounding them?

Observations: Detecting food is crucial to success of planaria. They must "taste" the water for compounds released from nearby food. Some classes of macromolecules (fats, carbohydrates, proteins, etc.) dissolve and disperse more readily than others.

Question: What types of organic molecules does *Dugesia* respond to most readily?

a. Establish a working lab group and obtain Inquiry-Based Learning Worksheet 37 from your instructor.

b. Patiently observe *Dugesia* movement and its responsiveness. Discuss with your group a well-defined question relevant to planarian response to dissolved molecules from their prey. Record the question on Worksheet 37.

c. Translate your question into a testable hypothesis and record it.

d. Devise a procedure to determine if *Dugesia* respond to some classes of dissolved nutrients more than other classes.

e. Outline on Worksheet 37 your experimental design and supplies needed to test your hypothesis. Ask your instructor to review your proposed investigation.

f. Conduct your procedures.

g. Record your data, answer your question, and make relevant comments.

h. Discuss with your instructor any revisions to your questions, hypotheses, or procedures. Repeat your work as needed.

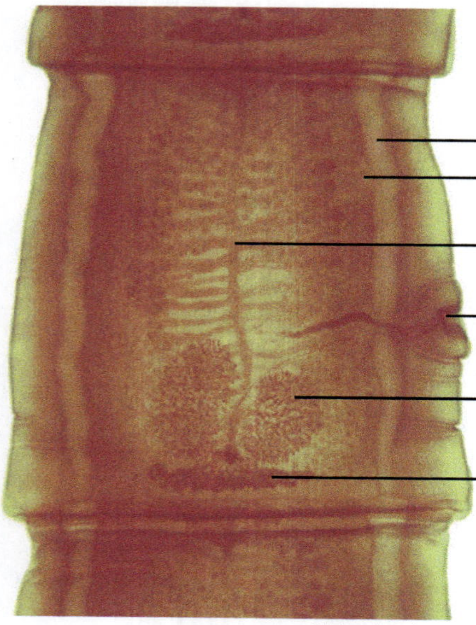

Figure 37.11 Mature proglottid of *Taenia pisiformis,* the dog tapeworm. Portions of two other proglottids are also visible at the top and bottom of the image (30×).

Question 5

a. How does a scolex compare in size to a proglottid near the scolex? Near the posterior end?

b. Tapeworms have no digestive system or mouth. How, then, do they obtain food?

c. Examine figure 37.10 and its caption carefully. Which proglottids, mature or gravid, occur closest to the scolex?

d. What is the difference between a mature and a gravid proglottid?

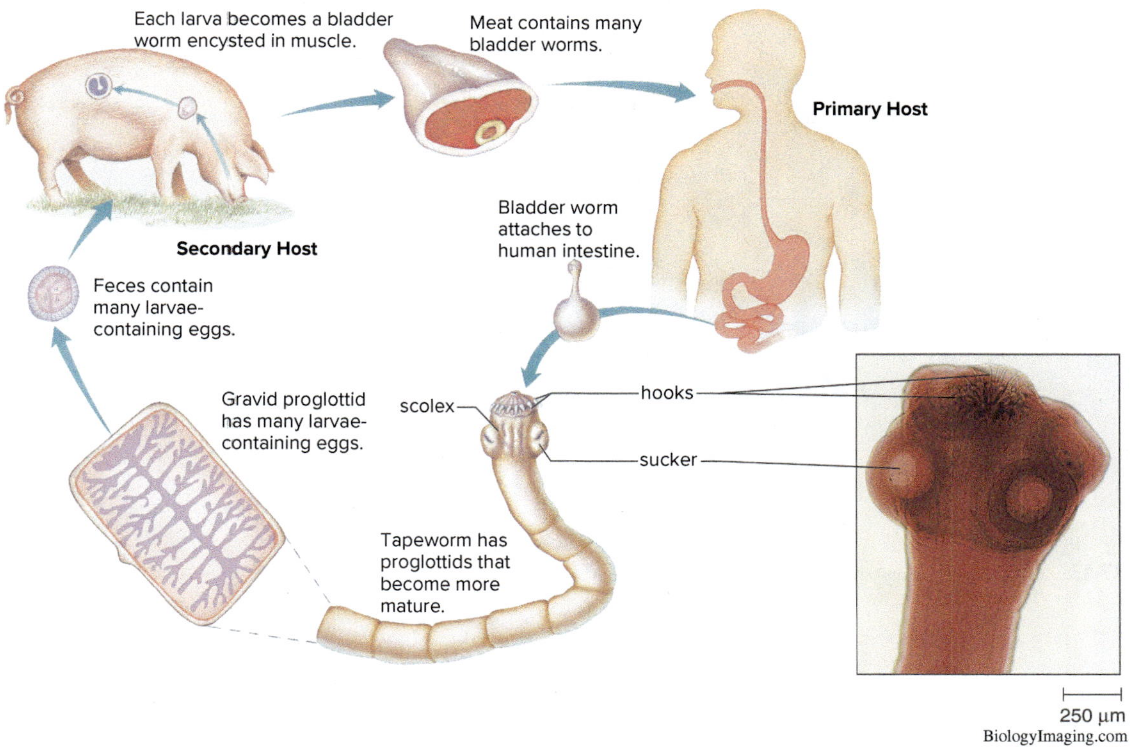

Figure 37.12 Life cycle of a tapeworm, *Taenia.* The life cycle includes a human (primary host) and a pig (secondary host). The adult worm is modified for its parasitic way of life. It consists of a scolex and many proglottids, which become bags of eggs.

e. Consider Learning Objective 1 listed at the beginning of this exercise. Is having a scolex significant to a fundamental process for tapeworms? For which process, and how is it significant?

f. Tapeworms are specialized. Would you expect specialists to have a unique morphology? Why?

PHYLUM MOLLUSCA

Mollusks such as snails, clams, octopuses, and squids are soft-bodied animals with a specialized layer of epidermal cells called a **mantle** that secretes a **shell** (fig. 37.13). They have evolved a remarkable diversity of forms built on a consistent ancestral body plan (fig. 37.13). Molluskan forms are so diverse that a clam and a squid belong to the same phylum!

Mollusks produce many kinds of external shells (fig. 37.14*a,b*)—some mollusks have only a remnant internal shell (fig. 37.14*c*), whereas others have no shell at all. This phylum's diversity of 110,000 species is surpassed only by arthropods and probably nematodes. Although mollusks are coelomate, their coelom is often reduced to a small chamber surrounding the heart. The circulatory system is open (except in cephalopods), meaning that blood pools in sinuses and bathes the organs directly. **Open circulatory systems** have a few large vessels and a heart but no smaller vessels and capillaries. Mollusks have a **complete digestive tract** with a mouth and anus. This allows one one-way movement of food and specialized digestive processes along the tract.

The basic body-plan of a mollusk shows little segmentation but consistently includes (1) a **visceral mass** of organ systems (digestion, excretion, and reproduction) and

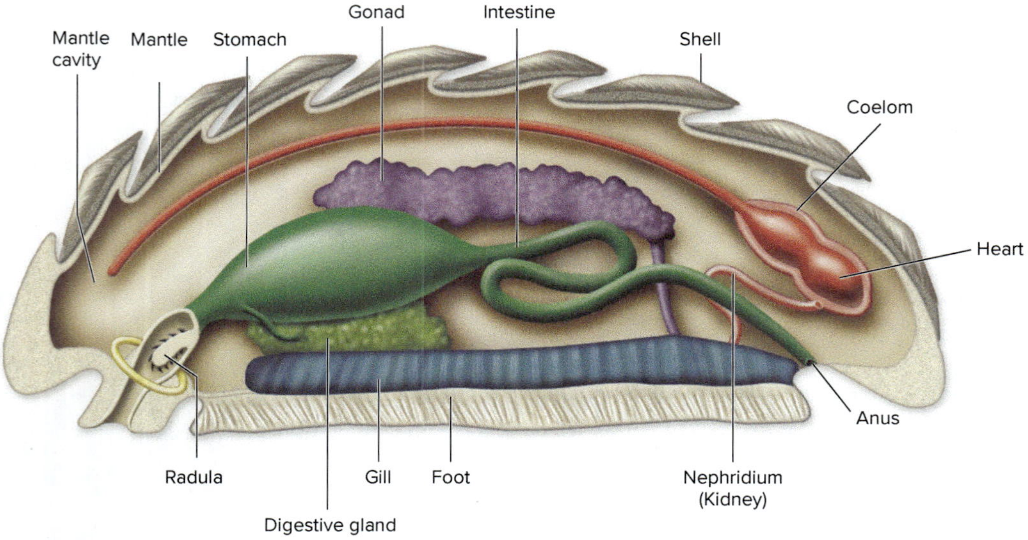

Figure 37.13 Generalized molluskan body plan.

(a) BiologyImaging.com (b) Stephen Frink/Photodisc/Getty Images (c) Rich Carey/Shutterstock

Figure 37.14 The body plans of the three major classes of mollusks: (*a*) Gastropoda, (*b*) Bivalvia, and (*c*) Cephalopoda. A mollusk's shell has an outer layer overlying layers of densely packed crystals of calcium carbonate. A visceral mass includes organs of digestion, excretion, and reproduction and extends as a muscular foot adapted for locomotion, attachment, or food capture (in squids and octopuses). The radula is a scraping and feeding organ characteristic of mollusks (except bivalves, which obtain their food by filter feeding). Folds of tissue called the mantle arise from the dorsal body wall, line the shell, and enclose a cavity adjacent to the visceral mass. Within the mantle cavity are gills or lungs.

sensory structures; (2) a ventral, muscular, and often highly modified **foot** used for locomotion; (3) a calcium-based shell, though occasionally absent; and (4) a mantle that secretes the shell and may aid in respiration and locomotion in some species (fig. 37.14). Some mollusks also have a differentiated **head.**

Class Polyplacophora

Obtain a preserved chiton and examine its external features. Polyplacophorans (*poly* = many, *placo* = plate, *phora* = move), commonly called chitons, are exclusively marine and have a primitive molluskan structure (fig. 37.15). The dorsal shell is divided into eight plates embedded in the mantle. The ventral foot is a broad oval muscle used to propel chitons slowly over the surface of rocks. The **radula,** a horny-toothed organ in the mouth, scrapes food (algae) from rocks.

Class Gastropoda

Most gastropods (*gastro* = stomach, *poda* = foot) (snails) have a single shell that is often coiled and elaborate. Some gastropods, such as marine nudibranchs and the common garden slug, do not produce a shell (fig. 37.16). Most gastropod species are marine, but freshwater snails and land snails are common. Gastropods commonly feed with a rasping band of teeth called a radula (fig. 37.17)

Examine preserved gastropods and aquarium snails and compare the relative spiraling of the shells of different species. This spiral growth is common in mollusks and results from unequal growth of the two halves of the larva and mantle.

Class Bivalvia

Clams, oysters, scallops, and mussels are bivalves (*bi* = two, *valve* = door or shell) having a dorsally hinged shell in two parts (figs. 37.14*b*, 37.18). The mantles of the left and right valves join posteriorly to form a ventral **incurrent siphon** and a dorsal **excurrent siphon** that direct water through the clam (fig. 37.19*a*). When a clam burrows in sediment, the siphons extend to the water. The space between the

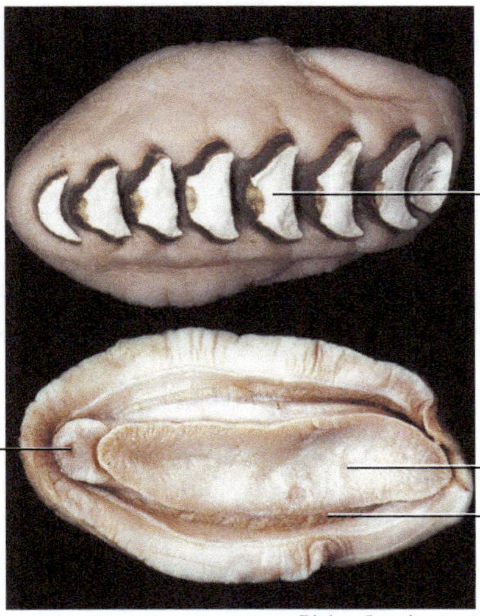

BiologyImaging.com

Figure 37.15 Chiton, *Chiton* (class Polyplacophora). Dorsal view (*above*) and ventral view (*below*) (1×). The shell of chitons has eight plates atop a mantle; visceral mass; and large, oval foot.

CK Ma/Shutterstock.com

Figure 37.16 Colorful nudibranchs, such as *Flaballina iodinae*, are gastropod mollusks without a shell. They have a rather mysterious defensive strategy—nudibranchs use the weapons of their prey. They eat sea jellies, hydrozoans, and corals, all of which have stinging structures called nematocysts. When a nudibranch attacks and eats sea jellies, it can swallow and digest these nematocysts without discharging them. The stingers pass through the digestive tract and are stored in feathery projections on the dorsal surface of the nudibranch A predator taking a mouthful of nudibranch gets a nasty taste.

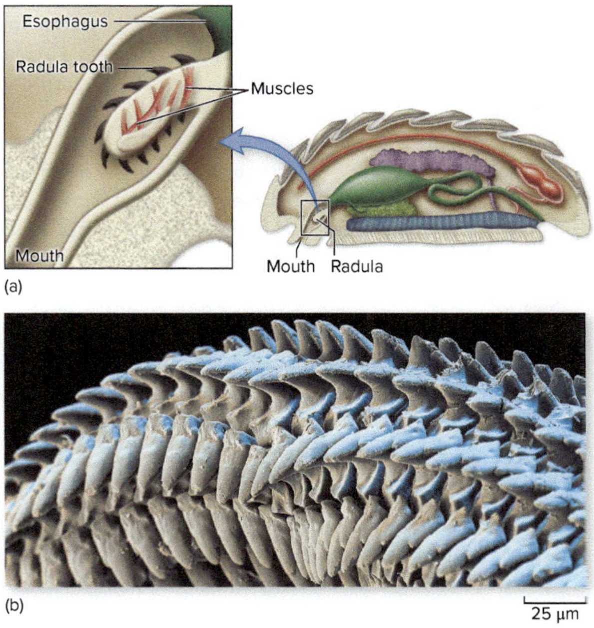

(a) McGraw Hill (b) Eye of Science/Science Source

Figure 37.17 Radula of mollusks. (a) A radula is a unique rasping tongue of mollusks with rows of chitinous teeth covering its surface. (b) The membranous radula is stretched over a rigid cartilaginous rod pressed against a surface to be scraped. Snails use their radula to scrape algae from rocks or tear pieces from plant leaves.

BiologyImaging.com

Figure 37.18 A giant clam (class Bivalvia, *Tridacna* sp.) of the Indopacific may weigh up to 900 lb (400 kg). Its mantle is often intensely colored by blue, green, and brown pigments, and it harbors rich and colorful colonies of symbiotic algae in the blood sinuses of the tissue. Photosynthesis by these algae provides extra food for the clam. Some clams have lenslike structures that focus light deep within their tissues and promote photosynthesis by the algae.

mantle and visceral mass is the **mantle cavity.** Water flows into the mantle cavity through the incurrent siphon, over the gills, and then dorsally into a space between the visceral mass where gills attach to the mantle. Water exits this chamber posteriorly through the excurrent siphon. As water flows over gills, suspended food particles are filtered by cilia and swept along grooves on the gills to the ventral edge. Cilia at the ventral edge move food to the labial palps that surround and direct food to the **mouth** (fig. 37.19*b*).

SAFETY FIRST Before coming to lab, you were asked to read this exercise so you would know what to do and be aware of safety issues. In the space below, briefly list the safety issues associated with today's procedures. If you have questions about these issues, contact your laboratory assistant before starting work.

Procedure 37.2 Examine bivalve anatomy

1. Obtain a preserved freshwater bivalve, such as *Anodonta*. The valves should be slightly separated by a wooden peg. Rinse the specimen thoroughly before dissection.
2. Locate the two valves and the anterior, posterior, dorsal, and ventral regions of the clam. The **hinge** and the **umbo** are on the dorsal surface (fig. 37.19*b*). The umbo is toward the anterior end.
3. Place the organism vertically and dorsal side (umbo) down on a dissecting pan. To further separate and loosen the valves, slip the flat end of a dull kitchen knife between the valves and rotate the knife. Be careful. Notice that a layer of mantle lies against the inside of each of the two valves.
4. Review figure 37.19 to find the general position of the **adductor muscles.**
5. When the valves are loose enough to be held slightly open (about 1.0 cm), slip a scalpel blade between the mantle tissue and shell of one side. When you have felt resistance of an adductor muscle against the tip of the blade, draw the blade snugly along the shell to cut the posterior and anterior adductor muscles. Try to cut them as close to the inner shell surface as possible. Always cut down toward the dissecting pan.
6. Pull the valves apart so that the organism lies on one side and the other valve lifts away. The mantle that once lined the removed shell now covers the visceral mass. Cut the mantle away to expose the visceral mass.
7. On the separated and empty valve notice the muscle's scars, hinge teeth, mantle line, and the inner, iridescent surface of the shell called **mother of pearl.**
8. Locate the **gills, labial palps,** tough muscular **foot,** and the softer, more dorsal visceral mass. The labial palps channel food to the open mouth.
9. To expose the internal organs of the visceral mass, slice the foot and visceral mass longitudinally (anterior-posterior plane) from the ventral surface toward the dorsal surface. This will produce left and right halves. Make the cut slightly off center to avoid cutting through all of the tough foot tissue.
10. Locate evidence of the coiled intestine, the green **digestive gland,** brown **gonad** tissue, and the dorsal **heart.** This sac is thought to be a remnant of the ancestral coelomic cavity. The intestine runs through it, and the sac contains the heart tissue.
11. When you finish your observations, dispose of the preserved tissue and shells in appropriate containers.

Question 6

a. What is the texture of the mantle of *Anodonta?*

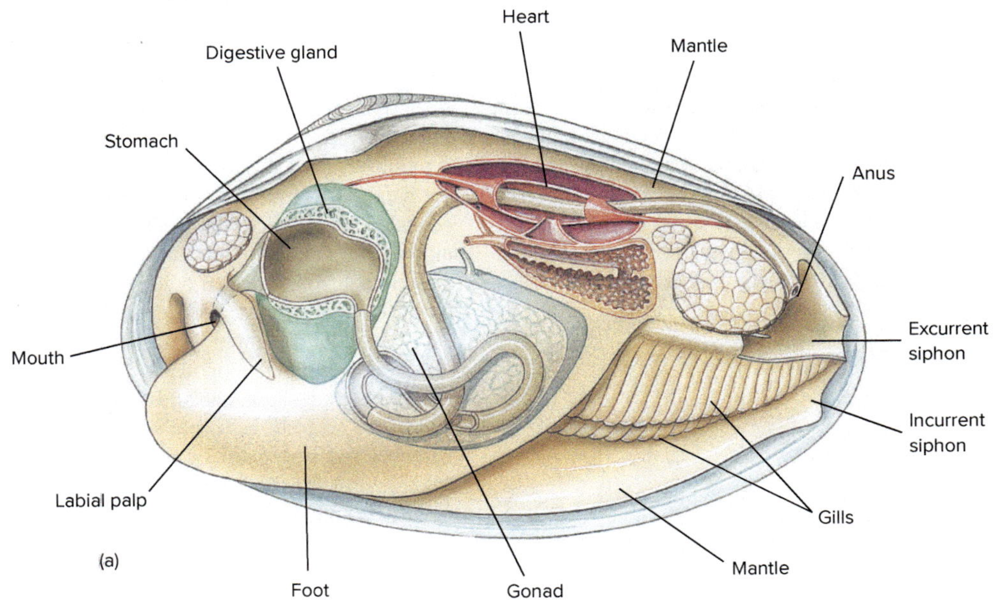

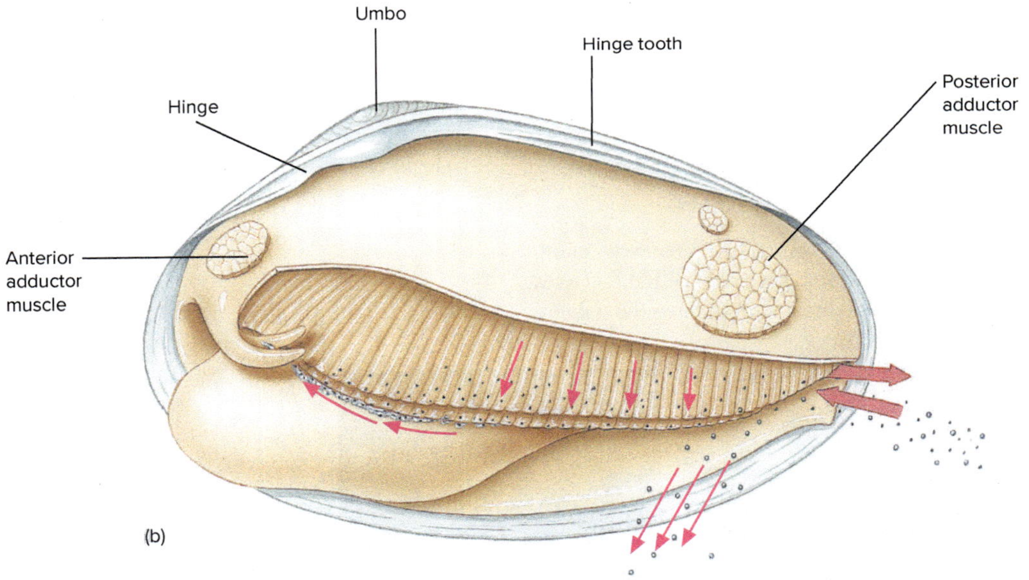

Figure 37.19 (*a*) Clam (*Anodonta*) anatomy. The left valve and mantle are removed. (*b*) During filter feeding, water enters the mantle cavity posteriorly and is drawn forward by ciliary action to the gills and palps. As water enters the tiny openings of the gills, food particles are filtered out and caught in strings of mucus carried by cilia to the palps and directed to the mouth. Arrows show the path of food particles being moved toward the mouth by cilia on the gills. Sand and debris drop into the mantle cavity and are removed by cilia.

b. Are siphons in a preserved specimen of *Anodonta* as obvious as they are in figure 37.19?

c. Consider Learning Objective 1 listed at the beginning of this exercise. In what ways would having a shell contribute to the survival and reproductive success of mollusks in their environment?

Examine a cluster of oyster shells and note how they are stuck together. A mature oyster is attached permanently to a substrate. Most bivalves produce pearls, but the finest pearls are made by *Pinctada* in the warm waters of the Pacific. Natural pearls form in oysters when an irritant such as a grain of sand lodges between the mantle and the shell. The mantle responds by surrounding the irritant with layers of the same crystalline material used for the shell. Cultured pearls are produced by oysters about 3 years after aquaculturists introduce an irritant to their mantle.

Question 7

a. How does the foot of a bivalve differ from that of snails or chitons?

b. How could predators attack an animal closed "tight as a clam"?

c. Is immobility a problem for filter feeders such as oysters? Why or why not?

Class Cephalopoda

Examine a preserved specimen of the common squid, *Loligo*. Note the number and arrangement of the tentacles and the pointed, hardened beak surrounding the mouth. Most cephalopods (*cephalo* = head, *poda* = foot) (e.g., squids, octopuses, nautilus, and cuttlefish) do not resemble their molluskan relatives because cephalopod shells may be absent or reduced to an internal remnant (fig. 37.20*a,b*). For example, a cuttlebone is actually the internal shell of a mollusk called a cuttlefish. (You may have offered your pet parakeet a cuttlebone to sharpen its beak.) The foot of a cephalopod is modified into tentacles. Squids and other cephalopods are predatory, and their external features are appropriately adapted. Unusual among mollusks is the relatively closed circulatory system of cephalopod arteries and veins.

Examine the eyes of *Loligo* closely. Also examine other displayed cephalopods such as the chambered nautilus; its external shell is beautiful and unusual for a cephalopod (fig. 37.20). Review mollusks by completing table 37.2 with brief descriptions of basic molluskan features. The eyes are probably the most surprising feature of a squid. The largest eyes in kingdom Animalia belong to the giant squid; they resemble mammalian eyes and use a lens to form clear images.

Question 8

a. What features of squids and octopuses are adaptations for predation?

b. Do all of the tentacles of a squid have suckers?

(a) oceanbounddb/Getty Images

(b) ©atese/Getty Images

(c) bluehand/Shutterstock

Figure 37.20 Cephalopod diversity. (*a*) An octopus. Octopuses generally move slowly along the bottom of the sea. (*b*) A squid. Squids are active predators, competing effectively with fish for prey. (*c*) Pearly nautilus, *Nautilus pompilius*.

Table 37.2

A Comparison of Major Characteristics of Four Classes of Mollusks

	Polyplacophora	Gastropoda	Bivalvia	Cephalopoda
Shell				
Mantle				
Foot				
Locomotion				
Feeding				
Sensory structures				

c. What are some functions of suckers?

d. Find the mouth at the base of the tentacles. What are the shape and consistency of the jaws? You may need to make an incision to expose the mouth and jaws.

e. Why are sensory organs more prominent in cephalopods than in other classes of mollusks?

f. Consider Learning Objective 1 listed at the beginning of this exercise. In what ways are image-forming eyes significant to fundamental processes for cephalopods?

Questions for Further Study and Inquiry

1. Flatworms are the first organisms we have discussed with an anterior-posterior orientation. How does this affect their movement compared to the movement of more primitive organisms?

2. What are the disadvantages of a flatworm's digestive system having only one opening?

3. The complete digestive tract of mollusks and other phyla allows functional specialization. What specializations are common in the digestive tract of higher organisms such as humans?

4. Prepare a simple table of all of the taxonomic groups covered so far, their common names, their distinguishing characteristics, and all representative genera covered in this exercise. Keep this table with your study notes.

5. What is the advantage of radial symmetry for sessile animals such as hydras and bilateral symmetry for mobile animals such as planaria? What major evolutionary trends accompany bilateral symmetry?

6. Mollusks exhibit a variety of feeding methods. List at least four and discuss adaptations and examples for each type.

7. A snail shell is quite different from the familiar bony skeleton of a mammal. In what ways does a shell function as a skeleton?

8. Some land snails have formed a lunglike structure from a major layer of tissue. What is that layer?

9. Cephalopods are considered by many to be the most distinctive class of mollusks. What makes them seem almost out of place?

Survey of the Animal Kingdom

Phyla Annelida and Nematoda

EXERCISE 38

Learning Objectives

By the end of this exercise you should be able to:

1. Describe how structures specific to annelids and nematodes help them survive in their environment and promote their evolutionary persistence.
2. Describe the general morphology of organisms of phylum Annelida and phylum Nematoda.
3. List the characteristics that phyla Annelida and Nematoda have in common with phyla Platyhelminthes and Mollusca.
4. Discuss those characteristics of annelids and nematodes that are newly derived from the ancestors of these two phyla.
5. List examples of each of the major classes of mollusks and annelids.
6. List examples of roundworms and examples of each major class of annelids.
7. Understand the differences between acoelomates, pseudocoelomates, and coelomates and know which phyla are associated with each.

Please visit connect.mheducation.com to review online resources tailored to this lab.

Annelids of phylum Annelida, commonly called segmented worms, share a close common ancestry with mollusks and platyhelminthes (see Exercise 37). They are coelomate protostomes with a free-living larval stage called a trochophore (see fig. 37.3). This shared type of larval stage indicates a close evolutionary relationship between annelids, mollusks, and platyhelminthes.

In contrast, nematodes of phylum Nematoda have ancestry more closely related to that of arthropods (see Exercise 40). Both of these phyla molt their outer covering. Nematodes molt their outer cuticle, just as arthropods molt their exoskeleton. As a nematode grows larger it forms a new, expandable cuticle underneath its existing one. Periodically the body swells, cracks, and sheds the old cuticle and replaces it with a new, expanded covering allowing for further growth. This periodic process of molting and shedding an outer covering is ecdysis.

Nematodes have a coelomic cavity, but contrary to that of mollusks and annelids it is not completely lined with mesoderm. This is why nematodes are called pseudocoelomates (fig 37.2).

PHYLUM ANNELIDA

Annelids include earthworms, leeches, and many less familiar marine and freshwater species (Table 38.1). The most distinctive characteristic of this phylum is **segmentation.** The body of an annelid is divided into repetitive **segments** arranged on a longitudinal axis and divided by **septa.** Each segment contains parts of the circulatory, digestive, nervous, and excretory systems. The circulatory system of annelids is closed, with blood always retained in vessels. Annelids also have **setae,** small, bristlelike appendages often occurring in pairs on lateral and ventral surfaces. The degree of setal development is distinctive for each of the three classes of annelids.

Class Polychaeta

Most polychaetes, such as the clam worm *Nereis,* are marine worms living in sediment (fig. 38.1). *Nereis* is distinctly segmented and each segment bears a pair of fleshy appendages called **parapodia.** These appendages have a large surface area, are highly vascularized with blood vessels, and help the polychaete move and respire. Protruding from the fleshy parapodia are many setae from which the class derives its name (*poly* = many, *chaeta* = setae). In some species the brittle, tubular setae are filled with poison and used for defense; in others the setae help filter food from the water.

Table 38.1

Phyla Nematoda and Annelida

Phylum	Typical Examples	Key Characteristics	Approximate Number of Named Species
Nematoda (roundworms)	*Ascaris*, pinworms, hookworms, *Filaria* 	Pseudocoelomate, unsegmented, bilaterally symmetrical worms; tubular digestive tract passing from mouth to anus; tiny; without cilia; live in great numbers in soil and aquatic sediments; some are important animal parasites	20,000+
Annelida (segmented worms)	Earthworms, polychaetes, beach tube worms, leeches 	Coelomate, serially segmented, bilaterally symmetrical worms; complete digestive tract; most have bristles called setae on each segment that anchor them during crawling.	20,000+

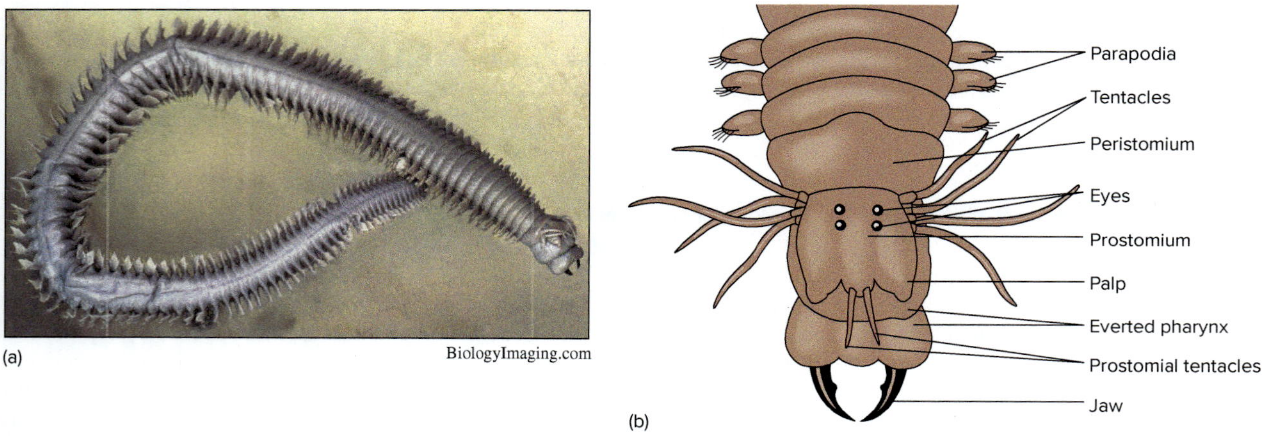

Figure 38.1 *Nereis virens,* a common polychaete (phylum Annelida). (*a*) External structure. (*b*) Anatomy of the anterior end.

Procedure 38.1 Examine polychaetes

1. Examine a preserved *Nereis* and a prepared slide of a parapodium. Sketch the parapodium. Note the tufts of bristles and lobes of tissue.
2. Examine the well-developed head of *Nereis* and locate the features shown in figure 38.1*b*. The mouth and jaws of polychaetes are retractile, so you may have to pull out the jaws with a pair of forceps.
3. Examine other displayed polychaetes such as *Aphrodita*, the sea mouse, and *Chaetopterus,* the parchment worm. These worms have many setae and highly modified parapodia. The parchment worm gets its common name from the paperlike tube it builds.

Question 1

a. The common name of the sea mouse refers to what external feature characteristic of polychaetes?

b. List several functions of parapodia and setae.

c. What is the probable function of the tentacles shown in figure 38.1*b*?

d. Consider Learning Objective 1 listed at the beginning of this exercise. Are parapodia significant to fundamental processes for polychaetes? In what ways?

e. What features of *Nereis* indicate that it is an annelid?

Class Oligochaeta

A common oligochaete is *Lumbricus terrestris,* the familiar earthworm.

Earthworm Locomotion

Movement of oligochaetes in their natural soil environment is not undulatory—instead, movement involves extension, anchoring, and contraction. These motions occur by alternating contractions of circular and longitudinal muscles. These contractions pull against a **hydrostatic** (water pressure) support system rather than against a rigid skeleton (fig. 38.2).

Procedure 38.2 **Examine locomotion in earthworms**

1. Watch a living earthworm move on the hard surface of a pan, then compare and contrast this motion with that of a snake and a vinegar eel.
2. Place the worm on some loose soil, and describe its burrowing motion.

Question 2

a. Does the earthworm move randomly?

b. What do you suppose the worm is seeking or avoiding?

c. What muscles allow the worm to change its length and thickness?

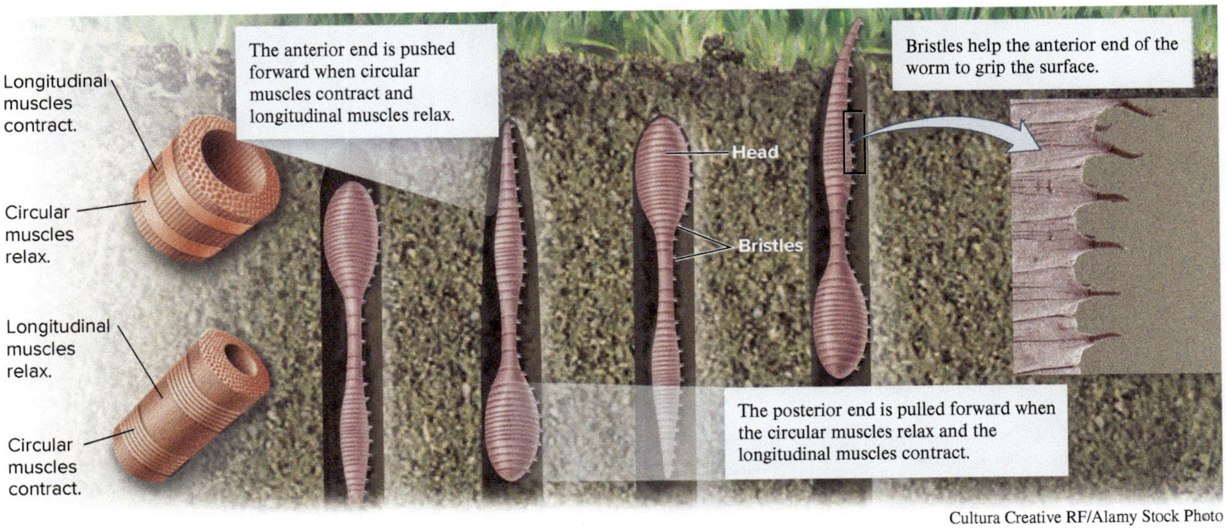

Cultura Creative RF/Alamy Stock Photo

Figure 38.2 Hydrostatic skeleton in a worm. By alternately contracting and relaxing circular and longitudinal muscles, earthworms use hydrostatic pressure to achieve locomotion. Clinging bristles along the body surface help prevent backsliding.

d. How does an earthworm's motion differ from that of a snake and nematode?

e. What features of *Lumbricus* indicate that it is an annelid?

External Anatomy of Earthworms

Oligochaetes (*oligo* = few or small, *chaeta* = setae), include many freshwater species besides the common earthworm. They lack parapodia and have few setae. The anus is on the terminal segment, but the mouth is preceded by a fleshy lobe called the **prostomium** (*pro* = before, *stoma* = mouth). Posterior to the mouth is the first body segment, the **peristomium** (*peri* = around, *stoma* = mouth) (fig. 38.3).

The most obvious external feature of an earthworm is the **clitellum,** a series of swollen segments at the anterior third of the body. Copulating worms attach at their clitella and exchange sperm (fig. 38.4). Earthworms are hermaphroditic, and each of the copulating worms produces egg and sperm cells. Sperm mature in **seminal vesicles** and exit the worm through **male gonopores** on segment 15 (fig. 38.3). Sperm then pass to the adjacent worm and move along its body surface to openings of the **seminal receptacles** (each side of segment 10), where they are stored temporarily. After copulation, the worms separate. A few days later the clitellum secretes a mucous band that slides anteriorly and picks up eggs from the **female gonopores** (segment 14) and stored sperm from seminal receptacles. After the eggs are fertilized in the mucous band, the worm releases it as a **cocoon.**

Procedure 38.3 **Examine the external features of an earthworm**

1. Examine the external features of a preserved *Lumbricus,* and notice its segmentation.
2. Although the body lacks parapodia, if you touch the smooth epidermis and cuticle you can feel short setae on the ventral surface.
3. Use a dissecting microscope to determine the number and arrangement of these setae.
4. Locate the mouth and anus of the earthworm.
5. Use a dissecting microscope to locate the male gonopores and seminal receptacle openings on your specimen (fig. 38.3).
6. Locate the female gonopores on your specimen.

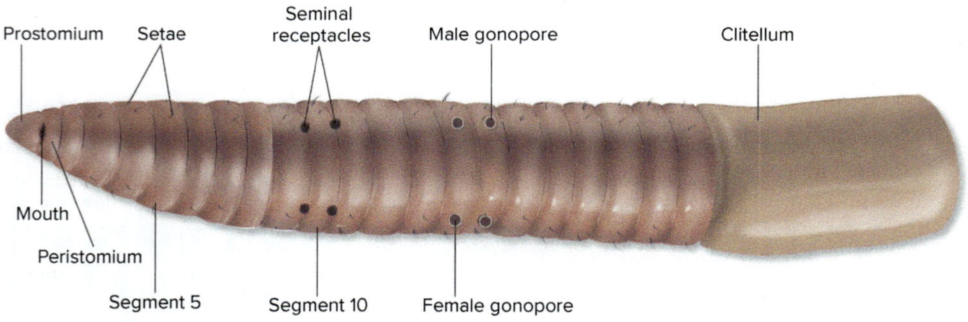

Figure 38.3 External features of an earthworm's ventral surface.

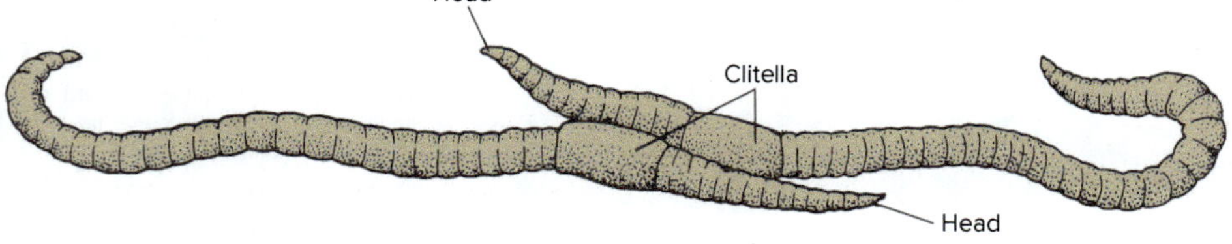

Figure 38.4 Position for copulation and transfer of sperm in earthworms (class Oligochaeta).

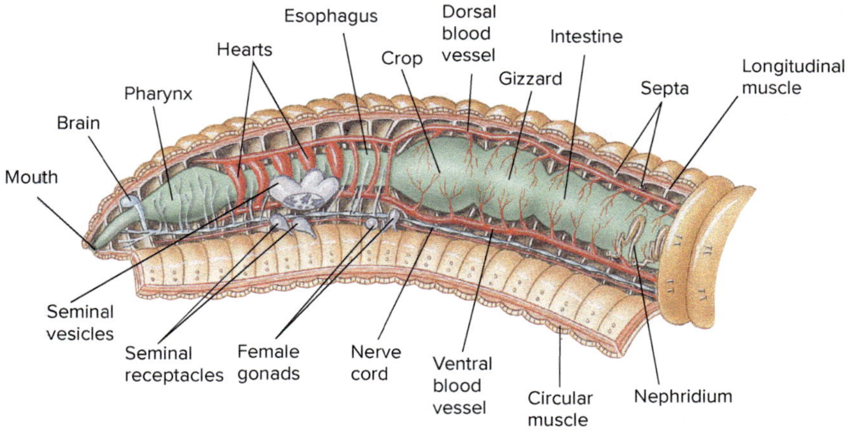

Figure 38.5 Internal anatomy of an earthworm.

Question 3

a. How many setae are on each segment of the earthworm?

b. Are they paired?

Internal Anatomy of Earthworms

Segmentation is best appreciated by looking at internal anatomy (fig. 38.5). Some of the repeating segments are not identical. Segments of the digestive tract are fused and specialized to form a muscular **pharynx** (for suction and ingestion of food), **esophagus** (for transport of food), **crop** (for food storage and some digestion), **gizzard** (for maceration or crushing of food), and **intestine** (for absorption of nutrients). Recall that nematodes also have a linear digestive tract, but it lacks the specialization of that of annelids. Reproductive organs cluster around the anterior segments. Locating small structures such as the female gonads and seminal receptacles will require careful work. A rudimentary **brain** is just anterior and dorsal to the pharynx and is continuous with the **ventral nerve cord.**

Some structures such as lateral branches of the ventral nerve cord and paired **nephridia** occur in each segment (fig. 38.6). Nephridia, which are small, white, convoluted tubes, are found on the inner surface of each segment. Nephridia function like kidneys—they collect and release excretory wastes. Ciliated, funnel-shaped **nephrostomes** on the ends of the nephridia gather waste products from the coelom. The waste is released through external pores called **nephridiopores.**

The **dorsal blood vessel** and **ventral blood vessel** (above and below the intestine) are the main vessels of the closed circulatory system. They are connected by five lateral **"hearts."** However, these hearts promote blood circulation no more than the dorsal and ventral vessels.

The digestive tract has an internal fold of tissue called the **typhlosole** arising from the dorsal wall. This creates a U-shaped intestinal lumen and doubles the surface area for absorption.

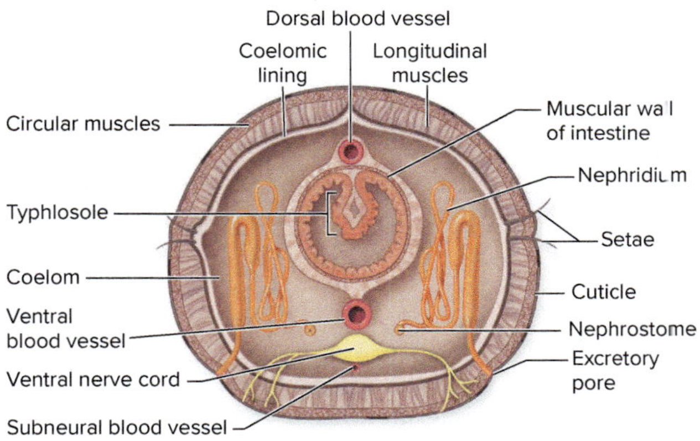

Figure 38.6 Cross section and internal anatomy of an earthworm.

Procedure 38.4 Examine the internal features of an earthworm

1. Pin the terminal segments of a preserved *Lumbricus,* dorsal side up, near the edge of a dissecting pan.
2. Open the body with a shallow, longitudinal incision. Use sharp-pointed scissors, and be careful not to cut too deeply. Hold the skin up with forceps to prevent damage to internal organs.
3. Pin the skin back on both sides of the incision, and expose the organs lying in the coelomic cavity. You may have to cut each septum so the body wall lies flat when pinned.
4. When you have completed the dissection, pour a small amount of water into the pan to cover the exposed worm.
5. Locate the structures shown in figure 38.5. You may need to use a dissecting microscope for close examination.
6. Locate the hearts, dorsal blood vessel, and ventral blood vessel.
7. Locate the organs of the digestive system. Slit open the digestive tract and observe its contents.
8. Locate a nephridium in each segment of your dissected specimen.
9. When you have completed your observations, dispose of your specimen in an appropriate container.
10. Examine a prepared slide of a cross section of an earthworm; locate the structures in figure 38.6.

Question 4

a. How many segments of an earthworm have a heart?

b. Does the ventral nerve cord traverse the entire length of the body?

c. Is the inside of the digestive tract the same from the pharynx to the end of the intestine? Explain.

d. How could absorption in the intestine be increased without increasing the intestine's length?

e. How do the layers of musculature in an earthworm differ from those of a nematode?

f. List two or three features of an earthworm cross section that distinguish the dorsal and ventral surfaces.

Class Hirudinea

Hirudineans include leeches, which are primarily freshwater ectoparasites (fig. 38.7). Leeches are not segmented as distinctly as are other annelids. Leeches lack setae, are dorsoventrally flattened, and have anterior and posterior suckers that hold prey. Blood sucking leeches eat infrequently, but species such as *Hirudo medicinal* can quickly consume five to ten times their body weight in blood. Many years ago physicians used these leeches for bloodletting from sick patients thought to have "too much bad blood." Today, laboratory-grown leeches are occasionally used to extract fluid that has accumulated around injuries and surgical incisions to enhance healing. Leeches extract fluid more efficiently and with less damage than does hypodermic suction (fig. 38.8).

Leeches reproduce sexually and individuals are hermaphroditic. Two leeches intertwine, and species with copulatory organs inject a packet of sperm called a **spermatophore** into the female gonopore. Other species have no copulatory organ; these leeches copulate by injecting a spermatophore directly through the epidermis of their partner (the spermatophore may have tissue-dissolving enzymes to aid penetration). The introduced spermatophore releases sperm into the coelomic cavity, and the cells move to the ovaries to fertilize the eggs. In two days to many months after copulation, the leech secretes a nutrient-rich cocoon to protect the eggs, and extrudes eggs into the cocoon. The cocoon is brooded by the leech or attached to submerged objects or vegetation.

Procedure 38.5 Observe leeches

1. Observe living leeches; compare their movement with that of earthworms.
2. Examine the external anatomy of a preserved leech and locate the two suckers.
3. Open the body with a pair of scissors and look for signs of segmentation. A dissected leech may be on demonstration.

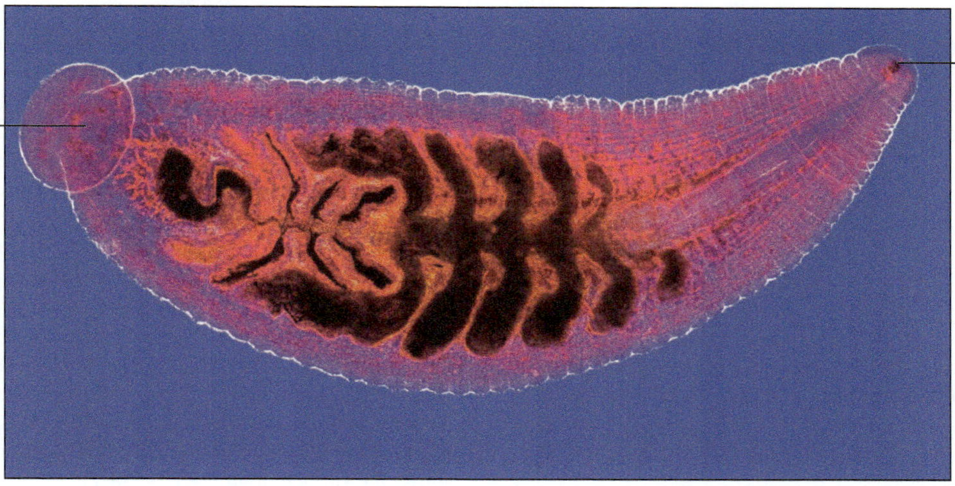

M. I. Walker/Science Source

Figure 38.7 External view of a leech (20×). Leeches are primarily freshwater ectoparasites. The mouth is within the anterior sucker.

BiologyImaging.com

Figure 38.8 Leech. These advanced annelids have external signs of segmentation but have little repetition of internal parts or internal compartments.

Question 5

a. What is the difference in general body shape of leeches compared to oligochaetes or polychaetes?

b. What function other than feeding do suckers serve?

c. Are setae visible on a leech?

d. Is internal segmentation of a leech as distinct as that of an oligochaete?

e. Consider Learning Objective 1 listed at the beginning of this exercise. How could production of a packetlike spermatophore contribute to the evolutionary success of leeches in their environment?

Question 6

a. Now that you have examined the unifying characteristics of mollusks and annelids, can you list three or four characteristics that they share with flatworms and nematodes as their close ancestors?

b. Draw and label three coelomic body plans and list which of the phyla in Exercises 37 and 38 are associated with each plan.

PHYLUM NEMATODA

Nematodes (fig. 38.9), commonly called roundworms, are everywhere and often occur in great numbers—a single decomposing apple may contain 100,000 nematodes of different species. If we removed everything but nematodes from the environment, we would still see a ghostly outline of our entire biosphere. Estimates of the global number of species living in all aquatic, terrestrial, and parasitic environments often exceed one million, most of which have not yet been formally described and named. Almost all feeding types (parasitic, predatory, etc.) are represented. Reproductive morphologies include dimorphic species, hermaphroditic species, and even species with males as well as hermaphrodites. Nematode diversity is extraordinary.

Many nematodes cause diseases in humans, other animals, and plants. One of the most serious of these diseases is elephantiasis, the grotesque swelling of an arm or leg resulting from nematodes (commonly *Filaria*) clogging the lymphatic system that drains the host's appendage (fig. 38.10). Fluid accumulates, and the appendage swells. Another parasitic nematode is the eye worm, *Loa loa,* which lives under the skin of humans and occasionally crawls across the surface of an eye (fig. 38.11).

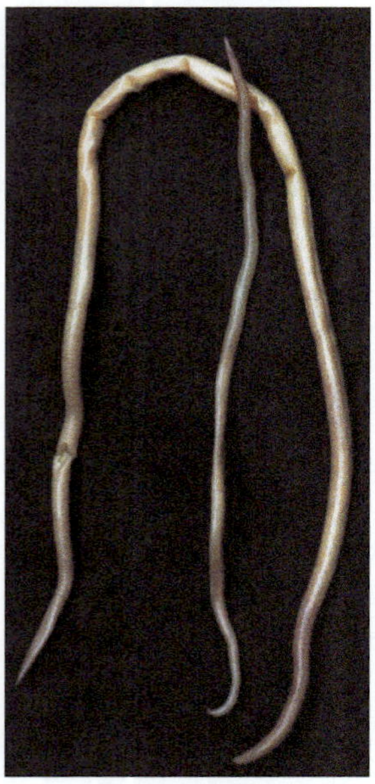

BiologyImaging.com

Figure 38.9 *Ascaris lumbricoides,* a common roundworm, inhabits the intestines of pigs and humans (0.5×). Male ascarid worms are smaller than females and have a curved posterior end.

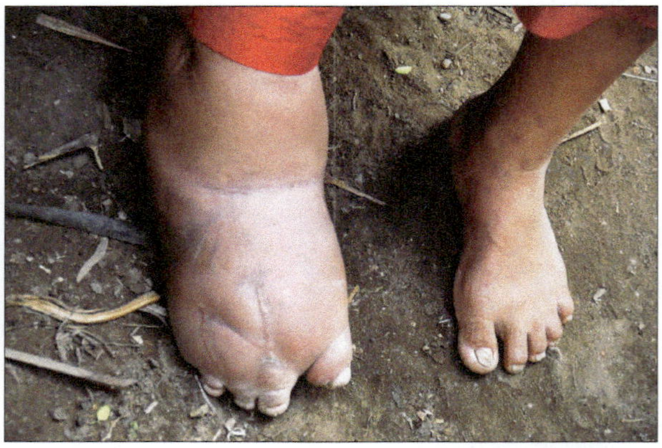

Rungtip Chatadee/age fotostock

Figure 38.10 Elephantiasis of a leg caused by adult filarial worms that live in lymph passages and block the flow of lymph. Tiny juveniles, called microfilariae, are picked up in a blood meal of a mosquito, where they develop to the infective stage and are transmitted to a new host.

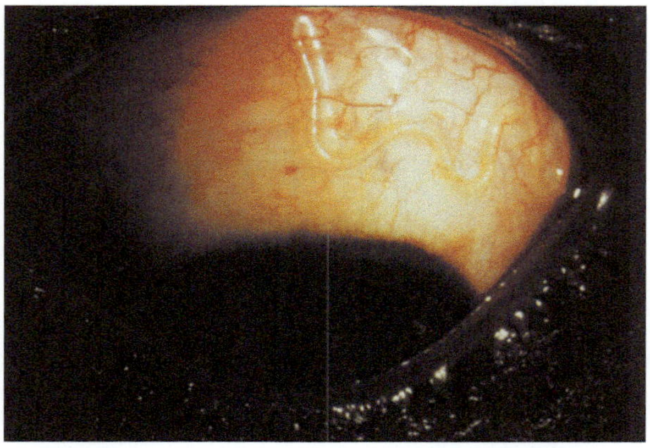

Sue Ford/Science Source

Figure 38.11 The nematode *Loa loa* appears in the eye of its host. This one is being surgically removed from someone's eye.

Nematodes are slender and long with a rather featureless exterior. They lack flagella and cilia and are covered with a somewhat flexible and chemically complex **cuticle.** The cuticle resists digestive enzymes and is permeable only to water, dissolved gases, and some ions. Nematodes have a **complete digestive tract** with a mouth and anus. In addition to their digestive cavity, roundworms have a body cavity called a **pseudocoelom** consisting of a fluid-filled space between the body wall and digestive tract (see fig. 37.2). Internal organs are suspended in this cavity.

Question 7

a. How does the number of body cavities of nematodes compare with that of flatworms?

b. What are the advantages of a digestive tract having a separate entrance and exit?

c. Female *Ascaris* are more numerous than males. Why might this be adaptive?

Rhabditus and *Turbatrix*

Procedure 38.6 Examine living nematodes

1. Examine a culture of living *Rhabditus, Turbatrix* (the vinegar eel), or *Cephalobus*.
2. Using a toothpick, put a small piece of the culture medium on a slide with a drop of water and coverslip, and focus with low magnification of your microscope.
3. Watch the nematodes move.

The flexible cuticle and hydrostatic pressure of fluid in a nematode's pseudocoel aid locomotion by resisting antagonistic muscle contraction. Nematodes have only longitudinal muscles and lack circular or diagonal muscles. This combination of features produces a characteristic motion.

Question 8

a. How would you describe the motion of a nematode?

b. How is this movement related to the movement of its muscle layers?

Ascaris

Ascaris is a large nematode that infects the intestinal tract of humans and other vertebrates (fig. 38.9). Males are smaller than females and have a hooked posterior end. The opening in the posterior end is the anus.

Procedure 38.7 Dissect *Ascaris*

1. Obtain a preserved *Ascaris lumbricoides* and examine its external features using a dissecting microscope. Compare the external features of males and females.
2. At the anterior end, locate the mouth surrounded by three lobes of tissue.
3. Prick the cuticle with a dissecting probe (teasing needle) and determine its consistency.
4. Pin the ends of a female specimen near the edge of a dissecting pan to permit viewing with a dissecting microscope. Slit the body wall longitudinally with a dissecting needle or sharp-pointed scissors.
5. Pin the body wall open and locate the internal organs shown in figure 38.12. The excretory pore may be small and difficult to find. A dissection of a male may be on demonstration.
6. Examine a prepared slide of a cross section of both a male and female *Ascaris*.

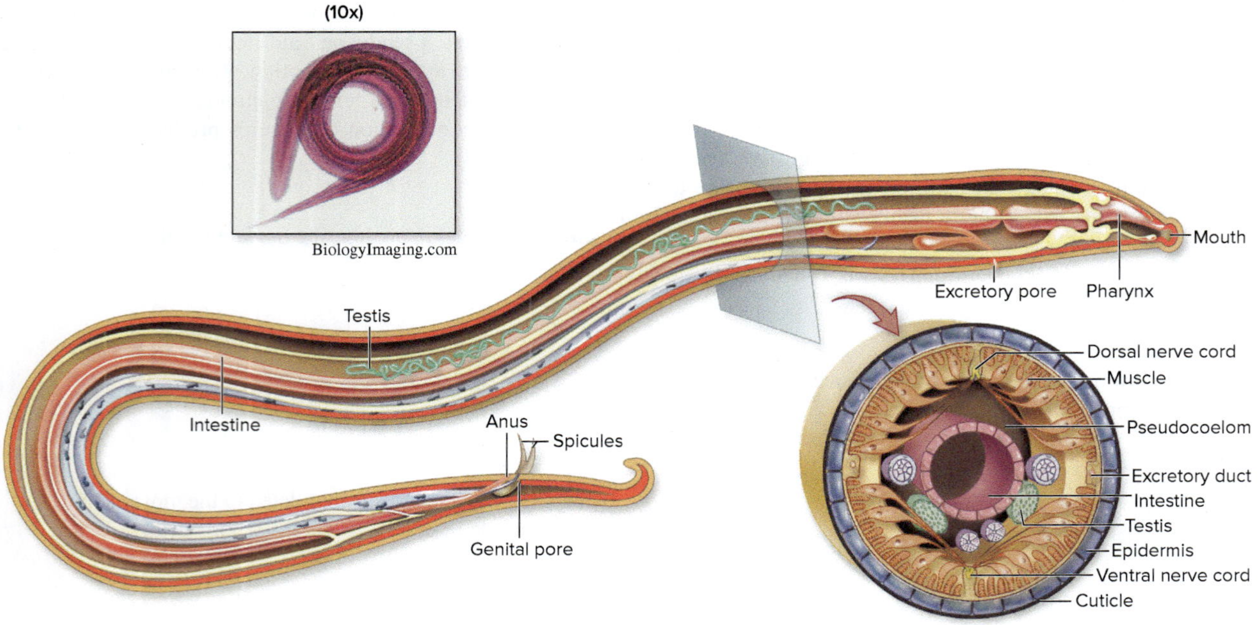

Figure 38.12 Internal anatomy of nematodes, which are commonly called roundworms.

7. Locate the features shown in figure 38.13 and determine from where along the length of your dissected specimen this section was taken.

8. Examine cross sections from different areas of the body if appropriate slides are available. Gametes mature as they move along the length of the tubular reproductive organs.

9. Dispose of waste in labeled containers.

Question 9

a. The cuticle of *Ascaris* is flaky and tough. Consider Learning Objective 1 listed at the beginning of this exercise. What might be an adaptive advantage of a thick and tough cuticle?

b. Where do the internal organs of *Ascaris* attach to the body wall?

c. How does the diameter of the female reproductive tract change?

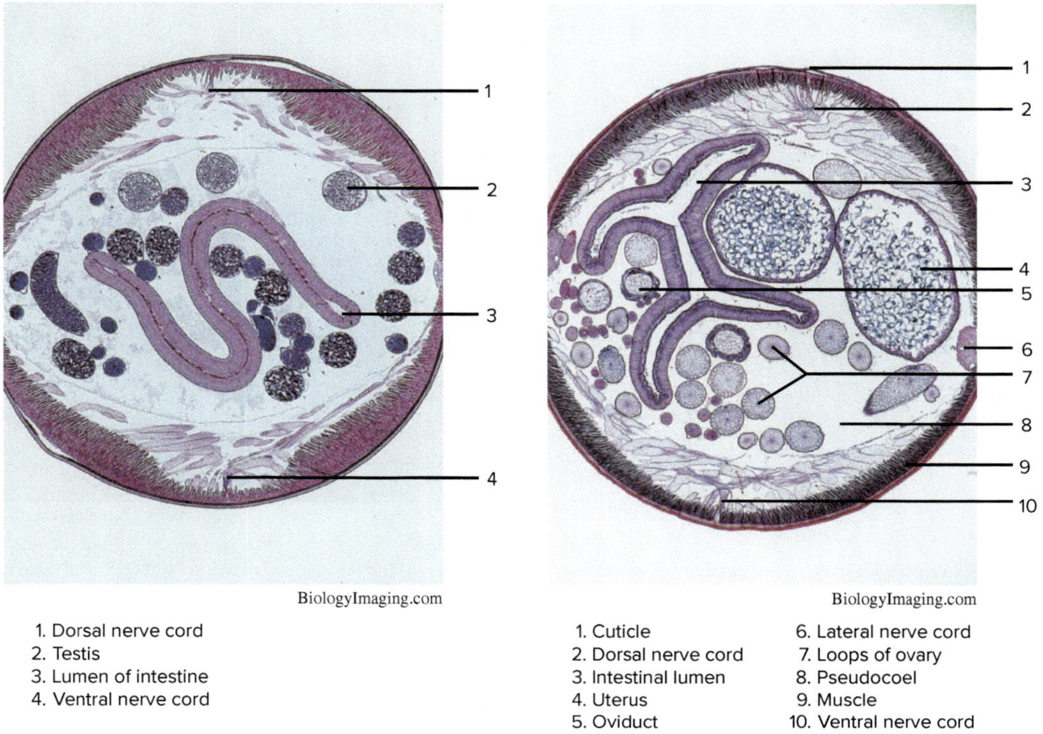

Figure 38.13 Cross sections of male (left) and female (right) *Ascaris,* a large nematode that infects the intestinal tract of a variety of vertebrates (30×).

d. Are any sensory organs evident in *Ascaris?* Why would this be adaptive?

Trichinella

Trichinella spiralis causes the disease trichinosis. Adult females of *Trichinella* live in the intestine of their host and release larvae. These larvae migrate through the body to striated muscles, especially in the diaphragm and tongue, where they form painful calcified cysts (fig. 38.14). The larvae remain encysted until the muscle tissue is eaten by another host, where the larvae mature. Humans infect themselves by eating poorly cooked pork containing encysted larvae. However, the occurrence of trichinosis is decreasing. In the United States fewer than 50 cases of human trichinosis are reported annually to the Centers for Disease Control and Prevention, compared with 500 cases a year in the 1940s.

Procedure 38.8 **Examine *Trichinella, Necator, Enterobius,* and *Dirofilaria***

1. Examine a slide of muscle tissue containing encysted larvae of *Trichinella.*
2. If slides are available, examine specimens of *Necator* (hookworm), *Enterobius* (pinworm), and *Dirofilaria* (heartworm).
3. Note variation in the morphology of the anterior and posterior ends of the organisms.
4. Sketch and note the body sizes of *Trichinella, Necator, Enterobius,* and larval *Dirofilaria.*

Hookworms cause anemia in infected livestock, which causes great economic losses. Pinworms, although less dangerous, irritate many children because the pinworms infect the intestine and inflame the anus (fig. 38.15). Children

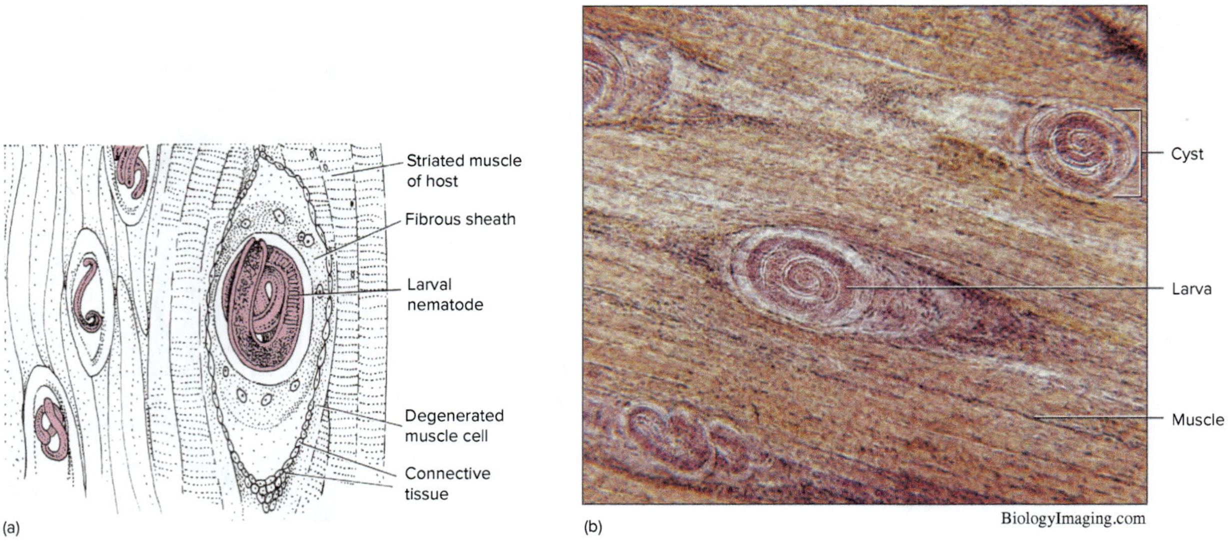

Figure 38.14 *Trichinella spiralis.* (*a*) Anatomy of larvae encysted in muscle. (*b*) Larvae in muscle section (400×). The disease trichinosis is acquired by eating poorly cooked meat that contains encysted larvae.

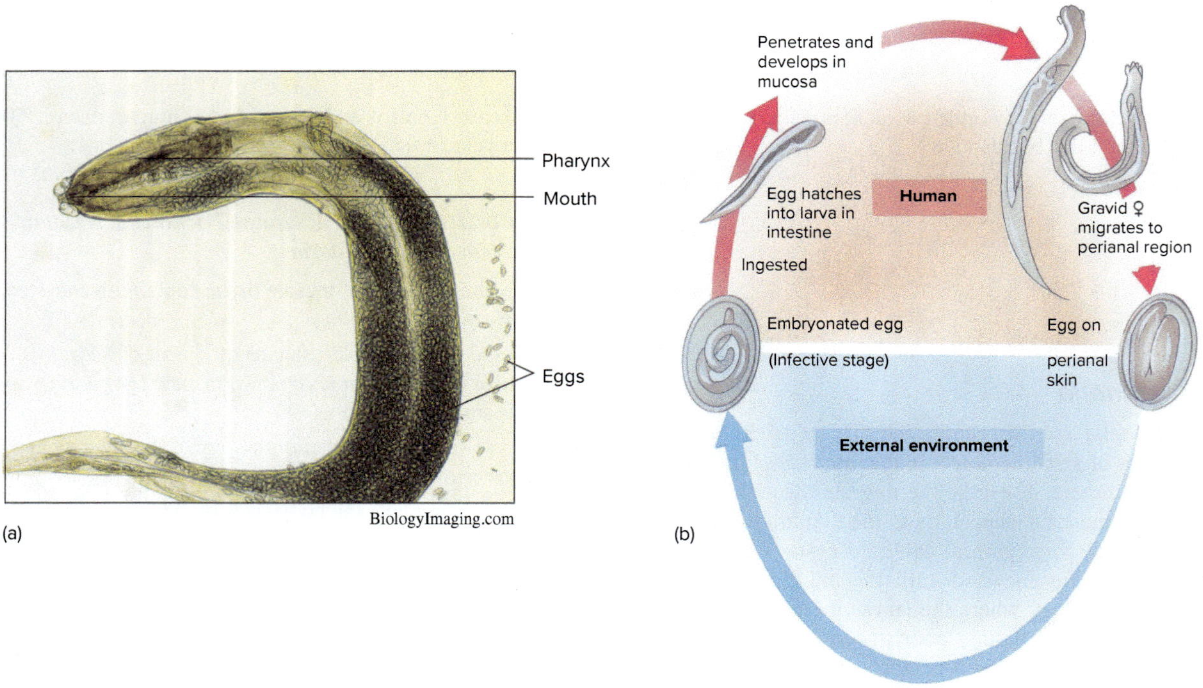

Figure 38.15 (*a*) *Enterobius*, a pinworm (60×). Pinworms are the most common parasite in the United States—30% of children and 16% of adults are infected. (*b*) The life cycle of the pinworm, *Enterobius vermicularis.* The adult crawls from the intestine and lays eggs on the skin just outside the anus. The host, usually a human child, is irritated by the pinworms and scratches. Eggs are then unknowingly eaten and the larvae hatch in the intestine to complete the life cycle.

often scratch themselves, put their fingers in their mouths, and thereby reinfect themselves. Pinworms infect about 30% of children and 16% of adults in the United States. Heartworms (*Dirofilaria immitis*) are among the most significant parasites of dogs and cats. The larvae are transmitted by mosquitoes and large, mature stages (9 to 16 in. long) reside in the heart and lungs, where they disrupt blood flow.

INQUIRY-BASED LEARNING

Of what use could leeches be to medicine?

Observations: Many years ago, physicians used leeches for "bloodletting" from sick patients thought to have "too much bad blood." Today, laboratory-grown leeches are occasionally used to enhance healing by extracting fluid that accumulates around incisions following microsurgery and injuries. Leeches extract fluid more efficiently and with less damage than hypodermic suction (fig. 38.16). Medicine is a business as well as a science that must be effective and delivered safely to large populations.

Question: Are leeches appropriate as a viable medical treatment?

a. Establish a literature search group according to your instructor's guidelines, and obtain Inquiry-Based Learning Worksheet 38 from your instructor.

b. Discuss among your group and record on Worksheet 38 the requirements of effective use of leeches in medicine. Include necessary characteristics of the organism and the logistics of delivering treatment.

c. Determine with your group the specific questions that doctors must answer concerning leeches and their effective, safe use. Record these questions on Worksheet 38.

d. Research the current literature and outline information about the use of leeches in medicine.

e. Record your findings on Worksheet 38.

f. Discuss your findings with your instructor and with other students.

Questions for Further Study and Inquiry

1. You have examined at least three phyla commonly referred to as "worms." How would you define this term?

2. Of what economic importance are earthworms?

3. Earthworms have no lungs or gills. Do they "breathe"? If not, how do they obtain oxygen to survive?

4. Do you suppose bloodletting by leeches was a good technique to cure psychosomatic illnesses? Why or why not?

5. Prepare a simple table of all of the taxonomic groups covered so far, their common names, their distinguishing characteristics, and all representative genera covered in this exercise. Keep this table with your study notes.

WRITING TO LEARN BIOLOGY

Are leeches predators or parasites? What reasons would you give for both possibilities?

Survey of the Animal Kingdom
Phylum Arthropoda

EXERCISE 39

Learning Objectives

By the end of this exercise you should be able to:
1. Describe the structures that contribute significantly to the survival of arthropods in their environments.
2. Describe the general morphology of organisms of phylum Arthropoda.
3. List characteristics that arthropods share with the phyla discussed previously.
4. Discuss those characteristics of arthropods that were newly derived from those of their ancestral phyla.
5. List examples of the major classes of arthropods.
6. Describe modifications of the exoskeleton and paired appendages of arthropods.

Please visit connect.mheducation.com to review online resources tailored to this lab.

The success and predominance of the arthropod body plan in marine, freshwater, and terrestrial environments is extraordinary. **Phylum Arthropoda** is the most diverse as well as most abundant phylum of animals—estimates range from 2–10 million species. Their tremendous success in all major habitats is due mainly to the adaptive value of their rigid external skeleton with **jointed appendages** (*arthro* = jointed, *poda* = foot or appendage) (table 39.1, fig. 39.1). Their appendages are extensions of the main body and highly adapted for locomotion, feeding, reproduction, defense, and sensing the environment. Bodies of arthropods are segmented, as are annelids (fig. 39.2), although segments of arthropods often fuse during development. The classification of arthropods is typically based on the arrangement of these segments and the structure of their appendages. Arthropods are coelomate, their circulatory systems are open, and all organ systems are well developed.

The **exoskeleton** of arthropods is made of **chitin,** a long chain of nitrogen-containing sugar molecules arranged in strong fibers. Chitin may be as soft as the body of a butterfly or, if impregnated with calcium carbonate, as hard as the shell of a lobster. This exoskeleton provides protection, a moisture barrier, and a place for muscle attachment. Although this tough covering limits growth, arthropods periodically shed their exoskeleton (ecdysis) and quickly enlarge before the new exoskeleton hardens.

Each body segment of ancestral arthropods had a single pair of appendages. Recall that this condition of paired appendages also occurs in annelids. Appendages of modern arthropods are covered by the rigid exoskeleton and usually have flexible joints. This provides arthropods with tremendous strength and flexibility. The variation in structure and function of appendages is enormous.

BiologyImaging.com

Figure 39.1 Scorpions are the oldest known terrestrial arthropods. Typical of all arthropods, scorpion bodies are segmented, have jointed appendages, and are covered with a hard, chitinous exoskeleton. Shown here is the world's largest scorpion (18 cm), the Emperor scorpion (*Pandinus emperator*) from west Africa. Some extinct species were five-times larger than the Emperor. The sting and neurotoxic venom of most species are equivalent to those of a hornet sting. The most notorious stingers are *Androctonus* of North Africa and species of *Centruroides* in Mexico, Arizona, and New Mexico. Children and weak adults may die from the neurotoxin of a scorpion in 6–7 hours by paralysis of respiratory muscles or cardiac failure. Scorpions locate their prey by detecting vibrations, often through sand rather than air. Desert species can locate and dig out a burrowing cockroach in just a few seconds. Desert species can also withstand temperatures of 46°C (115°F) and tolerate water loss equivalent to 40% of their body weight.

Table 39.1

Major Groups within the Traditional Classification of the Phylum Arthropoda

Group		Characteristics	Members	Approximate number of named species
Crustaceans	 U.S. Fish and Wildlife Service	Mouthparts are mandibles (biting jaws); appendages are biramous ("two-branched").	Lobsters, crabs, shrimp, isopods, barnacles	67,000
Chelicerates		Mouthparts are chelicerae (pincers or fangs).	Spiders, mites, ticks scorpions, daddy long-legs, horseshoe crabs	78,300
Centipedes and millipedes		Mouthparts are mandibles: bodies consist of a head and numerous body segments bearing paired uniramous ("single-branched") appendages.	Centipedes, millipedes	20,000
Insects		Mouthparts are mandibles; appendages are uniramous.	Beetles, bees, flies, fleas, true bugs, grasshoppers, butterflies, termites	900,000

Figure 39.2 Annelids, arthropods, and chordates all exhibit segmentation.

During the evolution of arthropods, their prominent characteristics (e.g., an exoskeleton and jointed appendages) were highly adaptive and probably allowed them to replace other species that weren't equipped to deal as well with their environment. These adaptations allowed primitive arthropods to exploit new environments, microhabitats, and niches. Taxonomists currently recognize four extant classes of arthropods: Chelicerata, Crustacea, Myriapoda, and Hexapoda. Review the introductory information in this exercise (as well as in your textbook) and complete table 39.2 for six or more arthropod characteristics.

Who Cares about Horseshoe Crabs?

Blood from the horseshoe crab, *Limulus polyphemus,* is more valuable than you might think! It annually is used for tests worth $50–60 million to the medical industry. Amoebocytes circulating in the blood of horseshoe crabs have an ancient immune response that releases coagulogen from granules in the cells in response to bacterial toxins. The reaction forms a "clot" around a site of invading bacteria. The medical industry catches the crabs, bleeds about one-third of their blood, then releases them. Mortality is about 7.5% to 10%. Limulus Amoebocyte Lysate (LAL) is extracted from the blood, marketed, and used to detect any trace of Gram-negative bacterial endotoxin present on newly manufactured equipment, needles, vaccines, prosthetic limbs, or even blood samples from patients. As required by the Food and Drug Administration (FDA), injectable drugs and vaccines must pass a Limulus Test with LAL before they are approved. There is no synthetic substitute for LAL, so conservation and management of horseshoe crab populations is critical—and may save your life.

Courtesy U.S. Fish and Wildlife Service

Table 39.2

The Major Characteristics of Arthropods and Their Adaptive Advantage

Characteristic	Adaptive Advantage

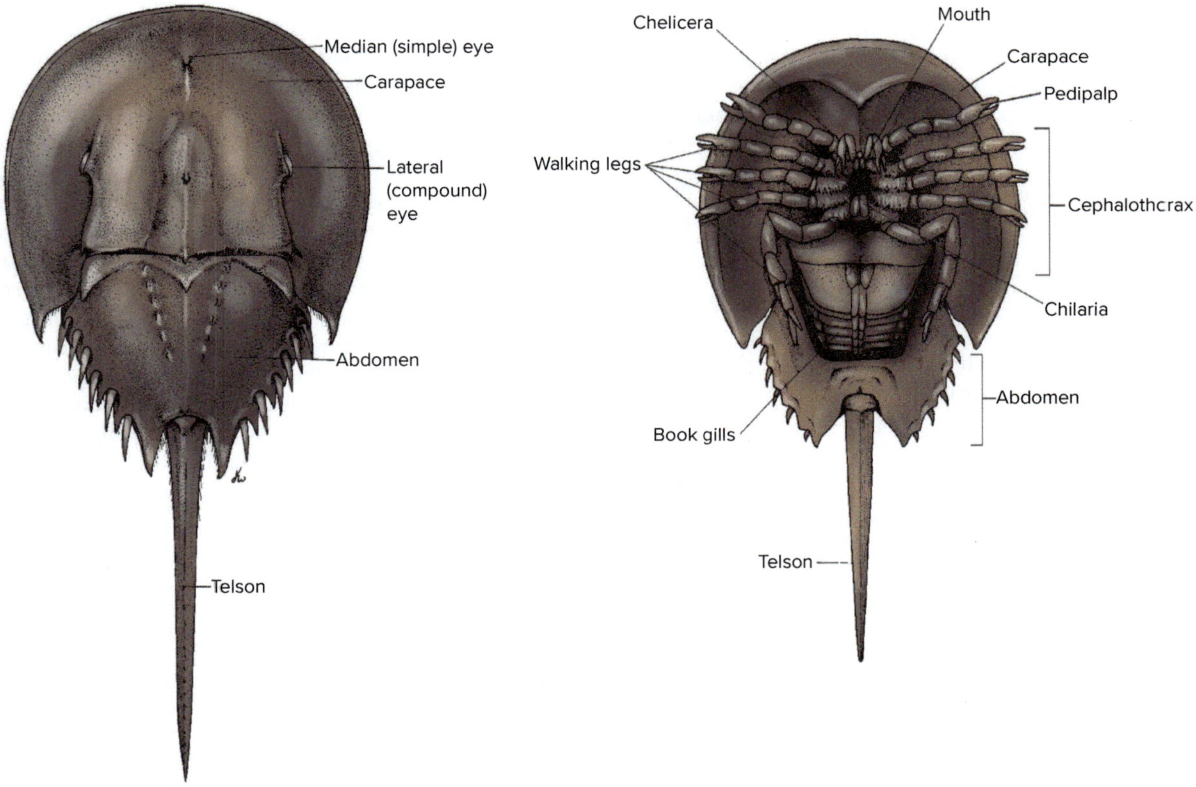

Figure 39.3 Dorsal and ventral views of a horseshoe crab, *Limulus* sp. Horseshoe crabs are ancient chelicerates abundant on the Atlantic Coast and Gulf of Mexico.

CLASS CHELICERATA

Chelicerates such as spiders, mites, scorpions, and horseshoe crabs are arthropods with the appendages of their most anterior segment modified into feeding structures called **chelicerae.** The second pair of appendages are **pedipalps** and are modified for capturing prey, sensing the environment, or copulating. Body segments of chelicerates are fused into two body regions: a **cephalothorax** consisting of a fused head and thoracic segments and an **abdomen** as the most posterior body region. Chelicerates lack antennae.

Horseshoe Crabs

Horseshoe crabs, commonly *Limulus,* are ancient marine chelicerates existing since the Cambrian period (500 million years ago). They are abundant on the Atlantic Coast and Gulf of Mexico.

Examine a preserved *Limulus* and find the structures illustrated in figure 39.3. A horseshoe-shaped **carapace** covers the cephalothorax and a flexible joint seen easily from a dorsal view separates the cephalothorax from the abdomen. The ventral surface includes five pairs of appendages

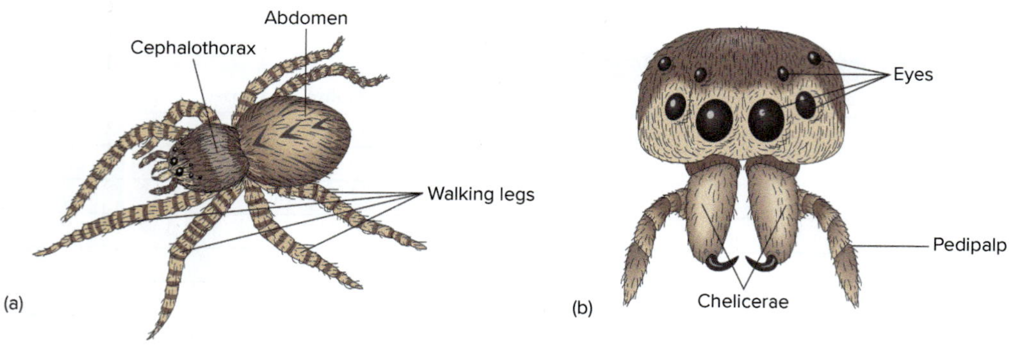

Figure 39.4 Anatomy of a jumping spider. (*a*) External view. (*b*) Anterior view of head.

modified as walking legs. The most anterior pair of legs are modified pedipalps. Anterior to the pedipalps are the chelicerae. Posterior to the walking legs are a pair of degenerated legs called **chilaria.** The appendages of the abdominal segments are modified as **book gills** and promote gas exchange.

Question 1

a. Why do we call the external chitinous covering of an arthropod a skeleton even though it is not made of bone?

b. What external features of a horseshoe crab are used for burrowing and scavenging in sand?

c. Why are appendages of abdominal segments called "book" gills?

Scorpions, Ticks, Mites, Daddy Longlegs, Spiders

These diverse groups, commonly referred to as arachnids, are diverse (85,000 species) and mostly terrestrial. The cephalothorax has chelicerae modified as fangs to pierce prey, has pedipalps to manipulate food and sense the environment, and has four pairs of walking legs.

Examine a preserved scorpion, noting the many distinct segments of its thorax and abdomen. Compare the segmentation of scorpions with that of other available preserved chelicerates such as mites, ticks, and daddy longlegs. Scorpions were the first terrestrial arthropods and have been around since the Silurian period (425 million years ago). Their lack of fused segments is unusual for chelicerates and indicates the scorpion's ancient origins (fig. 39.1). These secretive carnivores are 1–8 cm long. Stings of *Centruroides,* a common scorpion in Mexico, Arizona, and New Mexico, have killed many humans, mostly children. Its venom is neurotoxic and causes convulsions, paralysis of respiratory muscles, and heart failure. However, most scorpion stings are not fatal.

Examine the external features of a preserved spider and compare them with the features shown in figure 39.4. Spiders are the most familiar arachnids and comprise 30,000 species. They are terrestrial and prey mostly on insects and other small invertebrates (figs. 39.5, 39.6). Spiders are often confused with insects, but spiders have two body regions and eight legs. Insects have three body regions and six legs. The spinnerets of spiders are independently moving nozzles that release silk from internal silk glands. Web silk is made primarily of polypeptides of the amino acids glycine, alanine, and serine. When this fluid is emitted it hardens, not from exposure to air but from polypeptide cross-linkages that form during release. The fluid is not forced out under pressure; rather, it is drawn out by the hind legs or by the weight of the body falling through the air. Spiders can produce silk as strong as nylon, highly elastic, and dry or sticky depending on the construction of the web.

BiologyImaging.com

Figure 39.5 Tarantulas, sometimes called bird spiders, may have a 25-cm leg span. They live on the ground, dwell in trees, and sometimes burrow. The hairy body of this brown tarantula is extremely sensitive to vibration. Many hairs contain chemoreceptors as well as mechanoreceptors, and hairs on the tips of the feet are often iridescent. Captive tarantulas can live 20 years or more.

BiologyImaging.com

Figure 39.6 Black widow spiders (*Latrodectus mactans*) are shiny black with a red hourglass on the abdomen's ventral surface. As with other spiders, their chelicerae have poison glands opening at the tip of the fangs. Toxins and enzymes are injected into prey, liquefying the prey's tissue. The nutritious broth is then eaten. Black widows live in most parts of the world, and their venom causes nausea, muscular spasms, respiratory paralysis, and pain in the abdomen and legs. Human death from their bite is rare. Black widows, like other spiders, can withstand months of starvation by reducing their metabolic rate up to 40%. A male is only half the size of the female and is killed soon after mating.

Question 2

a. What external features make scorpions appear so menacing?

b. Which are larger, a scorpion's chelicerae or pedipalps?

c. What is the shape of a spider's chelicerae?

d. Do you see any evidence that a spider's body is segmented?

e. How many eyes do most spiders have? Are they paired and similar in size?

f. Many spiders are hairy. How might this feature be adaptive?

CLASS CRUSTACEA (CRAYFISH, CRABS, SHRIMPS)

Crustaceans (65,000 species) live in marine and freshwater. Only a few species are terrestrial. Crustaceans differ from the other classes because they have fundamentally **biramous,** or double-branched, appendages (fig. 39.7). Crustaceans have two pairs of antennae and usually have **compound eyes** with multiple lenses. Crustaceans along with insects are commonly called mandibulates because they have opposing mandibles derived from an anterior pair of appendages.

Crustacean Anatomy

The body of a crustacean such as a crayfish usually has two regions: a cephalothorax covered by a carapace and an abdomen (fig. 39.8). The five anterior pairs of crustacean appendages are modified into **first antennae, second antennae, mandibles, maxillae,** and **maxillipeds.**

Obtain and examine a preserved crayfish. Appendages 3 **(mandible)** through 9 **(cheliped)** are used for feeding. These appendages have different shapes. The shape of some appendages such as the cheliped indicate an obvious function.

Question 3

How might mouthparts of various shapes be adaptive for crayfish?

The four pairs of walking legs attach to the thorax at the ventral edge of the carapace. Abdominal appendages are much smaller than walking legs and are called **swimmerets** or **pleopods.** If you study a male crayfish, you'll notice that each of the first pair of swimmerets has an odd spatulate shape, modified to transfer packets of sperm to a female during reproduction. The most posterior pair of appendages are broad, flat **uropods.** They surround the terminal abdominal segment called the **telson.** All crustacean appendages are fundamentally biramous (i.e., double-branched).

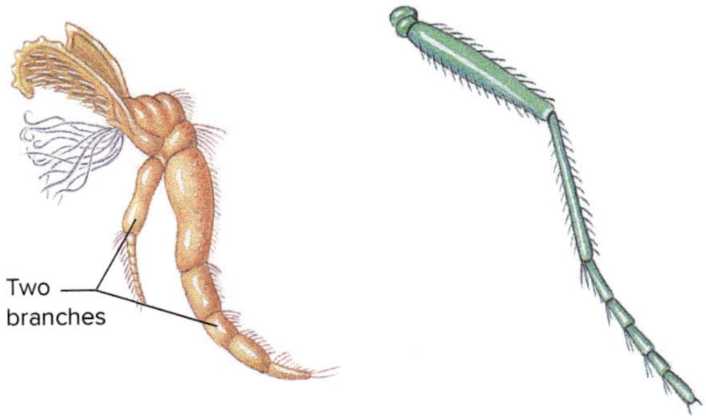

Figure 39.7 Branched and single appendages. A biramous leg in a crustacean (crayfish) and a uniramous leg in an insect.

By now you've probably guessed correctly that studying arthropods consists mostly of examining body segments and their paired appendages. Appendages of two different species may have similar functions but different embryological origin. Such structures are **analogous.** For example, chelicerae of a spider are analogous to mandibles of a crayfish because they have similar functions even though they are not derived from the same body segment. In contrast, **homologous** structures of two different species have similar developmental origin but may or may not serve the same function. Chelicerae of a spider are homologous to the most anterior pair of antennae of a crayfish.

Procedure 39.1 Study the external anatomy of a crayfish

1. Obtain a preserved *Cambarus* (the common crayfish) and locate the external features shown in figure 39.8.
2. To appreciate the variation in appendages in *Cambarus,* use forceps to remove one of the first antennae, a cheliped, a walking leg, an anterior swimmeret, a posterior swimmeret, and a uropod.
3. Arrange them in order in a dissecting pan or on a piece of paper and relate each one's structure to its function.
4. Examine the appendages you've removed and note which ones are biramous.
5. Observe a live crayfish (if one is available) and determine its two major methods of locomotion.
6. Examine other available crustaceans such as a crab. The cephalothorax of a crab is obvious, but the abdomen is highly modified.

Question 4

a. Which body region of a crayfish is most obviously segmented?

b. What structures are located just under the carapace and attached to each leg of a crayfish? What is the adaptive advantage of these structures being attached to legs?

c. How many legs are **chelate** (pincerlike with opposing claws)?

d. What is the function of the uropods and telson, and what feature indicates this function?

e. Are the anterior swimmerets different from the posterior pair? Is your crayfish a male or female?

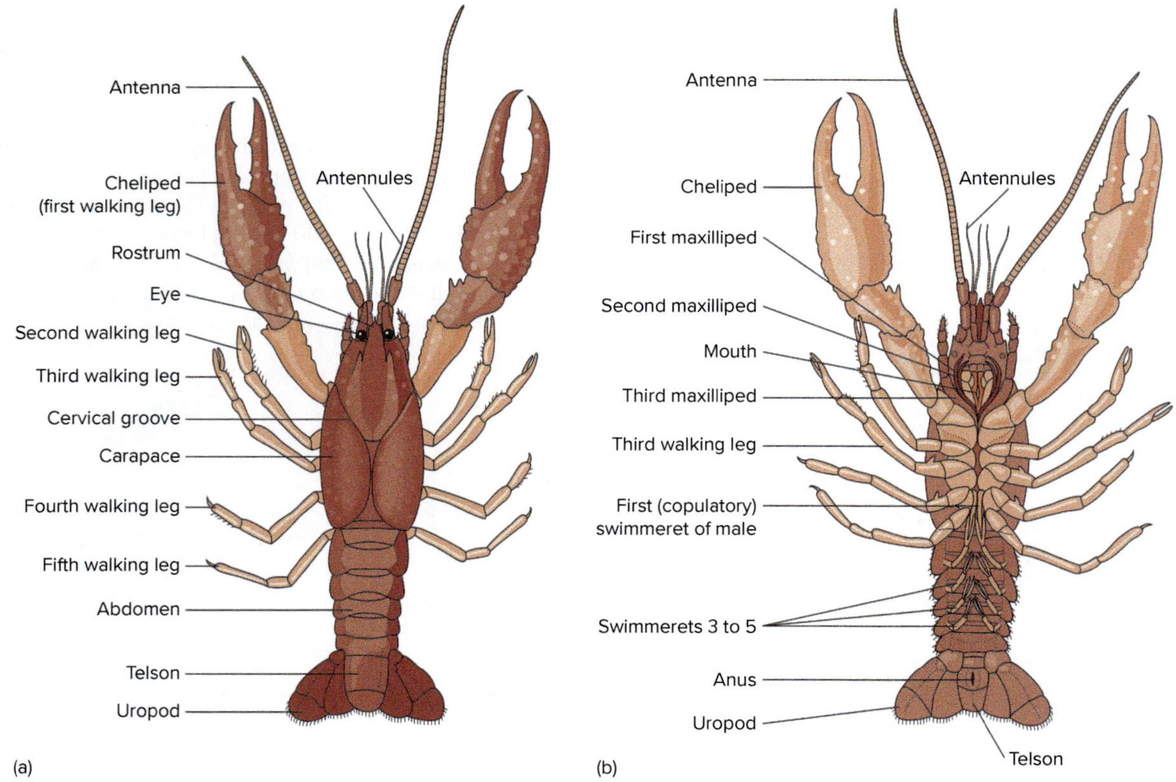

Figure 39.8 External structure of the crayfish. (*a*) Dorsal view. (*b*) Ventral view.

f. Which set of legs (swimmerets or walking legs) appears best adapted to carry an incubating egg mass delicately and in a protected place on the body?

g. Crayfish can walk. Can they also "swim"? How would the uropods and telson help them do this?

h. How does the shape of a crab's abdomen differ from that of a crayfish?

The internal anatomy of a crayfish features a diamond-shaped **heart** surrounded by a thin **pericardial sac.** The heart lies on the dorsal midline just anterior to the abdominal segments (fig. 39.9). In the open circulatory system of a crayfish, blood flows from the heart through large arteries to the gills and to sinuses surrounding and bathing the internal organs. Blood returns to the pericardial sac and enters the heart through small openings called **ostia.**

The **gonads** (testes or ovaries) are lateral and just anterior to the heart. Testes are usually white, and ovaries are orange. Sperm from testes exit the body through a pore at the base of the fifth pair of walking legs, and eggs are released at the base of the third pair of walking legs. During reproduction, males and females copulate and sperm cells are passed to the genital openings of the female. Fertilization is internal, and fertilized eggs are extruded and retained for maturation on swimmerets of the female.

The stomach is a continuous, membranous structure along the dorsal midline of the cephalothorax. It is surrounded by muscle, reinforced with ridges of tissue, and receives food from the **esophagus** and **mouth.** Food moves from the stomach, through the intestine, and out through the anus at the base of the telson. A large **digestive gland** that secretes enzymes and stores food lies just posterior and lateral to the stomach.

Beneath the internal organs lies the anterior part of the **ventral nerve cord.** A pair of nerves from the ventral nerve cord pass around the esophagus and come together anteriorly as a **brain** between and beneath the eyestalks. Posterior to the esophagus the nerve cord extends the length of the body as a series of swellings or **ganglia,** each of which controls organs in the immediate segment.

Procedure 39.2 **Study the internal anatomy of a crayfish**

1. Obtain a preserved crayfish.
2. Make an incision on each side of the cephalothorax. To make this incision, cut with scissors beginning at the posterior edge of the carapace halfway up the side (i.e , midlateral). Cut along the side toward the eye. Finish each lateral incision on each side of the rostrum.
3. Cut across the base of the rostrum and carefully remove the dorsal portion of the carapace.
4. Locate three pairs of ostia on the heart. Remove the heart.
5. Locate the stomach and digestive gland.
6. Carefully remove the internal organs (stomach, digestive gland, gonads). This exposes the end of the torn esophagus and the anterior part of the ventral nerve cord.
7. Remove a considerable amount of musculature on the floor of the abdomen (tail) to see the nerve cord more clearly.
8. Locate the **antennal glands.** They are excretory organs opening at the base of each antenna.
9. When you have finished examining your specimen, dispose of the material as directed by your instructor.

CLASS MYRIAPODA

Myriapods have one pair of antennae on the head and three pairs of appendages that are modified as mouth parts, including mandibles that act like jaws.

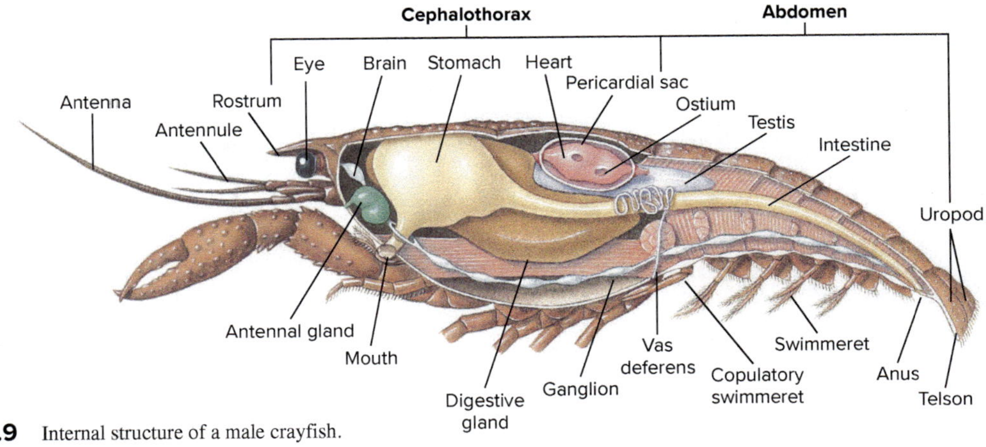

Figure 39.9 Internal structure of a male crayfish.

INQUIRY-BASED LEARNING

How can the same embryonic appendage pattern give rise to so many different appendage structures and functions?

Observations: Crustaceans are highly diverse but share fundamental similarities retained throughout their evolution. Much diversity among crustaceans stems from variation in their appendages, yet consistent patterns emerge in appendage morphology and body plan.

Question: What variations and similarities are apparent among appendages of crustaceans?

- **a.** Establish a working lab group and obtain Inquiry-Based Learning Worksheet 39 from your instructor.
- **b.** Discuss with your group and instructor specific observations to be made during dissection of two closely related arthropods relevant to the preceding observations and question. Record them on Worksheet 39.
- **c.** Translate your proposed observations into a testable hypothesis and record them on Worksheet 39.
- **d.** Obtain a large, whole shrimp and crayfish for dissection.
- **e.** Carefully remove appendages from one side of the specimens and tape the appendages side by side for comparison.
- **f.** Complete Worksheet 39.

Centipedes

Centipedes (3000 species) live in soil under logs and stones where they prey on small arthropods. Centipedes move rapidly and are dorsoventrally flattened. Some of the larger centipedes such as *Scolopendra* can inflict a painful bite but are not lethal to humans (fig. 39.10).

Examine a preserved centipede. Each body segment bears a pair of legs. The large fangs (sometimes called poison claws) on the head are not mandibles but are maxillipeds, which are appendages modified for feeding. The mandibles are smaller and lie between the maxillipeds.

Question 5

a. Do centipedes have 100 legs as their name suggests?

b. What structures of chelicerates are analogous to the antennae of chilopods (centipedes) and other mandibulates?

Millipedes

Examine a preserved millipede. Millipedes (8000 species) live in the same environment as centipedes but feed mainly on decaying plant material (fig. 39.11). Millipedes move slowly and are round in cross section. A disturbed millipede will frequently roll up to protect its soft underside. The number of legs on each segment distinguishes millipedes from centipedes.

Question 6

a. How many pairs of legs does a millipede have?

b. Each apparent segment is actually two fused segments. How many legs are on each apparent segment?

(a)

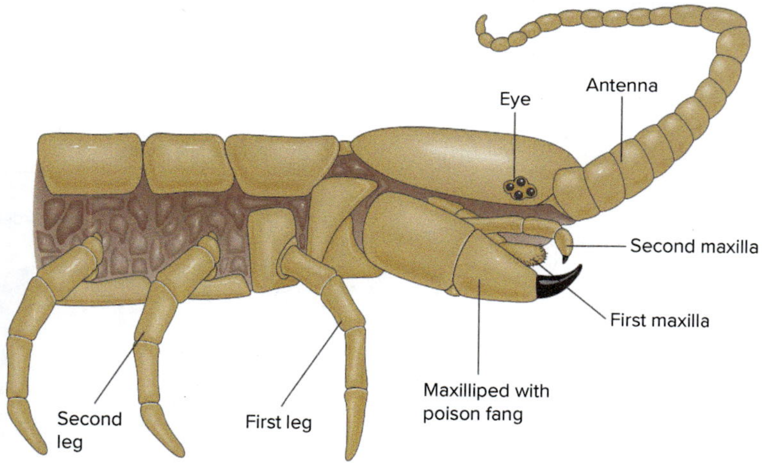

(b)

Figure 39.10 Centipede, *Scolopendra* sp. (*a*) Most segments have one pair of appendages each. The first segment bears a pair of poison claws (maxillipeds), which in some species can inflict serious wounds. Centipedes are carnivorous. (*b*) The head of a centipede.

Michael P. Gadomski/Science Source

Figure 39.11 Millipede, *Sigmoria*. Apparent body segments are actually fused pairs of body segments. Each apparent segment has two pairs of legs.

CLASS HEXAPODA (Insects)

Hexapods are six-legged arthropods. Most are insects, but there are a few earlier-diverging noninsect hexapods, including soil-dwelling groups representing a separate but related lineage.

Insects are by far the largest group of organisms on Earth and probably include 1 to 10 million species. At least 70% of all known species of animals are insects, and they dominate virtually all terrestrial habitats. Three separate body regions and six thoracic legs are the major diagnostic features of insects. Insects possess all the major characteristics of arthropods, but their ability to fly is probably the key to their great success. Although other groups of organisms can fly, insects were the first fliers. There are many advantages to flying, such as avoiding predators. Can you name some others?

Most insects have two pairs of wings. However, insect wings are not modified appendages. Rather, they are evaginations (outgrowths) of the thoracic exoskeleton.

The success of insects on land is further enhanced by an efficient system of respiratory tubes called **tracheae** that conduct air throughout the body.

External Anatomy of a Grasshopper

The grasshopper, *Romalea,* has characteristics typical of insects (fig. 39.12). Insects have three body regions: head, thorax, and abdomen; one pair of antennae; and six legs. The thorax is divided into the **pro-, meso-,** and **metathorax,** each having a pair of legs. Mouthparts are covered by the **labrum,** an extension of the head. Beneath the labrum are the **mandibles,** followed by a pair of **maxillae** with segmented extensions called **palps,** and then the **labium** with palps (fig. 39.13).

Romalea has 10 abdominal segments, each with a **spiracle** or breathing pore opening to the respiratory system of tracheal tubes (fig. 39.14). The terminal abdominal segment bears the reproductive genitalia. The terminal segment of males is blunt, whereas that of females is modified to lay eggs and is called an **ovipositor.**

Procedure 39.3 **Examine the external features of a grasshopper**

1. Obtain a preserved specimen and locate the external features shown in figure 39.12.
2. Examine the head and find simple eyes called ocelli as well as compound eyes (fig. 39.13).
3. Remove the labrum and locate the remaining mouthparts.
4. Use your probe to locate the stout mandibles. Press the mandibles with your probe to test their rigidity.

Question 7

a. How many ocelli does a grasshopper have?

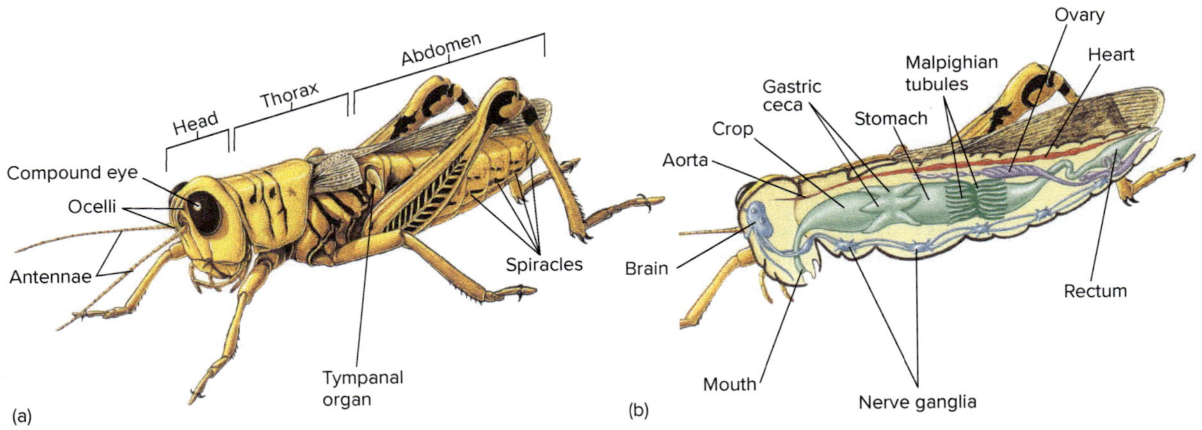

Figure 39.12 Grasshopper, *Romalea,* a member of class Insecta. (*a*) External anatomy. (*b*) Internal anatomy.

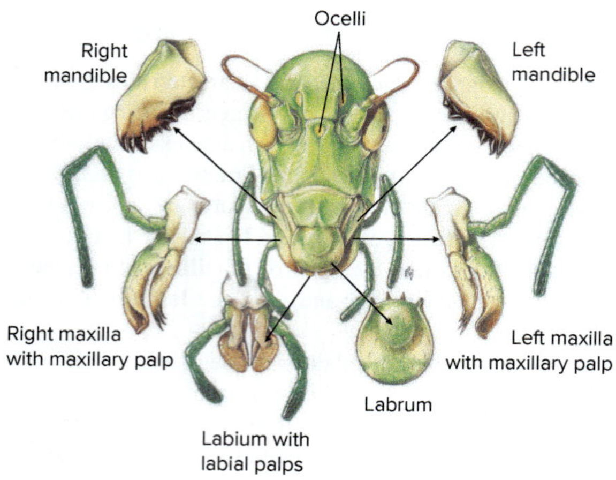

Figure 39.13 Head and mouthparts of a grasshopper.

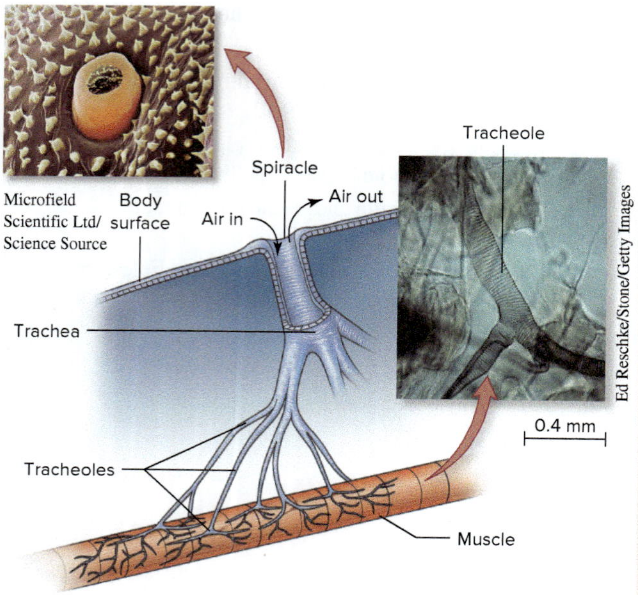

Figure 39.14 The tracheal system of an insect. Tracheae and tracheoles are connected to the exterior by specialized openings called spiracles and carry oxygen to all parts of a terrestrial insect's body.

b. What is the probable function of the maxillary and labial palps?

c. Wing morphology varies among species and orders of insects. Spread the wings of the grasshopper that you are examining. Do all of the wings have the same shape and consistency?

d. Do you suspect that the fore- and hindwings have different functions? Why or why not?

e. Consider Learning Objective 1 stated at the beginning of this exercise. Are wings significant to fundamental processes for arthropods? In what ways?

Procedure 39.4 **Examine the internal anatomy of a grasshopper**

1. Use scissors to clip the wings away from the grasshopper's body.
2. Make two lateral incisions (one on each side) from the posterior end. Cut the length of the abdomen and thorax along an incision line just dorsal to the spiracles.
3. Connect the two lateral incisions by cutting across the dorsal surface just behind the head. Then remove the dorsal strip of exoskeleton.
4. Pin the insect to the dissecting pan. Remove the legs if necessary.
5. Locate the heart and aorta just below the strip of exoskeleton that you removed (fig. 39.12*b*).
6. Notice the body cavity between the body wall and digestive tract. It is the hemocoel and is filled with hemolymph, a colorless blood.
7. The internal organs may be covered with a yellow fat body. Remove it and the heart to expose the digestive tract.
8. Search for tubules of the respiratory tracheal system. These tubules conduct air from the spiracles to the tissues.
9. Locate reproductive organs (ovaries and testes) that lie on either side of the digestive tract.
10. Examine the digestive tract and locate the major structures labeled in figure 39.12*b*.
11. Remove the digestive tract to examine the ventral nerve cord and its series of swollen nerve ganglia.

Question 8

a. How many abdominal segments does a grasshopper have? Is there a nerve ganglion for each segment?

b. Gastric ceca produce digestive enzymes and secrete them into the stomach. How many ceca are there?

c. The spiracles can be opened or closed. Why might this be adaptive?

d. The brain is in the dorsal part of the head. Where does the ventral nerve cord encircle the digestive tract and join with the brain?

The success of insects is also enhanced by a diversity of highly modified mouthparts (fig. 39.15). For example, mouthparts of grasshoppers grind coarse plant tissue, whereas mosquito mouthparts puncture and suck fluid from animals. Vision by insects is also remarkable. The structure of the arthropod compound eye allows extraordinary sensitivity and rapid detection of motion (fig. 39.16).

Examine all the insects available in the laboratory and note any differences from the basic anatomy of a grasshopper. If you find differences, ask your instructor to explain the variation among these insects. If you look carefully, you should keep your instructor busy for a long time discussing the most diverse animals in the world.

To review phylum Arthropoda, outline in table 39.3 the differences among the classes you have examined.

Question 9

a. Are the chelicerae of horseshoe crabs homologous or analogous to (1) scorpion pedipalps? (2) grasshopper mandibles? Explain.

b. Describe four specialized appendages of insects and their function.

c. Now that you have examined the unifying characteristics of arthropods, list three or four characteristics that they share with annelids, their most likely ancestors.

DICHOTOMOUS KEY

A common tool for identifying organisms is a **dichotomous key,** such as the one presented in Exercise 31. Dichotomous keys list and describe pairs of opposing traits, each of which leads to another pair of traits until a level of classification of the specimen being identified is reached. By using a key you'll learn the characteristics that distinguish each of the groups identified by the key.

The diversity of insects is immense. They are classified into 26 orders distinguished mainly by the structure of

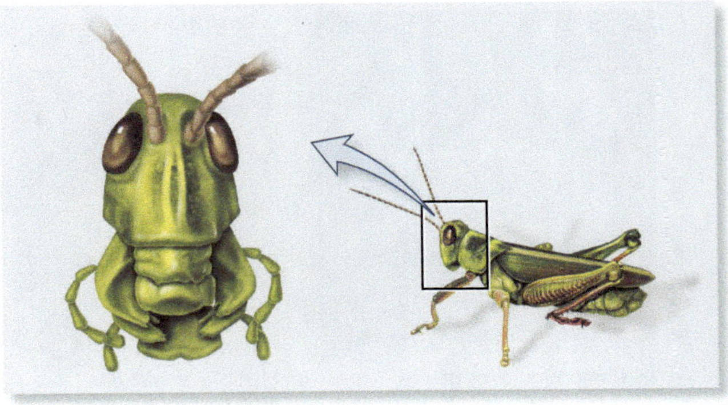

(a) Chewing (grasshopper)

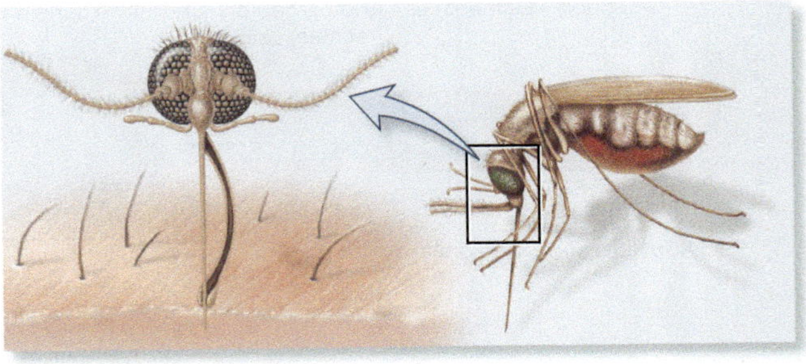

(b) Piercing and blood sucking (mosquito)

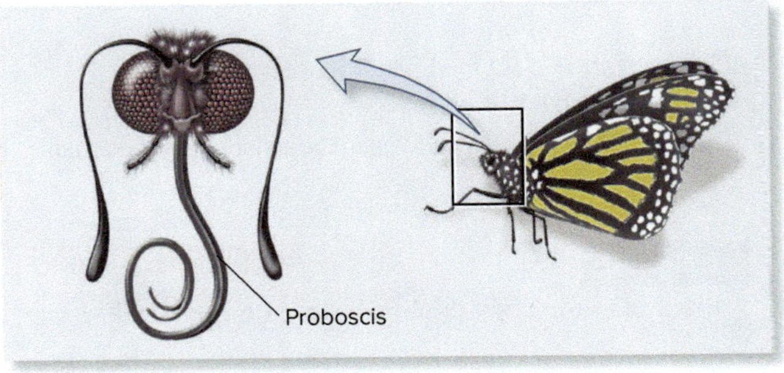

(c) Nectar sucking (butterfly)

(d) Sponging liquid (housefly)

Figure 39.15 Insect mouthparts have become modified in ways that allow insects to feed by a variety of methods, including (*a*) chewing (Orthoptera, Coleoptera, and others), (*b*) piercing and blood sucking (Diptera), (*c*) nectar sucking (Lepidoptera), and (*d*) sponging liquid (Diptera).

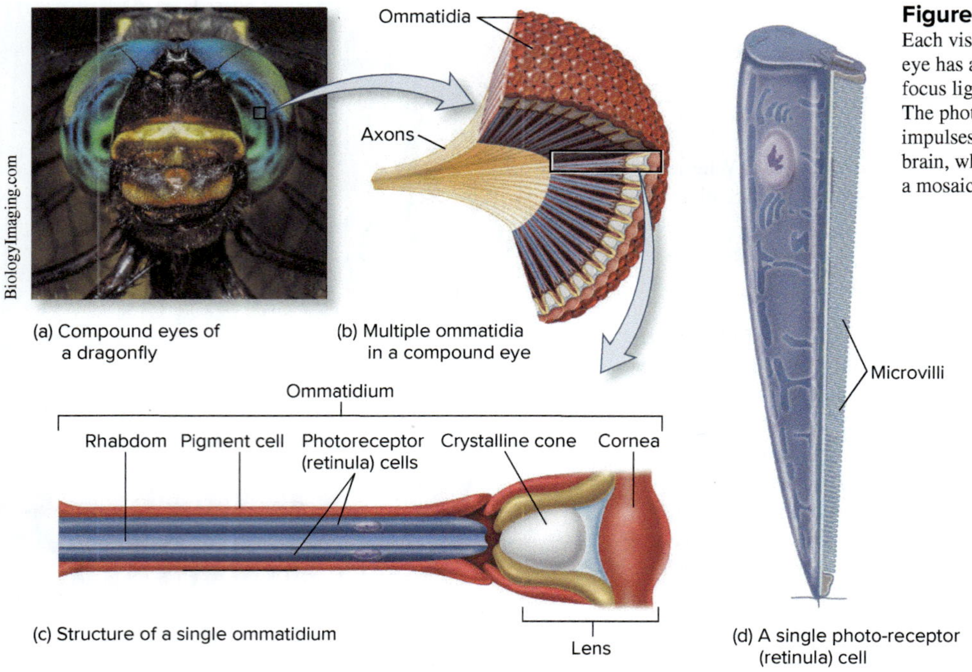

Figure 39.16 Compound eye. Each visual unit of a compound eye has a cornea and a lens that focus light onto photoreceptors. The photoreceptors generate nerve impulses that are transmitted to the brain, where interpretation produces a mosaic image.

wings, mouthparts, and antennae. You will be provided with five to eight preserved insects. Use this key to identify the order of each specimen.

Procedure 39.5 **Use a dichotomous key to identify the order of an insect**

The key will eventually lead you to the order of the specimen.

1. Select a specimen and read the first pair of characteristics.
2. Choose the one that best describes your specimen.
3. Proceed according to the number at the end of your choice to the next pair of characteristics.

Table 39.3

A Comparison of Major Characteristics of Organisms of Phylum Arthropoda

	Horseshoe crabs	Arachnids	Crustaceans	Centipedes	Millipedes	Insects
Names of body regions						
Number of legs						
Arrangement of legs						
Segmentation						
Number of antennae						
Names of major sensory organs						
Names of major mouthparts						

DICHOTOMOUS KEY TO SOME MAJOR ORDERS OF INSECTS

1. Insects with two wings ....**(flies) Diptera**
   Insects with four wings, a pair of forewings, and a pair of hindwings.. 2
2. Fore- and hindwings are not alike in texture and color. One pair may be hard and dense while the other may be light and transparent 3
   Fore- and hindwings similar, usually clear, thin, and transparent 5
3. Forewings thick and leatherlike at base, tips much thinner and may be transparent; mouthparts pointed and beaklike to puncture prey and suck body fluids ....**(bugs) Hemiptera**
   Forewings same texture throughout, biting mouthparts with opposing mandibles 4
4. Forewings leathery and with veins .... **(grasshoppers, crickets) Orthoptera**
   Forewings hard, without veins ....**(beetles) Coleoptera**
5. Wings of same length, antennae usually shorter than head 6
   Wings not of same length, antennae long or enlarged toward end 7
6. Large insects (usually > 3 cm), wings long, transparent, and with many strong veins; abdomen long and slender ....**(dragonflies) Odonata**
   Smaller insects, wing venation faint, wings extending posterior to the abdomen...................................**(termites) Isoptera**
7. Wings covered with fine, opaque scales; tubular, coiled, sucking mouthparts ....**(butterflies, moths) Lepidoptera**
   Wings thin, transparent, and not covered with scales; mandibles well developed ....**(ants, bees, wasps) Hymenoptera**

Questions for Further Study and Inquiry

1. Arthropods usually have a distinct head. How would you define a "head"? What are the advantages and disadvantages of having such a body region?
2. Does an insect's exoskeleton limit growth? Why or why not?
3. Diagram the arrangement of muscles necessary to bend a joint with an exoskeleton versus a joint supported by an endoskeleton.
4. Arthropod body segments are sometimes distinct, sometimes indistinct, and sometimes fused as groups to form body regions. Which groups of arthropods appear the most distinctly segmented? Which appear the least segmented?

5. What effect would 2.5 million spiders per acre have on the insect community?

6. Do you suspect that each eye of a spider provides the same sensory input to the brain? Why or why not?

7. What activities and body functions of arthropods require the most specialized appendages?

8. Do beetles have wings? If so, where are they?

9. What other group of organisms you have studied thus far has chitin as part of its outer covering?

10. What group of arthropods dominates the sea?

11. Prepare a simple table of all of the taxonomic groups, their common names, their distinguishing characteristics, and all representative genera covered in this exercise. Keep this table with your study notes.

12. Does a crayfish have an open or closed circulatory system? Review detailed references and summarize how the circulatory systems vary among animal phyla.

13. A major feature of annelids and arthropods is segmentation. Speculate on the adaptive advantages of segmentation.

WRITING TO LEARN BIOLOGY

Do you think that arthropods constitute a single phylum, or should they be divided into multiple phyla? Describe what divisions you would make and your reasons for them.

Survey of the Animal Kingdom

Phyla Echinodermata and Chordata

EXERCISE 40

Learning Objectives

By the end of this exercise you should be able to:

1. List echinoderm and chordate characteristics closely associated with successful food gathering and survival.
2. Describe the morphology of organisms of phyla Echinodermata and Chordata.
3. List characteristics that echinoderms and chordates share with phyla discussed previously.
4. Discuss characteristics of echinoderms and chordates that are unique or advanced compared to more primitive phyla.
5. Describe the water vascular system of echinoderms.
6. Discuss embryological characteristics that distinguish deuterostomes from protostomes.
7. Understand which phyla are protostomes and which are deuterostomes.

Please visit connect.mheducation.com to review online resources tailored to this lab.

Members of the two phyla remaining in our survey of animals—that is, Echinodermata and Chordata—are **deuterostomes.** Deuterostomes are a major departure from the phylogenetic line of **protostomes** such as annelids, mollusks, and arthropods. The fundamental basis for the separation of deuterostomes from protostomes involves different morphological patterns of embryonic development (fig. 40.1); most notably, the blastopore of deuterostomes gives rise to an anus rather than the mouth. The blastopore is the opening to the first cavity formed in a developing embryo and is discussed more in Exercise 50.

Examine figure 40.1 carefully. Refer to your textbook for a more detailed comparison of protostomes and deuterostomes.

Question 1

Describe three major differences between deuterostome and protostome development.

PHYLUM ECHINODERMATA

Echinoderms (6000 species) are marine bottom-dwellers and include sea stars, brittle stars, sea urchins, sand dollars, sea cucumbers, and sea lilies (table 40.1, fig. 40.2). These organisms are called echinoderms (*echino* = spiny, *derm* = skin) because their internal skeleton of calcareous plates, called **ossicles,** usually has spines protruding through a thin layer of skin. The five classes of echinoderms are distinguished primarily by the arrangement of their ossicles.

Adult echinoderms are radially symmetrical, and their bodies typically consist of a ring of five repetitive parts (i.e., they are pentaradial). In contrast, larvae of echinoderms are bilaterally symmetrical (fig. 40.3). This indicates that radial symmetry is secondarily derived and not directly related to the symmetry of more ancient phyla such as Cnidaria.

Echinoderms have a unique **water vascular system** consisting of a series of coelomic water-filled canals ending in hollow projections called **tube feet.** Muscle contractions and hydrostatic pressure in the water vascular system extend and move the tube feet and other parts of the system and thereby move the animal (fig. 40.4).

Class Asteroidea (Sea Stars)

Ossicles of class Asteroidea, including the common sea star *Asterias,* are arranged loosely under the skin, and spines are small and blunt. Arms of sea stars are continuous with the central disk. The mouth is at the center of the lower, **oral** surface, and the anus is on the upper, **aboral** surface. Surrounding the blunt spines of sea stars are **dermal gills** for respiration by diffusion and pincerlike **pedicellariae** used to remove debris from the surface (fig. 40.5). Also on the aboral surface is the **madreporite,** a sieve connecting the water vascular system with the environment.

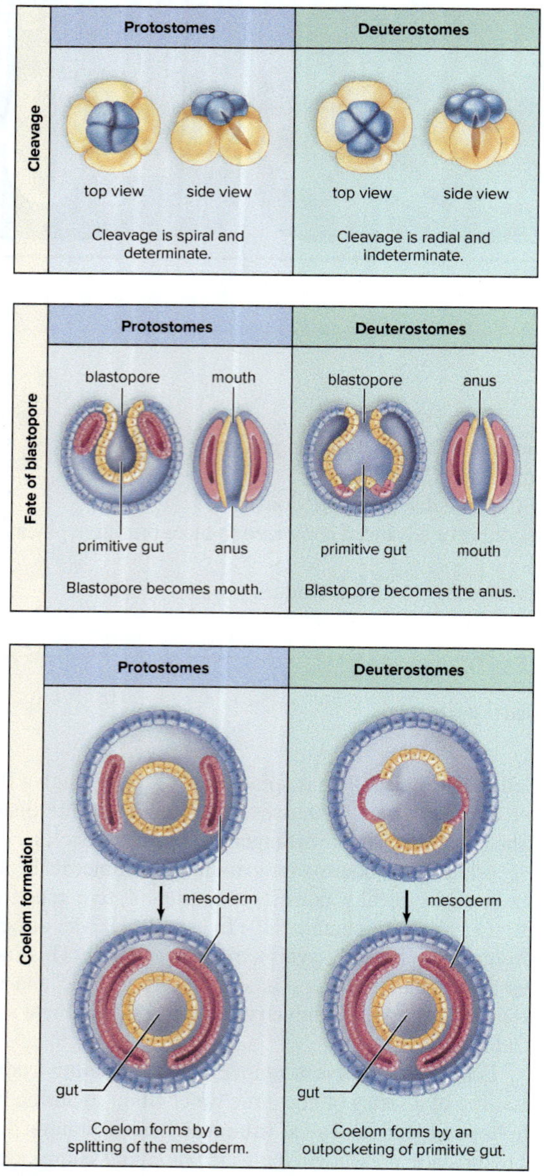

Figure 40.1 Patterns of embryonic development of protostomes compared to deuterostomes. *Left:* In the embryo of protostomes, cleavage is spiral and determinate—new cells form at an angle to old cells—and each cell has limited potential and cannot develop into a complete embryo. The blastospore is associated with the mouth, and the coelom, if present, develops by a splitting of the mesoderm. *Right:* In deuterostomes, cleavage is radial and indeterminate—new cells sit on top of old cells—and each one can develop into a complete embryo. The blastopore is associated with the anus, and the coelom, if present, develops by an outpocketing of the primitive gut.

Sea stars often prey on oysters and clams by using their arms and tube feet to grip the shell and persistently apply pressure to pry it open. Sea stars then evert their stomach inside the clam to digest and engulf the tissue.

SAFETY FIRST Before coming to lab, you were asked to read this exercise so you would know what to do and be aware of safety issues. In the space below, briefly list the safety issues associated with today's procedures. If you have questions about these issues, contact your laboratory assistant before starting work.

Procedure 40.1 Examine the external anatomy of a sea star

1. Examine a prepared slide of the bipinnaria larval stage of starfish development. Note the bilateral symmetry.
2. Examine a preserved sea star and locate the external and internal features shown in figure 40.6.
3. Examine the oral surface of the sea star, and locate the central mouth. Locate the tube feet protruding from the **ambulacral grooves.**
4. Use a dissecting microscope to examine the spines on the aboral surface (fig. 40.5).
5. Touch the madreporite with a probe, and note its consistency.
6. Examine figure 40.4 and trace the path of water from the madreporite to the tube feet.
7. If living sea stars are available, observe their locomotion.

Question 2

a. How many tube feet would you estimate are on the oral surface of a sea star?

b. Are sea star spines movable?

c. What is the consistency of the madreporite?

d. How fast does a living sea star move?

e. Do tube feet of a living sea star move in unison?

Table 40.1

Phyla Echinodermata and Chordata

Phylum	Typical Examples		Key Characteristics	Approximate Number of Named Species
Echinodermata (echinoderms)	Sea stars, sea urchins, sand dollars, sea cucumbers		Deuterostomes with radially symmetrical adult bodies; endoskeleton of calcium plates; five-part body plan and unique water vascular system with tube feet; able to regenerate lost body parts; marine	6000
Chordata (chordates)	Mammals, fish, reptiles, birds, amphibians		Segmented coelomates with a notochord; possess a dorsal nerve cord, pharyngeal slits, and a tail at some stage of life; in vertebrates, the notochord is replaced during development by the spinal column; 20,000 species are terrestrial; deuterostomes	42,500

Procedure 40.2 Examine the internal anatomy of a sea star

1. Obtain a preserved sea star, and cut off at least 2 cm of one arm.

2. Examine the severed end and compare the structures with those shown in figure 40.7.

3. Cut along both sides of the arm up to the central disk. Then cut across the aboral (upper) surface to join the two lateral incisions.

4. Remove the upper body wall and expose the coelomic cavity of the arm. Locate the internal organs shown in figure 40.8.

5. Remove one of the digestive glands so you can examine the gonad lying underneath.

6. Cut around the perimeter of the central disk. Then cut from the perimeter up to and around the madreporite. This incision will allow you to remove the upper body wall without tearing away the madreporite.

7. Remove the upper body wall. As you lift it, try to see the delicate connection between the anus and the surface of the thick-walled pyloric stomach below.

8. Locate the structures shown in figure 40.8.

Question 3

a. How many tube feet would you estimate are in one arm?

b. What part of the water vascular system extends into each arm?

c. What other phyla that you have examined rely on "hydraulics" as part of their locomotion system?

d. Does the stomach wall appear highly folded and extensible? How does that relate to the feeding method of most sea stars?

Class Ophiuroidea (Brittle Stars)

Examine a preserved brittle star such as *Ophioderma.* Brittle stars have slender, sometimes branched arms clearly demarcated from the central disk (fig. 40.2*d*). Ossicles of brittle stars are typically thick and have attached musculature. They may form "shields" on the surface. As the name "brittle" star implies, their arms detach easily, allowing escape from predators.

The ambulacral grooves are closed in brittle stars, and the reduced tube feet are not used for locomotion. The thin flexible arms of brittle stars allow them to crawl rapidly like an octopus rather than creep slowly like sea stars. Brittle stars eat suspended food particles captured with their tube feet and passed to their mouth.

(a) BiologyImaging.com

(b) BiologyImaging.com

(c) BiologyImaging.com

(d) Bill Ober

(e) BiologyImaging.com

Figure 40.2 Diversity in echinoderms (phylum Echinodermata). (*a*) Sea star, *Dermasterius imbricata* (class Asteroidea), in the Gulf of California. (*b*) California sea cucumber, *Parastichopus californicus* (class Holothuroidea). (*c*) Feather star, *Florometra serrastissima* (class Crinoidea), uses its highly branched arms in filter feeding. Although filter feeding probably reflects the original use of echinoderm appendages, most modern echinoderms use their arms for locomotion, capturing prey, and scavenging the substrate for food. (*d*) Brittle star, *Ophiothrix* (class Ophiuroidea). (*e*) Sea urchin, class Echinoidea. Urchins, such as this red sea urchin, *Strongylocentodus franciscanus,* defend themselves with long and sometimes barbed spines, often filled with toxins.

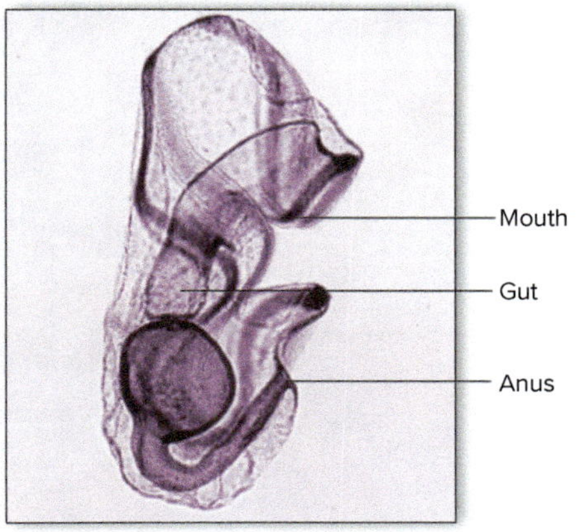

BiologyImaging.com

Figure 40.3 The free-swimming larva of the common sea star, *Asterias rubens.* Such bilaterally symmetrical larvae suggest that the ancestors of the echinoderms may not have been radially symmerical. Bilateral larva is shown in side view (100×).

Question 4

a. Between brittle stars and sea stars, which have the most apparent ossicles? Do they overlap?

b. Are tube feet visible in *Ophioderma?*

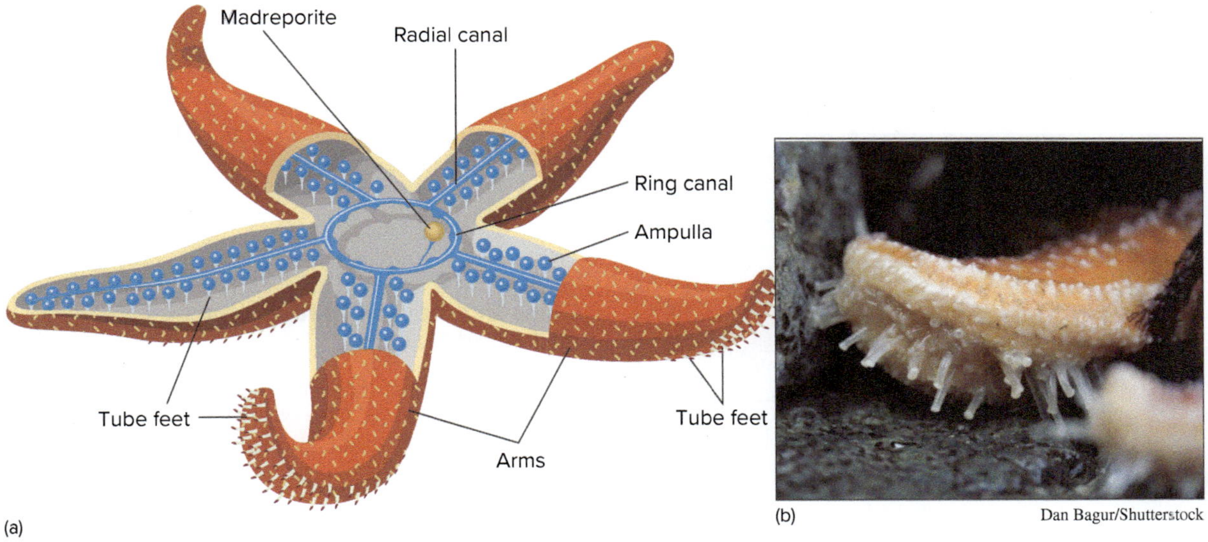

Dan Bagur/Shutterstock

Figure 40.4 Echinoderms. (*a*) The echinoderm body plan of a sea star, emphasizing the water vascular system. (*b*) The extended tube feet of a sea star, *Ludia magnifica* (10×). Tube feet are used for locomotion and for gripping and pulling apart prey such as clams.

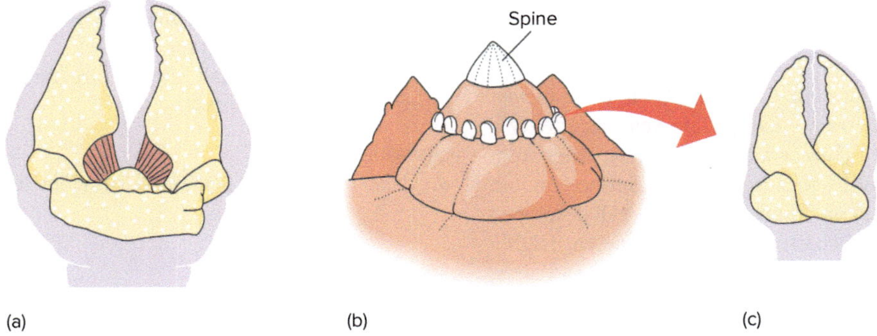

Figure 40.5 Pedicellariae of sea stars. (*a*) Forceps-type pedicellaria of *Asterias*. (*b*) Spine surrounded by (*c*) scissors-type pedicellaria of *Asterias*.

Figure 40.6 External features of sea stars.

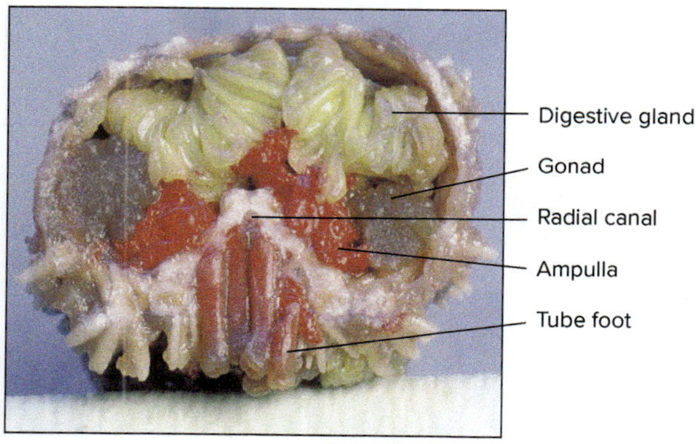

BiologyImaging.com

Figure 40.7 Cross section of an arm of a sea-star showing internal anatomy (5×).

Class Crinoidea (Sea Lilies and Feather Stars)

Examine a preserved crinoid. Crinoids are the most ancient echinoderms and flourished 450 million years ago; only a few genera live today (figs. 40.2*c*, 40.9). They differ from other living echinoderms because their oral surface (mouth and anus) usually faces up. Their ossicles are well developed and give the animal a coarse, jointed appearance. Highly branched and feathery arms surround the mouth and anus. Most ancient crinoids were attached to the substrate by a stalk and appeared to be plants. However, most modern species are not stalked or permanently attached. Crinoids filter feed by capturing food particles on the mucus of their tube feet.

Question 5

How does the position of the mouth and anus of a crinoid relate to a primitive sessile existence?

Class Echinoidea (Sea Urchins and Sand Dollars)

Examine an urchin such as *Arbacia* and locate the features shown in figure 40.10. Also examine a dissected Aristotle's lantern if one is available. In addition, examine a sand dollar and compare its test of fused ossicles to an urchin's test. Urchins lack distinct arms, and their ossicles are fused into a solid shell called a **test** (fig. 40.2*e*). Holes in the test allow long tube feet to protrude. Spines of sea urchins are jointed, movable, and longer than those of other classes of echinoderms. These spines and long tube feet control locomotion of urchins. The mouth contains five ossified plates, or teeth, used to scavenge and scrape surfaces of rocks and gather algae for food. This small internal structure of five teeth is called **Aristotle's lantern.**

Question 6

a. Is an urchin's test pentaradially symmetrical?

b. Urchins and sand dollars lack arms. How do they move?

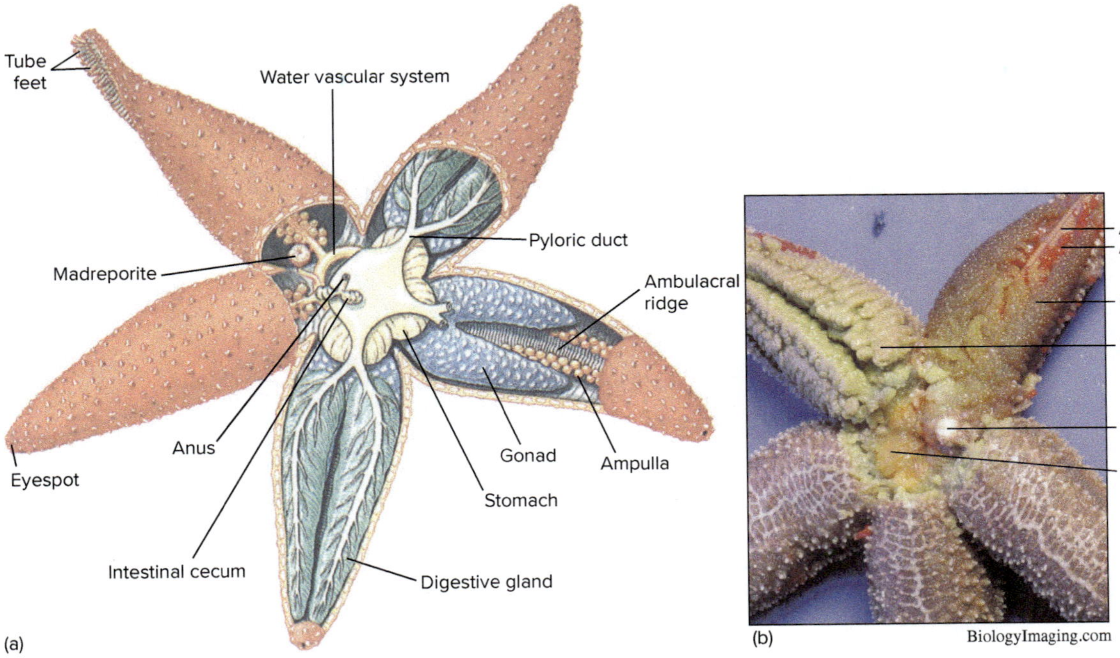

Figure 40.8 Sea star internal anatomy. (*a*) Schematic with aboral surface removed. (*b*) Partially dissected sea star.

Sea Urchins and RAGs

The vertebrate immune system is astoundingly complex, even though fewer than 1% of human genes encode antibodies that recognize invaders. Underlying this versatile immune system is the ability to shuffle and reshuffle DNA fragments that encode antibodies that recognize a virtually limitless variety of invading antigens. Surprisingly, scientists have learned much about the vertebrate immune system and its encoding system for antibodies by studying sea urchins.

By studying the genome of the purple sea urchin, *Strongylocentrotus purpuratus,* scientists have revised our understanding of the vertebrate immune system. In 2006, the purple sea urchin became the first echinoderm to have its DNA sequenced. When immunologist Jonathan Rast and his colleagues searched the echinoderm's DNA for evidence of genes required for vertebrate-style adaptive immunity, they found genes encoding a pair of enzymes that shuffle the echinoderm's DNA! The enzymes are called RAG proteins, for **r**ecombination **a**ctivating **g**ene. In vertebrates, two RAG proteins (called Rag1 and Rag2) must be present for lymphocytes to mature properly; these two proteins interact to recognize sites where a DNA fragment is to be cut and spliced. The origin of our vertebrate immune system may have begun taking shape in ancestors of our distant deuterostome cousins, the echinoderms!

gcammino/iStock/Getty Images

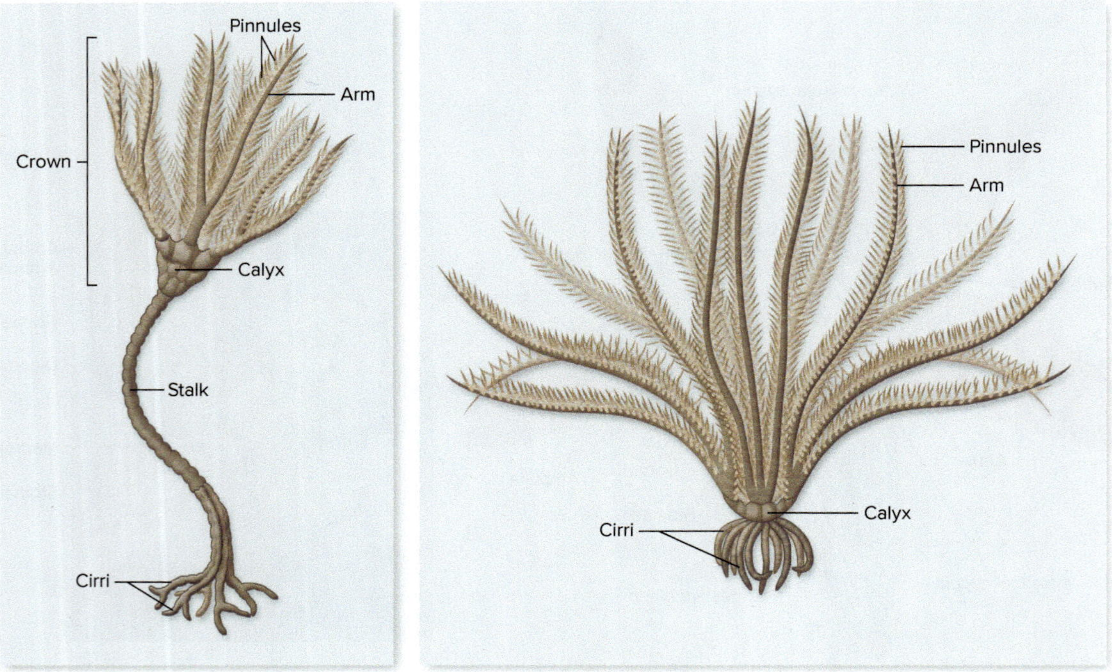

(a) (b)

Figure 40.9 Class Crinoidea. (*a*) A sea lily (*Ptilocrinus*). (*b*) A feather star (*Neometra*).

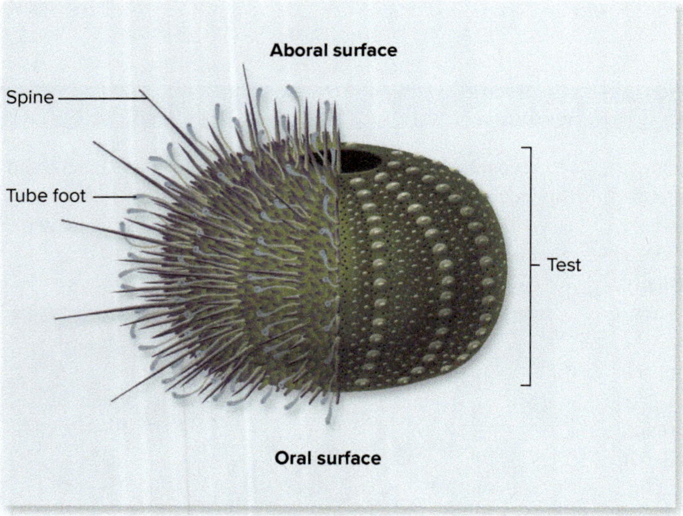

Figure 40.10 External anatomy of a sea urchin. Spines and tube feet are removed on the right half of the diagram to show the test.

Class Holothuroidea (Sea Cucumbers)

Examine a sea cucumber *Cucumaria* and determine the orientation of its radial symmetry. Find the mouth at one end; it is surrounded by modified tube feet called **tentacles.** Sea cucumbers look different from other echinoderms because they have soft bodies with reduced ossicles and few if any spines (figs. 40.2*b*, 40.11). Radial symmetry is less evident

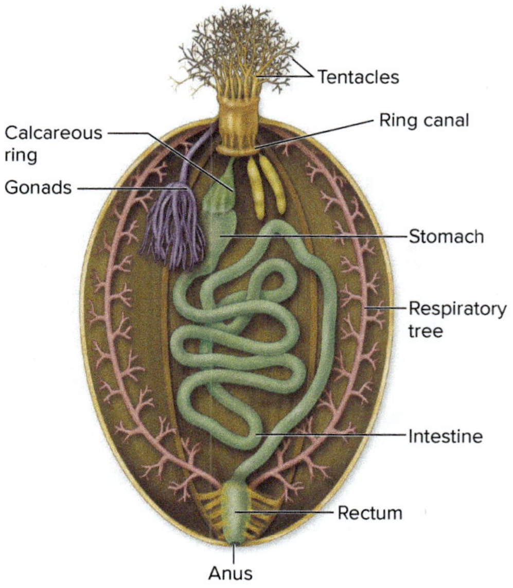

Figure 40.11 Internal structure of a sea cucumber, *Thyone.* The mouth leads to a stomach supported by a calcareous ring. The calcareous ring is also the attachment site for longitudinal retractor muscles of the body. Contractions of these muscles pull the tentacles into the anterior end of the body. The stomach leads to a looped intestine.

Table 40.2

A Comparison of the Major Characteristics of the Classes of Echinoderms

	Class				
	Asteroidea	**Ophiuroidea**	**Crinoidea**	**Echinoidea**	**Holothuroidea**
Shape of arms					
Development of tube feet					
Development of ossicles					
Feeding method					
Spine structure					

in sea cucumbers and their body axis is oriented horizontally. This orientation gives sea cucumbers a semblance of cephalization. The tentacles secrete a mucus that captures small floating organisms, which they eat. Interestingly, some sea cucumbers respond to stress by rupturing anteriorly and rapidly expelling their pharynx, digestive tract, and other organs. This process is called evisceration; because of it, the animal must regenerate the lost parts of the organs. Some gourmets consider sea cucumbers a delicacy.

Examine other preserved echinoderms and review in your textbook the major characteristics of each class. Then complete table 40.2.

Question 7

a. Are tube feet visible on the sea cucumber?

b. Hydras, octopuses, and sea cucumbers have tentacles. Do tentacles have a single universal function or varied functions? What functions are common?

c. Describe the functions of pedicellariae, madreporite, dermal gills, Aristotle's lantern, tube feet, water vascular system.

PHYLUM CHORDATA

Chordates include 42,500 species of fish, amphibians, reptiles, birds, and mammals. They all are characterized by (1) a **dorsal hollow nerve cord;** (2) a **notochord,** a cartilaginous rod that forms on the dorsal side of the gut in the embryo; (3) **pharyngeal slits,** openings in the throat that filter water that has entered through the mouth, and (4) a **postanal tail** (fig. 40.12). An internal, bony skeleton is also common and provides sites for muscle attachment for efficient movement.

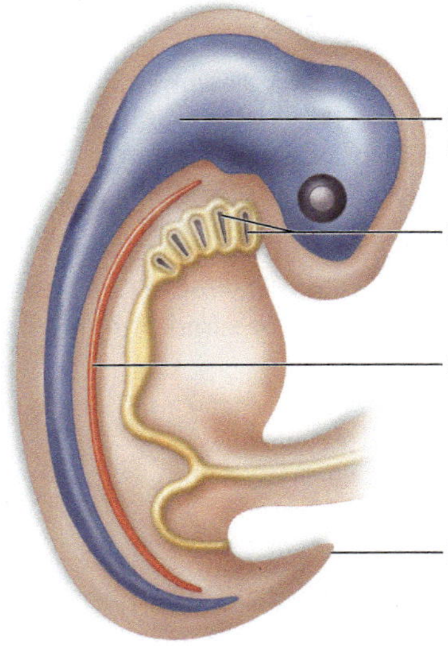

Figure 40.12 The four principal features of the chordates, shown in a generalized embryo.

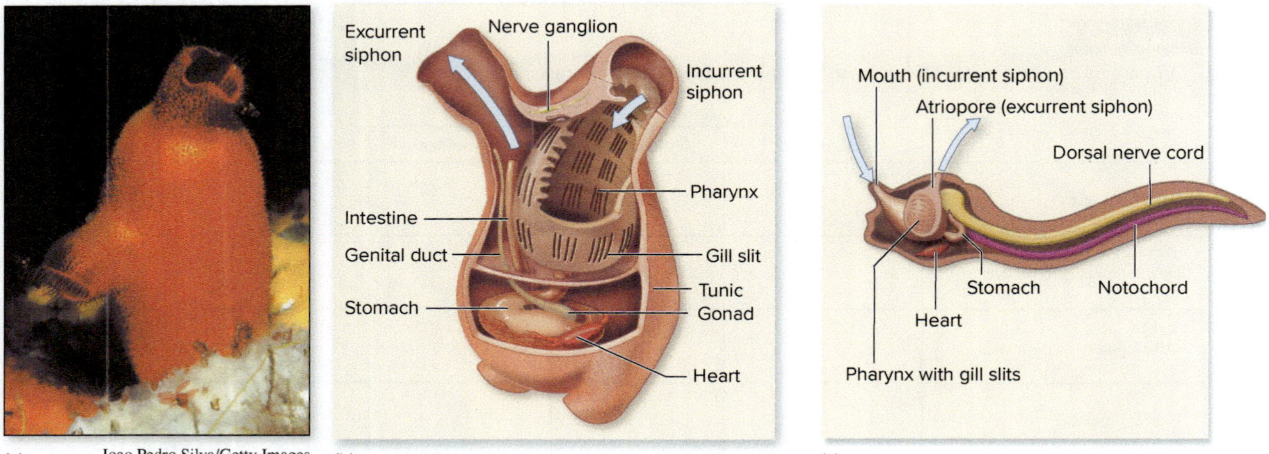

(a) Joao Pedro Silva/Getty Images (b) (c)

Figure 40.13 Tunicates (phylum Chordata, subphylum Urochordata). (*a*) Living adult. (*b*) Structure of adult tunicate. (*c*) Larval structure.

Subphylum Urochordata (Tunicates or Sea Squirts)

Examine a preserved adult tunicate. Urochordates, sometimes called tunicates, are sessile or planktonic marine organisms whose larvae possess the general chordate form—that is, they are elongated with a notochord and dorsal nerve cord (fig. 40.13). In contrast to larvae, the structure of an adult is highly modified to include a sievelike basket perforated with pharyngeal gill slits and surrounded by a cellulose sac called a **tunic.** Water enters through an incurrent siphon, is filtered by the pharyngeal basket, and exits through an excurrent siphon. Water is actively filtered; some tunicates only a few centimeters long can filter 170 liters of water per day. Food collected by mucus on the pharyngeal basket is moved by cilia to the stomach and intestine. The intestine empties into the body cavity near the excurrent siphon.

Examine a prepared slide of larval tunicates. A larval tunicate has bilateral symmetry, a dorsal nerve cord, a notochord, and a postanal tail but loses these features when it settles for adult life (fig. 40.13*c*).

Question 8
What other group of organisms has cellulose in its supporting structures? Does this shared feature surprise you?

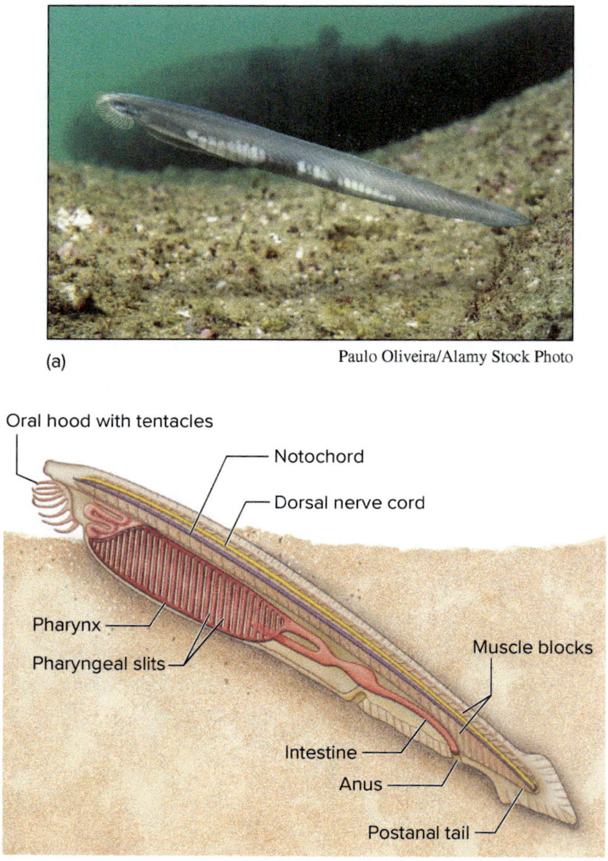

(a) Paulo Oliveira/Alamy Stock Photo

(b)

Figure 40.14 (*a*) A lancelet, *Branchiostoma lanceolatum* (phylum Chordata, subphylum Cephalochordata), protruding from shell gravel. The muscle segments are visible; the square, pale yellow objects along the side of the body are gonads, indicating that this is a male lancelet. (*b*) Internal structure of a lancelet. This bottom-dwelling cephalochordate has the four distinctive features of chordates: notochord, dorsal nerve cord, pharyngeal gill slits, and a postanal tail. The vertebrate ancestor probably had a similar body plan.

Subphylum Cephalochordata (Lancelets)

Examine either a preserved lancelet or a slide of a whole mount and compare the specimen with that shown in figure 40.14. Also examine a cross section through the pharynx and try to visualize the paths of food and water (fig. 40.15).

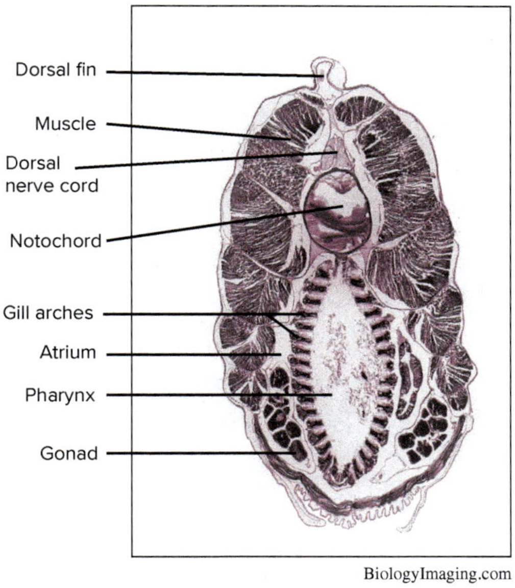

Figure 40.15 Cross section through the pharynx of a lancelet (20×).

Lancelets are small, fishlike, marine chordates that burrow in sand or mud. They are commonly called amphioxus, but the most common genus is *Branchiostoma*.

The dorsal nerve cord and notochord extend the length of the animal. The buccal cavity surrounds the mouth followed by a long pharynx with many gill slits (openings) separated by gill arches of reinforced tissue. As seawater enters the mouth and exits through the slits, it must pass over the surfaces of the arches that form the sides of the slits. As this occurs, food particles are caught on the arches and are eventually swept to the intestine. The anus is not terminal. Lancelets and vertebrates have a post-anal tail, another diagnostic trait of chordates. After water passes by the gill arches, it moves into a surrounding chamber called an **atrium** and then leaves the body through the **atriopore.**

Subphylum Vertebrata (Fish, Birds, Amphibians, Reptiles, and Mammals)

Vertebrates have a distinct head and a vertebral column that replaces the notochord in adults and surrounds the dorsal nerve cord (fig. 40.16). Three classes of vertebrates discussed here are fishes and four are terrestrial tetrapods.

Class Cephalaspidomorphi (Lampreys)

Examine a preserved lamprey, *Petromyzon* (fig. 40.17). Also examine a prepared slide of an **ammocoete,** the larva of a lamprey. Living lampreys (38 species) descended from representatives of the earliest stages in the evolution of vertebrates. They lack jaws typical of other vertebrates but have a cartilaginous endoskeleton and a notochord. Seven pharyngeal gill slits are evident near the head. The gill arches separating the gill slits are reinforced with cartilage. The mouth is at the center of the round **buccal funnel** and is armed with horny teeth and a rasping tongue. They attach their buccal funnel to the side of a fish, rasp a hole in the body with their tongue, and feed on the body fluids of the fish.

Question 9

Which closely related subphylum of chordates does an ammocoete resemble?

Class Chondrichthyes (Sharks, Skates, and Rays)

Sharks and their relatives (1000 species) are abundant in oceans as predators and scavengers. Their **endoskeleton** is cartilaginous and the anterior gill arches are modified into jaws. Jaws are a significant adaptation and modification of strong, anterior gill arches to process food (fig. 40.18). Like agnathans, their cartilaginous skeleton is not necessarily primitive but is probably derived secondarily from an ancestral bony skeleton.

Examine a preserved specimen of *Squalus,* the dogfish shark (fig. 40.19). Its external anatomy illustrates some advanced features appropriate for a predator. Fin structure

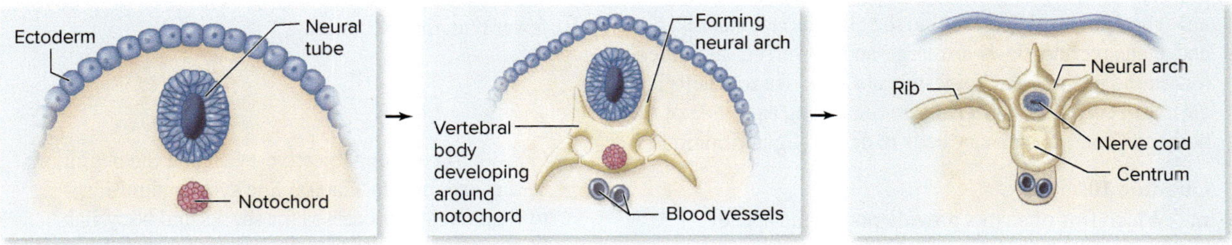

Figure 40.16 Embryonic development of a vertebra. During the course of evolution, or of development, the flexible notochord is surrounded and eventually replaced by a cartilaginous or bony covering, the centrum. The neural tube is protected by an arch above the centrum. The vertebral column is a strong, flexible rod that the muscles pull against when the animal swims or moves.

(a) Heather Angel/Natural Visions

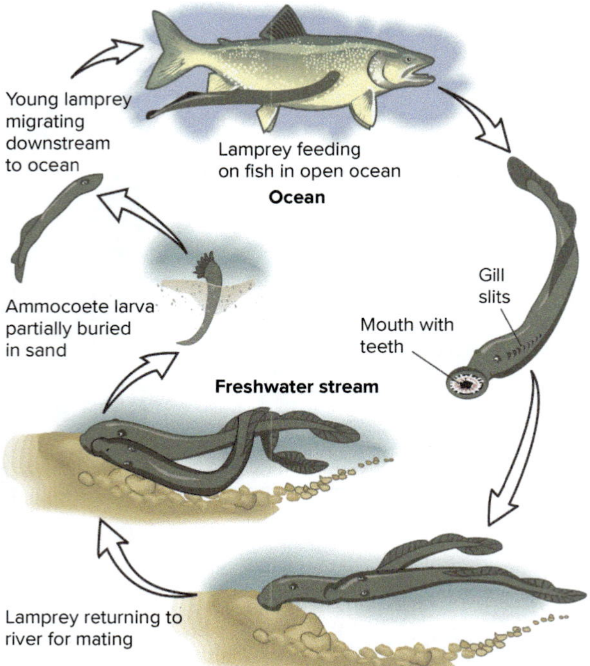

(b)

Figure 40.17 Lamprey (class Cephalaspidomorphi *Petromyzon marinus*). (*a*) Note the sucking mouth attached to aquarium glass and teeth used to feed on other fish. (*b*) External structure and life history of a sea lamprey. Sea lampreys feed in the open sea; toward the end of their lives lampreys migrate into freshwater streams, where they mate. Females deposit eggs in nests on the stream bottom, and the young larvae hatch 3 weeks later.

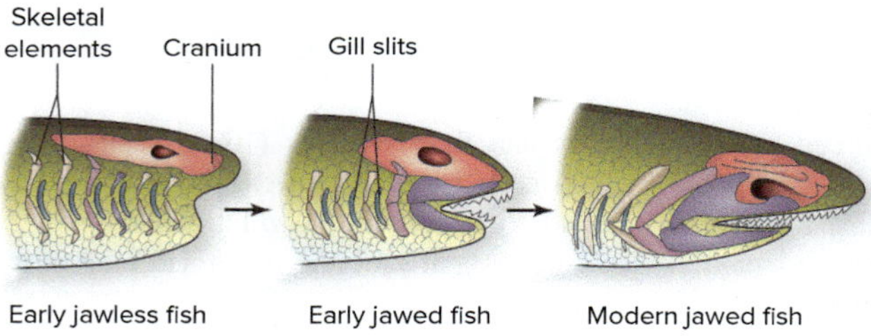

(a)

(b) Jeff Rotman Photography

Figure 40.18 (*a*) Jaws and support structures likely evolved from skeletal elements forming gill arches that separate gill slits of agnathans. (*b*) Head of sand tiger shark, *Carcharias* sp., showing a series of successional teeth on strong jaws.

includes paired **pelvic fins** (on the ventral surface near the anus) and **pectoral fins** (behind the gill slits) for stabilization and maneuvering. Jaws are large and powerful, and receptors in the nostrils and epidermis are sensitive to smells and electrical currents. A **lateral line** runs along each side of the body and contains sensory cells to detect slight vibrations.

Question 10

a. Which fins of sharks provide power and speed?

b. Why is the number of pharyngeal gill slits in sharks fewer than that in lampreys?

c. Consider Learning Objective 1 listed at the beginning of this exercise. Is a lateral line system significant to fundamental processes for sharks and bony fish? How so?

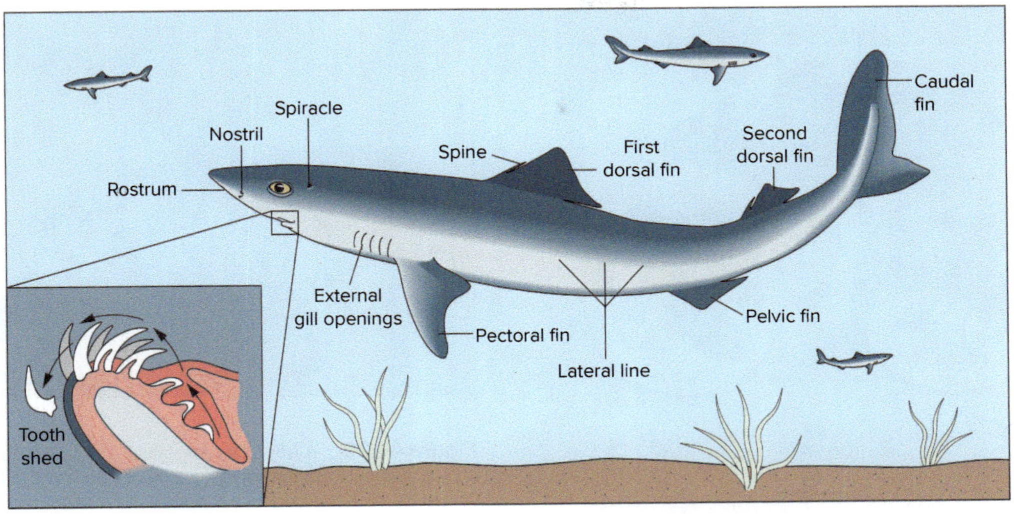

Figure 40.19 Sharks (class Chondrichthyes). Dogfish shark, *Squalus acanthias*. Section of lower jaw (inset) shows new teeth developing inside the jaw. These teeth move forward to replace lost teeth. The rate of replacement varies in different species.

Class Actinopterygii (Ray-finned Fish)

Procedure 40.3 **Examine the anatomy of a bony fish**

1. Examine the external anatomy of a preserved fish (fig. 40.20).
2. Although sharks are built for speed, the maneuverability of bony fishes is much greater. If living fish are available, observe their swimming and "breathing."
3. Examine a dissected perch and locate the structures shown in figure 40.20.

Ray-finned or "bony" fish are the most diverse class of vertebrates (30,000 species). Advanced features of ray-finned fish include a bony endoskeleton, modified gill arches, and internal air bladders for balance and buoyancy. Gills are protected by a movable gill cover called an **operculum.** Along each side and branching over the head of most fishes is a **lateral-line system** consisting of sensory pits in the skin. These pits detect water currents and predators or prey that may be moving near the fish.

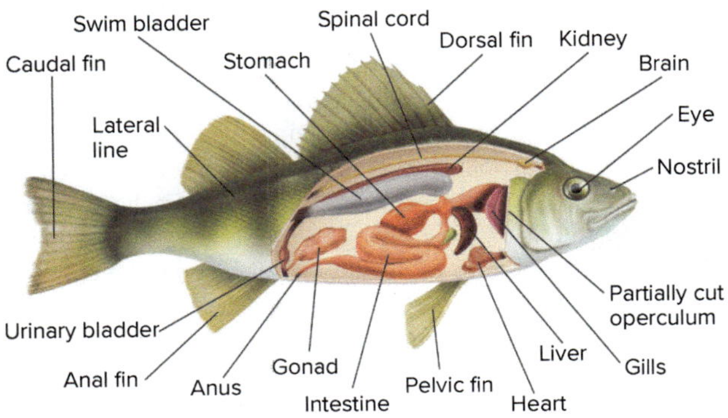

Figure 40.20 Anatomy of a bony fish, class Actinopterygii.

Question 11

a. How do the number and shape of fins of a ray-finned fish differ from those of a shark?

b. Fins of a ray-finned fish are flexible and diverse in shape. Describe the location of a fin present in ray-finned fish but not in sharks.

c. How does the symmetry of the tail of a fish compare with that of a shark?

d. Does most of the power for movement by a fish come from the tail or from other fins?

e. Can fish move water over their gills without moving through the water? What role does the operculum play in this movement?

f. How does the buoyancy of an air bladder affect the motion of a fish compared to that of a shark?

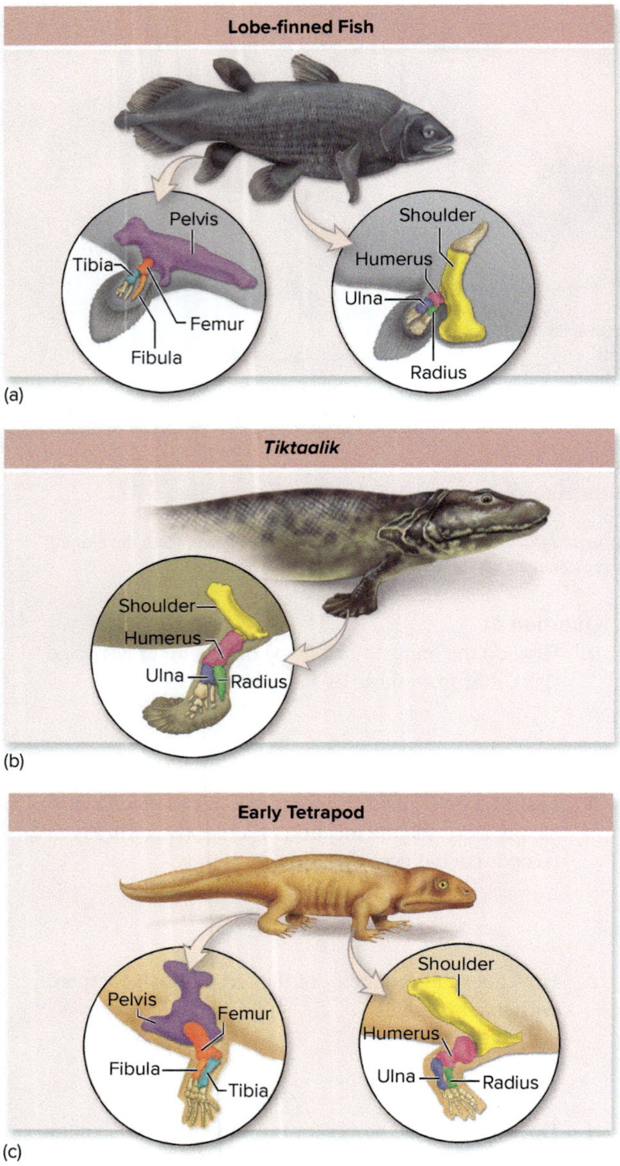

Figure 40.21 Transitional forms in the tetrapod lineage. This figure shows two early tetrapod ancestors, a lobe-finned fish and the transitional form *Tiktaalik roseae,* as well as a descendant, an early amphibian. Skeletal modifications show ancestral continuity as well as adaptation to brief periods out of the water by early amphibians.

Recent fossil evidence strongly supports our understanding of the evolutionary transition of ancestral bony fish to a tetrapod body plan of amphibians (fig. 40.21).

Class Amphibia (Frogs, Toads, and Salamanders)

Examine preserved amphibians on display. Amphibians were the first land vertebrates, arising from fish with stout, fleshy fins. Most amphibian adults are terrestrial, but they

BiologyImaging.com

Figure 40.22 A frog (class Amphibia). This poison arrow tree frog (*Dendrobates* sp.) exhibits strong coloration. These colors advertise its powerfully toxic secretions to predators, which quickly learn the frog's noxious taste. Natives of South America use the frog's toxins as a weapon; they kill the frog by piercing it with a sharp stick and holding it over a fire. The heat causes the cutaneous glands to secrete drops of venom, which are scraped into a container and allowed to ferment. Arrows dipped into the poison can paralyze birds and small monkeys. One poison arrow tree frog contains enough toxin to kill about 20,000 mice.

lay eggs in water (fig. 40.22). The eggs are fertilized externally and each hatches into an aquatic larval stage called a **tadpole.** Tadpoles undergo a dramatic metamorphosis of body shape as they become adults.

Development of legs and the development of lungs in amphibians were major evolutionary events. However, primitive lungs had already developed in some fish. In addition to lungs, the soft moist skin of some amphibians is highly vascularized and accounts for as much oxygen diffusion as the lungs.

Question 12

How are the legs of a frog different from the fins of a fish to enable movement on land?

Class Reptilia (Turtles, Snakes, and Lizards)

Examine preserved reptiles and note their morphological diversity (fig. 40.23). Reptiles, unlike their ancestors, are independent of aquatic environments and have developed structures for internal fertilization (fig. 40.24). Most reptiles also lay watertight eggs that contain a food source (the yolk) and a series of four membranes—the chorion, the amnion, the yolk sac, and the allantois (fig. 40.25). Each membrane plays a role in making the egg an independent life-support system. The outermost membrane of the egg, the **chorion,** allows oxygen to enter the porous shell but retains water within the egg. The **amnion** encases the developing embryo within a fluid-filled cavity. The **yolk sac** provides food from the yolk for the embryo via blood vessels connecting to the embryo's gut. The **allantois** surrounds a cavity into which waste products from the embryo are excreted.

BiologyImaging.com

Figure 40.23 Pit vipers are venomous reptiles; they have a pair of heat-detecting pit organs, one on each side of the head. Pit organs are visible between the eye and the nostril of this copperhead snake. These vipers can locate and strike a motionless warm animal in total darkness by sensing heat from its body. Pit organs are highly sensitive to infrared wavelengths and are especially sensitive to sudden changes of temperature. Pit organs can detect temperature differences of 0.2°C or less, allowing effective hunting of small animals at night.

Reptiles have a dry skin covered with scales that retard water loss. This dry skin does not aid respiration, but the lungs are well developed. Reptiles, fish, and amphibians are *poikilothermic,* meaning that their body temperature depends on the environment.

Question 13

a. What is the adaptive significance of internal fertilization and a watertight egg?

BiologyImaging.com

Figure 40.24 Internal fertilization. The male injects sperm-containing semen into the female's body during copulation. Ancestors of reptiles such as these lizards were the first terrestrial vertebrates to develop this form of reproduction, which is particularly suited to terrestrial existence.

b. How do the legs of different reptiles vary in number, size, and function?

c. Would you expect the legs of a terrestrial tetrapod to be more robust than those of an aquatic organism? Why or why not? Is this true for the reptiles and amphibians that you examined?

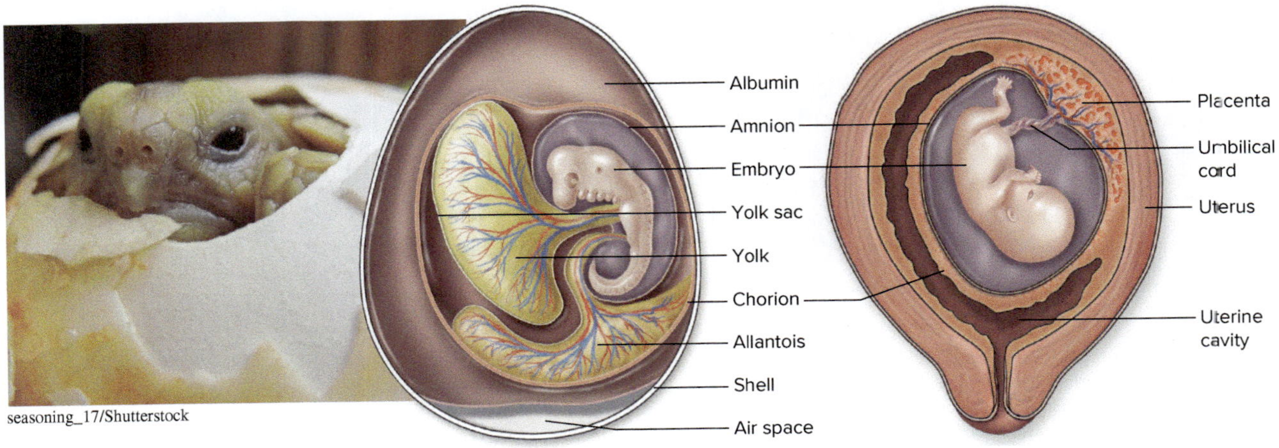

seasoning_17/Shutterstock

Figure 40.25 The watertight amniotic egg enables reptiles to live in a wide variety of habitats. This sea turtle is hatching from a typical reptilian egg. The amnion of vertebrate eggs is a sac that encloses the developing embryo of a reptile, bird, or mammal. In an amniotic egg, the embryo is encased in a hard, protective shell, and is supported internally by three membranes—the amnion, allantois, and chorion. Placental mammals also enclose their embryos in an amnion.

d. Consider Learning Objective 1 listed at the beginning of this exercise. How could poikilothermy contribute to the evolutionary success of reptiles in their environment?

Class Aves (Birds)

Examine a prepared slide and whole mount of a feather (fig. 40.26). Notice the interlocking structures. Also examine specimens of birds (fig. 40.27). Birds are the only animals with feathers, and they share the ability to fly with

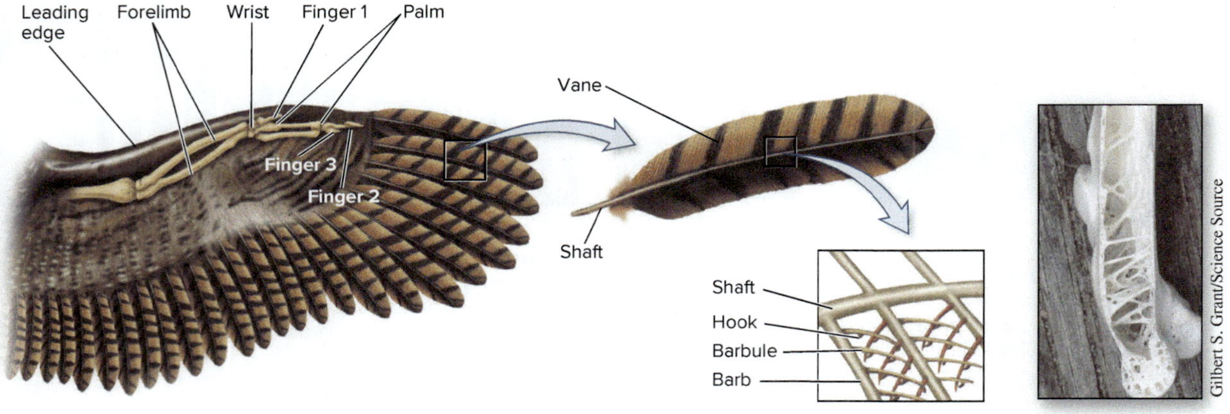

(a) Wing (b) Feather structure (c) Pelican bone (*Pelicanus occidentalis*)

Figure 40.26 Features of the bird wing and feather. (*a*) The wing is supported by an elongated and modified forelimb with extended fingers. (*b*) Each feather has a hollow shaft that supports many barbs, which in turn support barbules that interlock with hooks to give the feather its form. (*c*) The bones of a pelican (*Pelicanus occidentalis*) are hollow but crisscrossed with a honeycomb structure that provides added strength (8×).

(a) BiologyImaging.com (b) kojihirano/iStock 360/Getty Images

Figure 40.27 Birds (class Aves). (*a*) The flightless cormorant, *Phalacrocorax harrisi,* lives only on the Galápagos islands and is the only cormorant in the world (out of 30 species) that cannot fly. Ancestors arriving at the islands had no predators and little competition for their feeding niche of bottom-fish, eels, and octopuses. Flight was no longer adaptive. Over time, natural selection favored a streamlined body and strong legs for swimming. Flight muscles atrophied through the generations, and their sparsely feathered wings have become vestigial. (*b*) The California condor (*Gymnogyps californianus*) is the largest land bird in North America. Young condors acquire full adult plumage after 6 years and may live 50 years. They are efficiently adapted to soaring effortlessly in search of carrion. Their bald heads are adapted for reaching deep within the carcass and tearing pieces of meat. Unfortunately, they are in danger of becoming extinct. The remaining three or four wild individuals were captured in 1987. Offspring have been raised in captivity and have been periodically reintroduced into their dwindling habitat. Even efficient survival adaptations of the condor have not prevented a dramatic population decline. Condors are extremely sensitive to human disturbances, and we have steadily encroached on their habitat.

only a few groups. Eyes of birds are always prominent, and vision is one of their most highly developed senses. Birds are *homeothermic,* meaning that they maintain a constant body temperature. Other adaptations to flight include a high body temperature for high metabolism, a lightweight skeleton, an efficient respiratory system, and heavy musculature at the breast to move the wings. The evolutionary origin of feathers is hotly debated, but biologists agree that their lineage likely traces back to dinosaur ancestry (fig. 40.28).

Question 14

a. What are wings of flying animals other than birds made of?

b. Why might birds use keen vision more than reptiles or amphibians do?

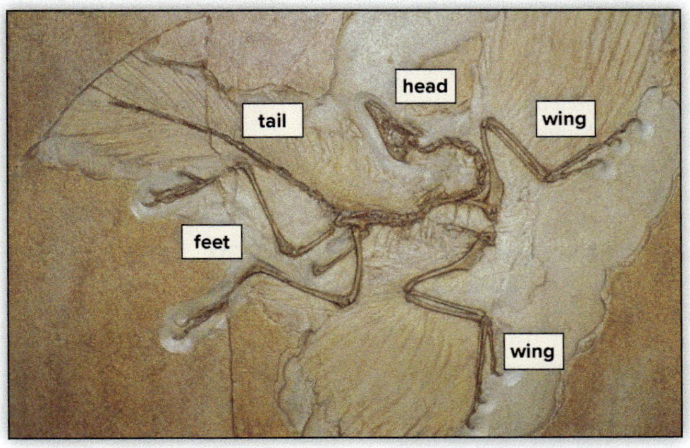

(a) *Archaeopteryx* fossil

(b)

(a) Jason Edwards/National Geographic/Getty Images; (b) Joe Tucciarone, Interstellar Illustrations

Figure 40.28 Transitional fossils. (*a*) *Archaeopteryx* was a transitional link between dinosaurs and birds. Fossils indicate that it had feathers and wings with claws, and teeth. Most likely, it was a poor flier. (*b*) *Archaeopteryx* also had a feather-covered, bony reptilian-type tail that shows up well in this artist's representation.

c. Consider what you have learned about enzymes in Exercise 11. What might be the adaptive advantage of homeothermy?

d. Describe six adaptations of birds to flight.

Class Mammalia

Examine some preserved mammals. Mammals are covered with insulating body fat and **hair** and maintain a constant body temperature as birds do (fig. 40.29). Mammals are active and have a well-developed circulatory system with a four-chambered heart. The circulatory system distributes oxygen, nutrients, and heat. Mammals nourish their young with milk produced by the mother's **mammary glands** (fig. 40.30).

Although you are already familiar with the external anatomy of *Homo sapiens*, you will study the anatomy of a rat, another representative mammal, in Exercises 47, 48, and 49. As you examine preserved and living mammals, search for common features such as hair distribution, body orientation, and structures for locomotion.

Magdalena Biskup Travel Photography/Getty Images

Figure 40.29 This tarsier (*Tarsius syrichia*) is a primitive primate (class Mammalia) found in the Philippines. It is the size of a rat, lives in a tree, and eats insects. The position of its eyes in the front of its head allows full stereoscopic vision. It has nails instead of claws, which indicates a common ancestry with higher primates. Its large eyes are efficient adaptations for nocturnal activity. The retinas lack cones for detecting color but are extra rich in rods for black/white sensitivity.

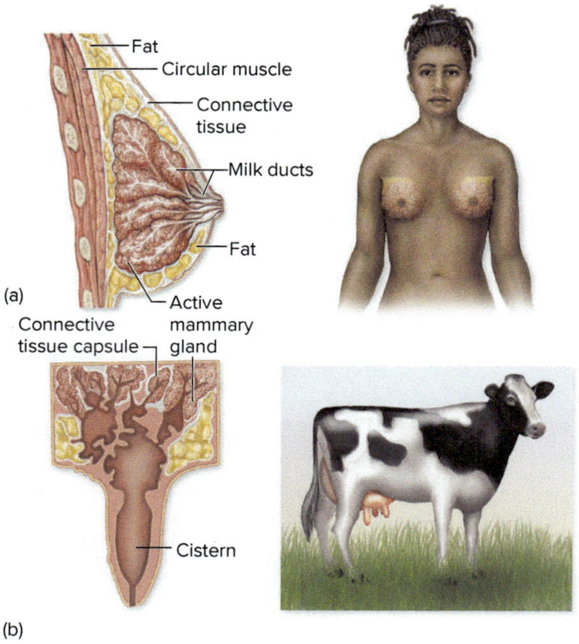

Figure 40.30 Mammary glands are specialized to secrete milk following the birth of young. (*a*) Many ducts lead from the glands to a nipple. Parts of the duct system are enlarged to store milk. Suckling by an infant initiates a hormonal response that causes the mammary glands to release milk. (*b*) Some mammals (e.g., cattle) have teats formed by the extension of a collar of skin around the opening of mammary ducts. Milk collects in a large cistern prior to its release.

Question 15

a. What factors govern the distribution of hair on species such as the human or rat?

b. How do the mammals that you are examining vary in body orientation (resting stance and position during movement)?

c. What characteristics of mammals help explain how they can occupy a variety of habitats?

Examine other preserved chordates and review in your textbook the major characteristics of each chordate group. Then complete table 40.3.

Table 40.3

A Comparison of the Characteristics of the Major Chordate Groups

	Descriptive Characteristics					
	Appendages description	Feeding methods	Skeletal elements	Diagnostic features	Gill slit development	Notochord/ spine condition
Tunicates						
Lancelets						
Lampreys						
Sharks						
Bony fish						
Amphibians						
Reptiles						
Avians						
Mammals						

INQUIRY-BASED LEARNING

Would you argue that skeletons of all major groups of vertebrates are similar? Different?

Observations: Natural selection shapes available genetic variation into adaptations that boost fitness. Over many generations, characteristics with no adaptive advantage for survival and reproduction may decrease in frequency, while those that confer reproductive advantages become increasingly frequent. External features interface an organism with its environment and are subject to strong selective pressures. External features serving multiple functions also vary a great deal among the various classes of vertebrates.

Question: To what extent are vertebrate external morphologies adapted for various functions?

a. Establish a working lab group and obtain Inquiry-Based Learning Worksheet 40 from your instructor.
b. Obtain and examine the external features of a preserved fish, amphibian, bird, reptile, and mammal.
c. Discuss with your group and instructor specific comparisons to make among these vertebrates. Record them on Worksheet 40.
d. The table on Worksheet 40 lists four broad functions of an adaptation. Can you think of others? List at least two other functions in the worksheet table.
e. Complete the investigation directed by Worksheet 40.

Questions for Further Study and Inquiry

1. Does it surprise you that echinoderms are more closely related to our own phylum (Chordata) than are other phyla? Why would you have thought otherwise?

2. Why are embryological features important for distinguishing the major groups of phyla?

3. Echinoderms lack cephalization. What characteristics of this group deemphasize the need for a head?

4. What problems were associated with colonizing land during the evolution of vertebrates?

5. Why do you suppose four rather than five or six appendages is the rule for vertebrates?

6. A cuticle occurs on the surface of organisms of many phyla and appears to be an advantageous feature. Why have higher organisms not retained this structure?

7. What is the difference in the developmental derivation of mandibles among insects, jaws of vertebrates, and the beak of an octopus?

8. Although other groups of vertebrates are more numerous and have existed longer than mammals, mammals are often called the most advanced form of life. Why?

9. Locomotion in mammals is varied. Do you believe that their powers of locomotion are superior to those of birds? Why or why not?

10. Compare the origin and function of reptile scales, bird feathers, and mammal hair. How are they similar? How do they differ?

11. Prepare a simple table of all of the taxonomic groups, their common names, their distinguishing characteristics, and all representative genera covered in this exercise. Keep this table with your study notes.

12. Vertebrates have a closed circulatory system, meaning that the blood is always enclosed within vessels and does not fill body cavities. Mollusks (Exercise 38) and arthropods (Exercise 39) have open circulatory systems, meaning that blood is pumped by a heart into body cavities, where tissues are surrounded by the blood. What are the advantages and disadvantages of each type of circulatory system?

13. Two classes of vertebrates (Aves and Mammalia) are *endothermic*. What is meant by this term? Hypothesize some evolutionary advantages of being endothermic. What are some of the costs?

WRITING TO LEARN BIOLOGY

What external anatomical features of amphibians are associated with their dual life on land and in water?

Credits

Online Supplements

Online Supplements

Connect Access for Biology

McGraw Hill Connect is a digital teaching and learning environment that improves performance over a variety of critical outcomes. With Connect, instructors can deliver assignments, quizzes and tests easily online. Students can practice important skills at their own pace and on their own schedule.

HOW TO REGISTER

Using a Print Book?
To register and activate your Connect account, simply follow these easy steps:

1. **Go to the Connect course web address provided by your instructor or visit the Connect link set up on your instructor's course within your campus learning management system.**
2. **Click on the link to register.**
3. **When prompted, enter the Connect code found on the inside back cover of your book and click Submit. Complete the brief registration form that follows to begin using Connect.**

Using an eBook?
To register and activate your Connect account, simply follow these easy steps:

1. **Upon purchase of your eBook, you will be granted automatic access to Connect.**
2. **Go to the Connect course web address provided by your instructor or visit the Connect link set up on your instructor's course within your campus learning management system.**
3. **Sign in using the same email address and password you used to register on the eBookstore. Complete your registration and begin using Connect.**

**Note: Access Code is for one use only. If you did not purchase this book new, the access code included in this book is no longer valid.*

Need help? Visit mhhe.com/support